Alternative Energies and Efficiency Evaluation

Edited by Muhammad Wakil Shahzad, Muhammad Sultan, Laurent Dala, Ben Bin Xu and Yinzhu Jiang

Published in London, United Kingdom

IntechOpen

Supporting open minds since 2005

Alternative Energies and Efficiency Evaluation
http://dx.doi.org/10.5772/intechopen.94657
Edited by Muhammad Wakil Shahzad, Muhammad Sultan, Laurent Dala, Ben Bin Xu and Yinzhu Jiang

Assistants to the Editors: Muhammad Ahmad and Nida Imtiaz

Contributors
Darko Goričanec, Danijela Urbancl, Cesar A.M. Abreu, Benjamim H.L. Silva, Godfrey Kabungo Gakingo, Tobias Muller Louw, Muhammad Wakil Shahzad, Kim Choon Ng, Ben Bin Xu, Muhammad Burhan, Chen Qian, Doskhan Ybyraiykul, M. Kumja, Muhammad Ahmad Jamil, Nida Imtiaz, Yuanlong Cui, Xuan Xue, Saffa Riffat, Zohreh Soleimani, Elmer Theo, Tugba Gurler, Laurentiu Constantin Lipan, Ali Alqahtani, Sami Almutairi, Zeeshan Farooq, Edwin Kwasi Tamakloe, Ngang Bassey Ngang, Homayoun Nikookar, Kailasnath B.B. Sutar, Ali Soofastaei, Milad Fouladgar, Vikram Rama Uttam Pandit, Mana Al-Owaidh, Abdulrahman Hazazi, Solomon Oji, Abdulaziz Al-Dulaijan, Patrick Sangouard

First published in London, United Kingdom, 2022 by IntechOpen
IntechOpen is the global imprint of INTECHOPEN LIMITED, registered in England and Wales, registration number: 11086078, 5 Princes Gate Court, London, SW7 2QJ, United Kingdom
Printed in Croatia

British Library Cataloguing-in-Publication Data
A catalogue record for this book is available from the British Library

Additional hard and PDF copies can be obtained from orders@intechopen.com

Alternative Energies and Efficiency Evaluation
Edited by Muhammad Wakil Shahzad, Muhammad Sultan, Laurent Dala, Ben Bin Xu and Yinzhu Jiang
p. cm.
Print ISBN 978-1-83969-827-9
Online ISBN 978-1-83969-828-6
eBook (PDF) ISBN 978-1-83969-829-3

For EU product safety concerns:
IN TECH d.o.o., Prolaz Marije Krucifikse Kozulić 3, 51000 Rijeka, Croatia, info@intechopen.com or visit our website at intechopen.com.

Meet the editors

Dr. Muhammad Wakil Shahzad is a senior lecturer in the Mechanical and Construction Engineering Department, Northumbria University (NU), Newcastle Upon Tyne, United Kingdom. He is an expert in renewable energy storage and its applications for water treatment, hybrid desalination processes, and life cycle cost analysis. He has won many international awards for his innovative desalination cycle, including the National Energy Globe Award Saudi Arabia 2021, Sustainability Medal 2020, Global Innovation Award 2020, National Energy Globe Award Saudi Arabia 2020 and 2019, Excellence and Leadership Award 2019, and IDA Environmental & Sustainability Award 2019. His research has been highlighted on Yahoo Business, Nature Middle East, Arab News, and many other national and international platforms. Dr. Shahzad has a Ph.D. in Mechanical Engineering from the National University of Singapore and research training from KAUST Saudi Arabia. He has extensive experience in intellectual property development and the commercialization of innovative technologies. He holds eleven international patents. To date, he has published two books, seventeen book chapters, more than seventy peer-reviewed journal papers, and more than 110 conference papers. He also received three best paper awards at international conferences. He is an editor of International Communications in Heat and Mass Transfer, an editorial board member of SN Applied Sciences, and a guest editor for topical collections. He is a Chartered Engineer and a mentor for the International Desalination Association's Young Leader Program (IDA-YLP). He is also a member of many professional organizations, including the Institute of Mechanical Engineers, International Desalination Association (IDA), International Water Association (IWA), and American Society of Mechanical Engineers (ASME).

Dr. Muhammad Sultan is an associate professor in the Department of Agricultural Engineering, Bahauddin Zakariya University, Multan, Pakistan. He has a BSc and MSc in Agricultural Engineering from the University of Agriculture, Faisalabad, Pakistan, and a Ph.D. in Energy and Environmental Engineering from Kyushu University, Japan. He did postdoctoral research in energy and environmental engineering at Kyushu University and in mechatronic systems engineering at Simon Fraser University, Canada. He has published more than 300 articles in international journals, conferences, books, and book chapters. He has been a reviewer and editor of numerous journals. His research focuses on developing energy-efficient temperature and humidity control systems for agricultural applications including greenhouses, fruits/vegetable storage, livestock, and poultry applications. His research keywords include adsorption heat pumps, desiccant air-conditioning, evaporative cooling, Maisotsenko cycle, adsorption desalination, energy recovery ventilator, atmospheric water harvesting, and wastewater treatment.

Dr. Ben Bin Xu is a Full Professor of Materials and Mechanics, at the Department of Mechanical and Construction Engineering (MCE), Northumbria University, United Kingdom and leads the research group of the Smart Materials and Surfaces Lab. He obtained his Ph.D. in Mechanical Engineering at Heriot-Watt University, Scotland, in 2011 and subsequently worked as a postdoc researcher at the University of Massachusetts Amherst (2011–2013). His research interests include advanced materials, smart surfaces, energy materials/systems, thin films, flexible electronics, and micro-engineering. He has published more than 120 journal articles and 3 book chapters. He has four patents to his credit. He has given more than sixty invited talks and won multiple awards, such as the Outstanding Paper Award at the 2018 IEEE International Flexible Electronics Technology Conference, 2016 Young Investigator Award from the International Polymer Networks Group, and the 3rd prize in the 2016 EPSRC photo competition, among others. Dr. Xu is a fellow of the Royal Society of Chemistry (RSC), Institute of Materials, Minerals and Mining (IoM3), and Royal Society for the Encouragement of Arts, Manufactures and Commerce (RSA). He is also a Chartered Scientist and Chartered Engineer. He has been an elected member in the Advisory Council and Nomination Committee of IoM3 since 2019, an elected member in the RSC Materials Chemistry Division Council since 2020, an elected member in the RSC Newcastle upon Tyne and North East Coast Committee since 2020, and the executive committee for the Chinese Society of Chemical Science and Technology (CSCST-UK) since 2016.

Yinzhu Jiang is a professor in the School of Materials Science and Engineering, Zhejiang University, China. He received his Ph.D. from the Department of Materials Science and Engineering, University of Science and Technology, China in 2007. He worked as a post-doctoral researcher at Heriot-Watt University, the UK from 2007 to 2008 and an Alexander von Humboldt Fellow at Bielefeld University, Germany from 2008 to 2010. His research interests focus mainly on energy-related materials and electrochemistry, including rechargeable batteries, metal anodes, and solid electrolytes. He has published more than 100 journal articles.

Professor Laurent Dala joined Northumbria University, UK in 2016 as a Full Professor of Aerospace Engineering, having previously held positions as Chief Scientist in Aerospace Engineering for the Council for Scientific and Industrial Research, South Africa, and chair of Aerospace Engineering at the University of Pretoria, South Africa. He was also the president of the Aeronautical Society of South Africa (AeSSA). He is a fellow of the Royal Aeronautical Society (FRAeS), a Chartered Engineer, and a Eur-Ing. He has more than 25 years of research experience in aerospace engineering. He has led various research projects around the world, particularly in flow control and multidisciplinary optimisation methods. He has been awarded as a Principal Investigator more than ZAR 3.0M and £250,000 since 2014. He has published more than 160 peer-reviewed papers/conferences. He has four patents with three in flow control. He has been invited more than sixty times as a guest speaker and is a winner of multiple awards. He is a member of the Specialist Committee for Aerodynamics at the Association Aeronautique et Astronautique de France and a visiting professor in France, South Africa, Russia, and China. He is a member of the EPSRC's College of Peer Reviewers, the European Research Council's College of Peer Reviewers, and the National Research Foundation's College of Peer Reviewers, South Africa. Dr. Dala is currently the head of the Department of Mechanical and Construction Engineering, Northumbria University.

Contents

Preface

In 2018, the global energy demand grew by 2.3%, representing a nearly two-fold growth since 2010. This greater energy demand was propelled by a global economy that expanded by 3.7% in 2018. The United States, China, and India together accounted for nearly 75% of the increase in global energy demand. Gas consumption increased 10% from the previous year in the United States alone. The exceptional change in winter and summer seasons was also responsible for a tremendous increase in global energy. Cold snaps drove demand for heating and, more significantly, hotter summer temperatures increased demand for cooling. As a result of greater energy consumption, CO_2 emissions also increased to 33.1 Gt, most of which are caused by coal-fired power generation.

Renewables contribute only 25%–30% of the growth in total primary energy demand. This was largely due to an expansion in electricity generation, where renewables accounted for 45% of the growth in 2018. The International Environmental Agency (IEA) set out the Net Zero by 2050 roadmap, according to which the electricity production sector needs to reduce global emissions by nearly three-quarters by 2025. To achieve this, the IEA recommends a more than 6% decrease in electricity generation from coal-fired power plants. Even though renewables installations are expanding quickly, there is not enough to satisfy a strong rebound in global electricity demand. This will result in a sharp rise in the use of fossil fuel electricity generation that risks pushing carbon dioxide emissions.

This book addresses key challenges related to energy production, efficiency evaluation, and CO_2 emissions. Section 1 discusses assorted renewable energy generation processes such as wave energy, hydrogen, and low-temperature sources. Section 2 covers efficiency evaluation frameworks, including a standard primary energy framework for efficiency evaluation and process optimization. It also discusses some aspects of industrial energy evaluation and optimization. Section 3 highlights greenhouse gas emissions, energy, and economic and environmental analysis. It also provides detail on energy efficiency technologies to reduce emissions and adverse environmental impacts.

Dr. Muhammad Wakil Shahzad
Department of Mechanical and Construction Engineering,
Northumbria University,
Newcastle upon Tyne, United Kingdom

Muhammad Sultan
Bahauddin Zakariya University,
Multan, Pakistan

Laurent Dala and Ben Bin Xu
Northumbria University,
Newcastle upon Tyne, United Kingdom

Yinzhu Jiang
Zhejiang University,
Hangzhou, China

Section 1

Energy Resources

Chapter 1

Waveform Design for Energy Efficient OFDM Transmission

Homayoun Nikookar

Abstract

In this chapter, a green radio transmission using the binary phase-shift keying (BPSK) modulated orthogonal frequency-division multiplexing (OFDM) signal is addressed. First, the OFDM transmission signal is clearly stated. For a specified performance of the system, the least transmit power occurs by the optimal OFDM shape, which is designed to minimize the average inter-carrier interference power taking into account the characteristic of the transmit antenna and the detection process at the receiver. The optimal waveform is obtained by applying the calculus of variations, which leads to a set of differential equations (known as Euler equations) with constraint and boundary conditions. Results show the transmission effectiveness of the proposed technique in the shaping of the signal, as well as its potential to be further applied to smart context-aware green wireless communications.

Keywords: green radio transmission, multicarrier, OFDM, signal design, optimization

1. Introduction

The advancements in the field of wireless communication have led to many exciting applications such as mobile internet access, health care and medical monitoring services, smart homes, combat radios, disaster management, automated highways, and factories. With each passing day, novel and advanced services are being launched even while existing ones continue to flourish. Wireless services have now found applicability in other sectors too including health care, transportation, security, logistics, education, and finance. Demand for wireless services is thus expected to grow in the foreseeable future. However, with the increasing popularity of wireless services (such as the 5G and the future 6G), the requirements on prime resources such as green transmission and radio spectrum are put to a great test. Recent studies have shown that the energy costs account for as much as half of a mobile service provider's annual operating expenses. Therefore, making the communication equipment more efficient in relation to its power consumption not only has implications with regard to environmental pollution and the level of CO2 emission, but also makes economic sense and can eventually reduce the cost of wireless services (for providers and users).

In this regard and given the 10% of the world's energy consumption due to the information and communication technology (ICT) industry [1], energy efficiency has become one of the key performance indicators (KPI) in the design and implementation of radio systems.

The theme of green radio communications is to design energy-efficient wireless communication techniques and protocols, which optimally utilizes available resources and minimize power consumption. Various strategies are employed for the design of energy-efficient wireless systems; among them are energy-efficient new radios [2, 3], minimization of interference [4], and optimal resource allocation [5]. In this chapter, we would like to design a signal for energy-efficient OFDM transmission. As can be seen from Section 4, we will reduce the power dissipation of the transmitter and consequently the level of CO2 emission of the radio system by optimal design of a waveform for transmission that minimizes inter-carrier interference of the OFDM system. In this way for reaching the same performance level of the provided communication services, a lower power level will be required for transmission, which leads to a green wireless radio system. As the reduction of the inter-carrier interference level is the basis for the optimization of OFDM signal, the approach explained in this chapter falls in the category of green radio by minimization of interference mentioned above.

The rest of the chapter is organized as follows. In Section 2, the basics of the OFDM transmission being provided. In Section 3, the data detection procedure in the receiver is mathematically explained. This procedure is further needed for the maximization of the detection performance of the system. Waveform design and the optimization procedure using calculus of variations and the Lagrange multiplier are detailed in Section 4, and the results are discussed. Conclusion remarks appear in Section 5.

2. OFDM transmission

Multicarrier technique, also called orthogonal frequency-division multiplexing (OFDM), used among others in the DAB (Digital Audio Broadcasting), DVB (Digital Video Broadcasting), and 5G wireless communication systems, is a modulation method that is used for the high-speed data communications. In this technique, transmission is carried out in parallel on different orthogonal frequencies (known as subcarriers). By orthogonality, we mean different frequencies—that are used for transmission—do not influence each other. Because of the orthogonality of subcarriers, data on different frequencies do not interfere with each other and subsequently, a higher performance can be achieved with this transmission technique. This technique is desirable for the transmission of digital data through multipath fading radio channels. Since by parallel transmission, the deleterious effect of fading is spread over many bits; therefore, instead of a few adjacent bits completely destroyed by the fading, it is more likely that several bits only be slightly affected by the channel. The other advantage of this technique is its spectral efficiency. In OFDM, the spectra of subchannels (subcarriers) overlap each other while satisfying orthogonality, giving rise to the spectral efficiency. Because of the parallel transmission, in the OFDM technique, the transmit symbol duration is increased. This has the added advantage of this method to work in the radio channels having impulsive noise characteristics. The other advantage of the OFDM method is its implementation with the fast Fourier transform (FFT) algorithm, which provides efficient full digital implementation of the modulator and demodulator. A detailed study of OFDM for wireless personal communications and its comparison with other modulation methods can be found in [6].

In the serial data transmission sequences of data are transmitted as a train of serial pulses. However, in the OFDM parallel transmission, each bit of a sequence of M bits modulates a subcarrier. A simple block diagram of the OFDM transmitter is shown in **Figure 1**. The input data with the rate R are divided into the M parallel

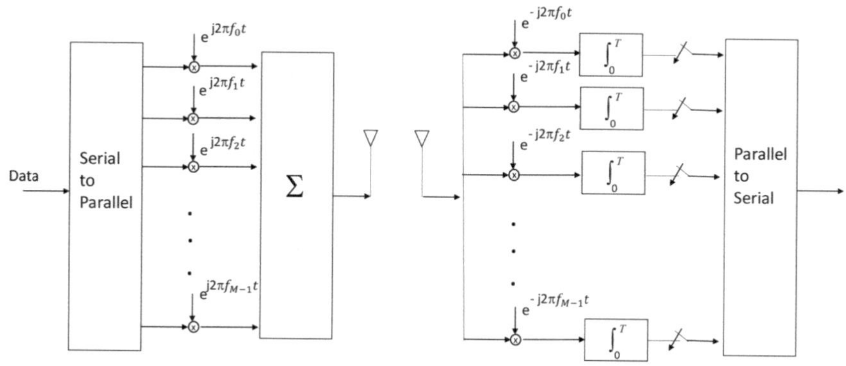

Figure 1.
Block diagram of the OFDM system.

information sequences with the rate R/M. Each sequence modulates a subcarrier. In the OFDM method, the frequency of the mth subcarrier is

$$f_m = f_0 + \frac{m}{T}, \qquad m = 0, 1, 2, \ldots, M-1 \tag{1}$$

where f_0 is the lowest frequency, which can be considered zero without loss of generality, M is the number of subcarriers, and T is the OFDM symbol duration. The OFDM transmitted signal is written as

$$s(t) = \frac{1}{\sqrt{M}} \sum_{i=-\infty}^{\infty} \sum_{m=0}^{M-1} b_m(i) e^{j2\pi f_m t} g(t - iT) \tag{2}$$

where $b_m(i)$ is the symbol of the mth subchannel at time interval iT, and for the BPSK modulation is ± 1 and $g(t)$ is the shape of the transmitter filter that is nonzero in $(0, T)$, which in this chapter we will try to optimize its shape. The factor $1/\sqrt{M}$ in (2) is used in order to keep the power of the OFDM signal constant disregarding the number of subcarriers. For the interval $(0, T)$, the OFDM signal can be expressed as follows:

$$s(t) = \frac{1}{\sqrt{M}} \sum_{m=0}^{M-1} b_m(0) e^{j2\pi f_m t} g(t), \qquad 0 \le t < T \tag{3}$$

In the transmitter, the abovementioned OFDM signal is sent to the antenna for transmission through the radio channel. Since the OFDM transmission is a very wideband transmission technique, the transmit antenna will influence this signal. The impact of the antenna in the transmission band can be modeled as a differentiator [7]. Accordingly, the transmitted signal is written as follows:

$$x(t) = \frac{d}{dt} s(t) = \frac{1}{\sqrt{M}} \sum_{m=0}^{M-1} b_m(0) \left[j2\pi f_m e^{j2\pi f_m t} g(t) + e^{j2\pi f_m t} \dot{g}(t) \right], \quad 0 \le t < T \tag{4}$$

where $\dot{g}(t) = \frac{d}{dt} g(t)$. The above Eq. (4) can be simplified as

$$x(t) = \frac{1}{\sqrt{M}} \sum_{m=0}^{M-1} b_m(0) e^{j2\pi f_m t} \left[j2\pi f_m g(t) + \dot{g}(t) \right], \qquad 0 \le t < T \tag{5}$$

3. Data detection procedure in the OFDM receiver

Now noting the block diagram of the OFDM system, **Figure 1** and assuming an ideal channel and no noise, in the receiver for the detection of the kth bit, the following operation is done:

$$z_k = \int_0^T x(t)e^{-j2\pi f_k t}dt = \int_0^T \frac{1}{\sqrt{M}}\sum_{m=0}^{M-1} b_m(0)e^{j2\pi f_m t}\left[j2\pi f_m g(t) + \dot{g}(t)\right]e^{-j2\pi f_k t}dt \quad (6)$$

The decision variable z_k can be written as follows:

$$\begin{aligned} z_k &= \frac{1}{\sqrt{M}}\int_0^T b_k(0)\left[j2\pi f_k g(t) + \dot{g}(t)\right]dt \\ &+ \frac{1}{\sqrt{M}}\sum_{m=0,\, m\neq k}^{M-1} b_m(0)e^{j2\pi(f_m - f_k)t}\left[j2\pi f_m g(t) + \dot{g}(t)\right]dt \end{aligned} \quad (7)$$

The first part of (7) is the useful signal for detection of bit b_k and the second term is the inter-carrier interference (ICI), that is,

$$ICI = \frac{1}{\sqrt{M}}\int_0^T \sum_{m=0,\, m\neq k}^{M-1} b_m(0)e^{j2\pi(f_m - f_k)t}\left[j2\pi f_m g(t) + \dot{g}(t)\right]dt \quad (8)$$

As our data bit modulation is BPSK, we are only interested in the real part of the first term of (7). Furthermore, by ignoring the ICI term the decision variable becomes

$$z_k = \frac{1}{\sqrt{M}}\int_0^T b_k(0)\dot{g}(t)dt \quad (9)$$

So, for the best detection performance, we would like to have

$$\int_0^T \dot{g}(t)dt = 1 \quad (10)$$

This is a normalized constraint that will be used later on in finding a solution to the optimization problem.

4. Waveform design and optimization procedure

In this chapter, we would like to design the waveform $g(t)$ to have the least transmit power while having the best data detection performance. This can be achieved by minimizing the power of interference. In this way, when the power of ICI is minimized the least transmit power will be needed to provide a specified data detection performance. In the following, the power of ICI is obtained. Since the mean of BPSK data is zero, $E(b_m) = 0$, using Eq. (8) the variance of the ICI interference is written as follows:

$$\begin{aligned} \sigma^2_{ICI} &= \frac{1}{M}\int_0^T\int_0^T \sum_{m=0,\, m\neq k}^{M-1}\sum_{n=0,\, n\neq k}^{M-1} \overline{b_m(0)b_n^*(0)}e^{j2\pi(f_m t - f_n u)}e^{j2\pi f_k(t-u)} \\ &\left[j2\pi f_m g(t) + \dot{g}(t)\right]\left[-j2\pi f_n g(u) + \dot{g}(u)\right]dtdu \end{aligned} \quad (11)$$

Since the BPSK-modulated data bits on different carriers are assumed to be independent, that is, $E[b_m b_n{}^*] = \delta_{m,n}$, and by using the properties of Dirac delta function, Eq. (11) reduces to

$$\sigma^2{}_{ICI} = \frac{1}{M}\int_0^T \sum_{m=0,m\neq k}^{M-1} \left(4\pi^2 f_m{}^2 g^2(t) + \dot{g}^2(t)\right)\, dt \tag{12}$$

By changing the order of integral and summation we have:

$$\sigma^2{}_{ICI} = \frac{1}{M}\sum_{m=0,m\neq k}^{M-1}\int_0^T \left(4\pi^2 f_m{}^2 g^2(t) + \dot{g}^2(t)\right)dt \tag{13}$$

Using Eq. (1) and by denoting

$$A_m = \frac{2\pi m}{T} \tag{14}$$

The power of ICI interference becomes

$$\sigma^2{}_{ICI} = \frac{1}{M}\sum_{m=0,m\neq k}^{M-1}\int_0^T \left(A_m{}^2 g^2(t) + \dot{g}^2(t)\right)dt \tag{15}$$

Therefore, our index function to minimize can be written as follows:

$$J_{min} = \int_0^T \left(A_m{}^2 g^2(t) + \dot{g}^2(t)\right)dt \tag{16}$$

We would like to find the optimal waveform by minimizing (16) and with respect to the constraint of Eq. (10).

By consideration of this restriction and using the Lagrange multiplier, the *augmented functional* for minimization is written as follows:

$$J_a(g(t),\dot{g}(t),p(t),t) = \int_0^T \left[A_m{}^2 g^2(t) + \dot{g}^2(t) + p(t)(\dot{g}(t) - \dot{\nu}(t))\right]dt \tag{17}$$

$$J_a(g(t),\dot{g}(t),p(t),t) = \int_0^T f_a(g(t),\dot{g}(t),p(t),\dot{\nu}(t),t)dt$$

$$f_a(g(t),\dot{g}(t),p(t),\dot{\nu}(t),t) = A_m{}^2 g^2(t) + \dot{g}^2(t) + p(t)(\dot{g}(t) - \dot{\nu}(t)) \tag{18}$$

where $p(t)$ is the Lagrange multiplier and according to (10)

$$\dot{\nu}(t) = \dot{g}(t) \tag{19}$$

The optimal waveform $g_*(t)$, (subscripts with * indicate the optimal waveforms), which is the extremal for the augmented functional J_a in (17), is the solution of the Euler differential Equation [8]:

$$\frac{\partial f_a}{\partial g}(g_*(t),\dot{g}_*(t),p_*(t),\dot{\nu}_*(t),t) - \frac{d}{dt}\left(\frac{\partial f_a}{\partial \dot{g}}(g_*(t),\dot{g}_*(t),p_*(t),\dot{\nu}_*(t),t)\right) = 0$$

$$\frac{\partial f_a}{\partial \nu}(g_*(t),\dot{g}_*(t),p_*(t),\dot{\nu}_*(t),t) - \frac{d}{dt}\left(\frac{\partial f_a}{\partial \dot{\nu}}(g_*(t),\dot{g}_*(t),p_*(t),\dot{\nu}_*(t),t)\right) = 0 \tag{20}$$

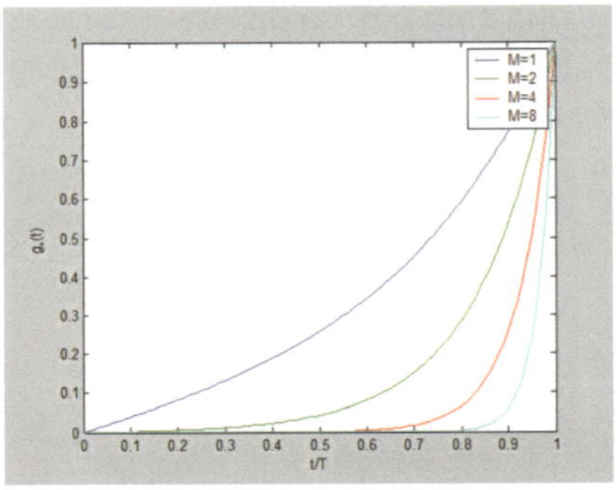

Figure 2.
The optimal energy efficient OFDM waveform for different values of M.

By calculation of the Euler equation, we obtain the following differential equations:

$$\ddot{g}_*(t) - A_m{}^2 g_*(t) = 0 \tag{21}$$

$$\frac{d}{dt}\left(\frac{\partial f_a}{\partial \dot{\nu}}(g_*(t), \dot{g}_*(t), p_*(t), \dot{\nu}_*(t), t)\right) = \dot{p}_*(t) = 0 \tag{22}$$

The solution for Eq. (21) is as follows:

$$g_*(t) = C_1 e^{A_m t} + C_2 e^{-A_m t} \tag{23}$$

By using the boundary conditions $g(0) = 0$, and $g(T) = 1$ we have

$$C_1 = -C_2 = \frac{1}{2\sinh(A_m T)} \tag{24}$$

Accordingly, the optimum waveform $g_*(t)$ for minimal interference, which leads to a minimal transmit power becomes the following:

$$g_*(t) = \frac{\sinh A_m t}{\sinh(A_m T)} \tag{25}$$

where A_m is a constant which is a function of the subcarrier number of the OFDM and the OFDM symbol duration T, see Eq. (14). Therefore, for a specified performance the *hyperbolic sine (sinh)* is the best shape for energy-efficient OFDM transmission. In **Figure 2**, the optimal waveform for different values of M is plotted.

5. Conclusion

In this chapter, we designed an optimal shape for energy-efficient OFDM transmission. We started with the formulating of the OFDM transmit signal and its shaping taking into account the behavior of the transmit antenna in the broad bandwidth of the OFDM signal and its detection at the receiver. The energy-efficient optimal waveform was obtained by minimizing the inter-carrier interference power level in the data detection process, which subsequently gives the best performance of the system. Results show that the *sinh* shape needs the least transmit energy to provide the specified performance. It has to be mentioned that the design framework presented in this chapter (i.e., minimization of ICI power that requires the least transmit power) can directly be applied to other design criteria such as security, spectral efficiency, performance, of OFDM wireless communication networks by merely changing the objective function. However, in this process, the desirable properties of the OFDM signal must be translated into realizable objective functions with constraints, a task that might be quite challenging.

Author details

Homayoun Nikookar
Netherlands Defence Academy, Den Helder, The Netherlands

*Address all correspondence to: h.nikookar@mindef.nl

References

[1] Usma M, Kautarci ME. A survey on recent trends and open issues in energy efficiency of 5G. Sensors. 2019;**19**(14): 3126. DOI: 10.3390/s19243126

[2] Nikookar H. Wavelet Radio: Adaptive and Reconfigurable Wireless Systems based on Wavelets. Cambridge: Cambridge University Press; 2013

[3] Nikookar H. Signal design for green radio transmission: A wavelet approach. IEEE International Symposium on Personal Indoor Mobile Radio Communications. Bologna, Italy: PIMRC; 2018

[4] Aydin O, Jorswieck EA, Aziz D, Zappone A. Energy-spectral efficiency tradeoffs in 5G multi-operator networks with heterogeneous constraints. IEEE Transactions on Wireless Communications. 2017;**16**:5869-5881

[5] Rehan S, Grace D. Efficient joint operation of advanced radio resource and topology management in energy-aware 5G networks. In: Proceedings of the 2015 IEEE 82nd Vehicular Technology Conference (VTC2015-Fall), Boston, MA, USA. Massachusetts, USA: IEEE. 2015. pp. 1–2

[6] van Nee R, Prasad R. OFDM for Wireless Multimedia Communications. London: Artech House; 1999

[7] Mireles FR. Signal detection for Ultra wideband communications in dense multipath. IEEE Transactions on Vehicular Technology. 2002;**51**(6): 1517-1521

[8] Kirk DE. Optimal Control Theory. New Jersey: Prentice Hall; 1970

Chapter 2

Hydrogen as a Clean Energy Source

Vikram Rama Uttam Pandit

Abstract

Sustainable development of the world is mainly dependent on the use of present energy resources, which primarily includes water, wind, solar, geothermal, and nuclear power. Hydrogen as a clean and green energy source can be the resolution of the energy challenge and may satisfy the demands of several upcoming generations. Hydrogen when used it does not produce any type of pollutant and this makes it a best candidate as a clean energy. Hydrogen energy can be generated from natural gas, oil, biomass, and fossil fuels using thermochemical, photocatalytic, microbiological and electrolysis processes. Large scale hydrogen production is also testified up to some extent with proper engineering for multi applications. Alas, storage and transportation of hydrogen are the main challenge amongst scientific community. Photocatalytic hydrogen production with good efficiencies and amount is well discussed. Till date, using a variety of metal oxide-sulfide, carbon-based materials, metal organic frameworks are utilized by doping or with their composites for enhance the hydrogen production. Main intents of this chapter are to introduce all the possible areas of hydrogen applications and main difficulties of hydrogen transportation, storage and achievements in the hydrogen generation with its applications.

Keywords: energy, hydrogen, photocatalysis, storage

1. Introduction

The fast and uncontrolled growing population is a result of urbanization and industrialization. States economic growth is dependent on the number of working industries in it. Also, to fulfill the demands of growing population textile, food, petrochemical, plastic, leather, and metal factories are emerging day by day [1–3]. Each of these industries helps to the mankind for their betterment and responsible for creating several pollutants, destroying agricultural lands and natural habitats. On the other hand, all these industries are working on non-renewable energy sources like wood, coal, natural gas, and petroleum products [4, 5]. Non-renewable energy sources are of one-time use, once used one cannot utilize them again. Two main problems are associated with non-renewable energy sources firstly when they burnt the produces harmful pollutants which is not acceptable [6, 7]. From past couple of decades scientific community, is engaged to solve the pollutants removal/degradation problems which are not economical. When fossil fuel gets extensively used, they produce global warming. Global warming is mainly caused by carbon dioxide emission. Carbon dioxide is when produce it remains in the atmosphere and absorbs the harmful infrared radiations which are reflected from earth surface. Secondly, after consumption these energy sources are going to be extinct as their availability is limited [8]. Geographically, non-renewable energy stock is located only to certain territories hence, it might be the reason for clashes between energy enriched and energy deficient countries which may leads to the world war again. Because of all these problems like pollution, environmental and health hazards the use of non-renewable energy sources for long period are not viable [9, 10].

Use of renewable energy sources like solar energy, hydro energy, wind energy, geothermal energy and nuclear energy are the only alternatives for all above said problems. Major advantages of renewable energy sources are that they when used does not create any type of pollution as well as one can use them in cyclic manner. Only one thing that we must keep in mind that these can not be used directly as an energy, but one must convert in the form of energy [11]. Also, sometimes for the energy conversion process complex, huge and expensive machinery may require. The ultimate solution for all above discussed problems is the use of hydrogen energy. The sustainable and better future is the motivation for the production and use of hydrogen energy.

Hydrogen gas is also known as a green fuel, as it when used does not produced any harmful and toxic pollutants. Hydrogen is a very first element from the periodic table with symbol H and atomic number 1. It is present in a S block, first A (IA) group with electronic configuration $1S^1$. Because of only one electron present in its outermost shell, it can form only a single sigma bond with many other atoms. It is a nontoxic, colorless, odorless, and combustible element. Hydrogen is a nonmetal, present in a gaseous form under normal conditions. Hydrogen is at the top list in terms of abundance in the universe. Many studies shows that hydrogen is not only present on the planet earth but also it is present on and other planets too. Stars are producing helium from hydrogen under high temperature and pressure at their core. Being a smallest and lightest element hydrogen atom consist of one proton and one electron without any neutron.

Hydrogen has three isotopes adds proton (H), deuterium (D) and tritium (T) along with these chemists succeeded in the synthesis of 4H and 7H in a laboratory, which are not occurring naturally. Arrhenius theory of acids and bases is very useful in chemical science is depends on the hydrogen. Most of the chemical compounds which are synthesized in lab or naturally occurring are bonded to hydrogen because of its high reactivity. Water is a most important source for living organisms which is also known as a universal solvent is composed of two hydrogens and one oxygen. In the present chapter, different sources of hydrogen as an energy with photocatalytic hydrogen production methods mechanism from water and toxic hydrogen sulfide is depicted along with respective photocatalytic reaction setup for the generation of hydrogen. Both water and H_2S splitting set ups are being reported in our latest work with the help of organic and inorganic semiconductor and composite photocatalyst materials.

2. Hydrogen production methods

After realizing the importance of hydrogen in clean energy and future fuel till date, many methods are well known for the generation of hydrogen.

Hydrogen can be produced mainly from biological, electrolytic, photocatalytic, steam reforming and thermochemical methods as shown in **Figure 1**. Microorganisms (algae and bacteria) can also be responsible for the generation of hydrogen using biological processes. Water can be spilt by using both electrolytic and sunlight (photo) into hydrogen and oxygen. Also, hydrogen sulfide can used as

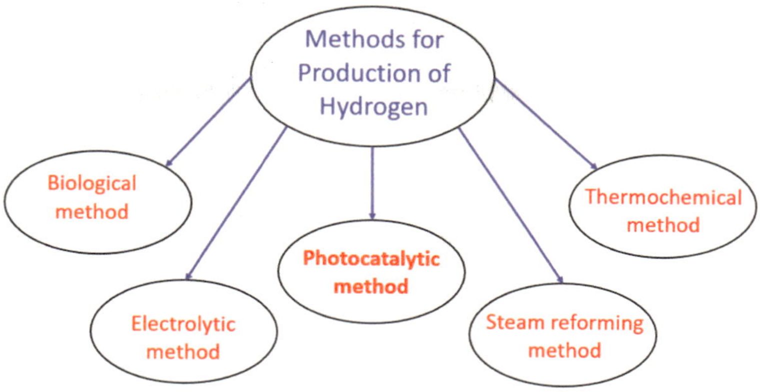

Figure 1.
Hydrogen production methods.

a rich source for hydrogen, which photocatalytically splits into hydrogen as a clean energy and sulfur for agriculture [12].

2.1 Biological method

Microorganisms such as algae and bacteria in absence of sunlight with organic matter can produce hydrogen using many biological reactions. In this method bacteria (microorganisms) break down the organic matter such as biomass, sugar, corn or waste and releases hydrogen gas. This method is also known as dark fermentation as no light is involved [13]. Biological method is under research and development stage as the efficiency of this process is not up to the mark.

2.2 Thermochemical method

Natural gas, coal, hydrocarbons and biomass is also rich with the hydrogen content. Thermal process like steam reforming is responsible for releasing hydrogen from these sources. Hydrogen when combines with carbon produces a hydrocarbon which are available naturally and are one of the main sources of hydrogen. Hydrocarbons like methane, ethane and propane (alkene and alkyne) are key compounds of hydrogen. More than 90% of hydrogen which we get nowadays is coming from hydrocarbons which has a fossil origin. These hydrocarbons are quite stable than the other sources of hydrogen and does not leave hydrogen easily unless catalysis process is used. Breaking of sigma bond present in carbon and hydrogen is most difficult but can be achieved by steam reforming in case of methane.

$$CH_4 + H_2O \rightarrow CO + 3H_2 \quad (1)$$

Generally, in steam reforming process water (steam) and methane are mixed in presence of catalyst (noble metals) inside a tube in appropriate proportion. Main advantage of using this process to produce hydrogen is that many companies/industries are already with all the equipment's setups are present. Perfect engineering and research and development in hydrogen generation from hydrocarbons field will fulfill the need of many upcoming generations [14].

2.3 Electrolytic method

This method uses electricity to get hydrogen and oxygen from decomposition of water. Now a days electrolytic water splitting is well developed and available in market commercially. This method is more efficient as compared to previous methods for production of hydrogen. Ultra-pure hydrogen can be produced by using electrolysis method. Also, bulk amount of hydrogen can be produced using this method but, the assembly required for this purpose is expensive [15]. Excess amount of energy is required in the form of overpotential to overcome activation barriers. If excess energy is not supplied the rate of hydrogen production is merely slow. Lately, hydrogen production using this method is not in demand as hydrogen can be generated in affordably amount from fossil fuels.

2.4 Photocatalytical method

In any chemical conversion reactants are converted into useful products either by photocatalytic or thermal reaction. The photocatalysis reaction method in which light/ photon is used to stimulate a photocatalyst substance (catalyst) which alters the rate of chemical reactions/process favorably without taking actual part in chemical process. Photocatalysis can also explained as a process in which light energy is used to activate the substance, which enhances the rate of a reaction without used in the chemical reaction [16]. Generally, electron-hole pairs are generated after absorption of light when the energy of the incident light/photons exact matches or exceeds the bandgap of that semiconductor photocatalyst material.

3. Mechanism of photocatalysis

In presence of light photocatalyst material shift the electron from valance band (VB) to the conduction band (CB). VB which has oxidative potential of +1.0 to +3.5 V versus the normal hydrogen electrode (NHE), and CB have a chemical potential of +0.5 to −1.5 V versus the NHE, hence they can act as reductants. **Figure 2**

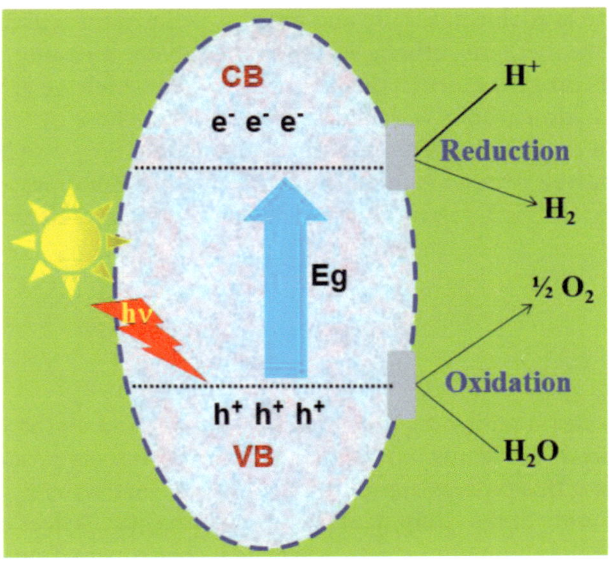

Figure 2.
Mechanism of photocatalytic H_2 production.

describes, photocatalysis method in three steps (1) migration of electrons (negative charge) from VB to CB, by leaving exact vacant holes (positive charge), (2) movement of excited holes and electrons to the surface and (3) excited electrons-holes they react with absorbed electron donors and electron acceptors for reduction and oxidation reactions. Along with proper band gap of semiconductor photocatalyst material its charge recombination rate, mobility and lifetime of electron and holes also plays a crucial role in overall hydrogen production [17].

3.1 Semiconductor photocatalyst material

Recently, charge recombination is delayed using either use of co-catalyst (platinum) on the surface of semiconductor photocatalyst which boosts the overall efficiency of hydrogen. Sacrificial agents such as methanol, EDTA, sulfides, sulfates and benzyl alcohol are also act as an oxygen scavenger in the process by enhancing the total yield of hydrogen production. After first report on TiO_2 for hydrogen production by Gratzel *et al.* the hydrogen generation field is considered as of immense importance [18]. Many researchers have published in the photocatalytic hydrogen generation field by changing catalysts, co-catalysts, combination of two catalysts (composites, coupled system). Till date, nano metal oxides, sulfides, niobates, tantalates and vanadates which contained the metal of d^0 or d^{10} electronic configurations such as In, Ga, Sb, Bi and Ag. Further, binary and ternary nano sulfides (CdS, ZnS, SnS_2, ZnI_2S_4, CdI_2S_4 and Sb_2S_3) nitrides oxynitrides, carbon based, and organic semiconductor materials has been reported as alternative photocatalysts for H_2 generation. Nanomaterial based semiconductor photocatalyst systems (TiO_2, SnO_2, WO_3, ZnO, Si_2O_7, ZrO_2, $SrTiO_3$, $LaCrO_3$, $BaTiO_3$) have many advantages as compared to their bulk counterparts and hence preferred in photocatalytic hydrogen generation [18–20]. These advantages are, high surface area, higher optical absorption, shorter charge migration length, higher solubility, tunable electronic structure, plasmonic resonance assisted charge injection and separation. All these plus points can be utilized to scale up photocatalytic hydrogen production using nanostructured semiconductor photocatalyst system. Following are the two photocatalytic hydrogen production setups are discussed with schematic representation using water and hydrogen sulfide as a source of hydrogen.

4. Photocatalytic hydrogen generation setups

4.1 Water splitting

Many attempts were taken place for the construction and engineering of setup for photocatalytic hydrogen generation [16]. Depending on the light source used (in sunlight or in a lab) reaction setup is modified, and the simplest form is described here with schematic representation **Figure 3**.

Generally, photocatalytic hydrogen production experiments were carried out in a wooden cupboard box which is known as a photocatalytic water splitting setup. This box is having the dimensions around 30 × 30 × 30 inch with closed wooden/stainless steel chamber with observing window as shown in schematic. **Figure 3** represents the schematic of setup and can be divided into three main parts as shown in figure. Firstly, the light source fitted vertically in quartz tube having water circulation (a) arrangement for cooling purpose. 400–600 W lamp with emission wavelength 200–800 nm can be used depending upon the photocatalyst. Secondly, inside this box a two neck 250 mL borosilicate round bottom flask (b) act as a photoreactor. In photoreactor water, co-catalyst, magnetic needle and scavenger

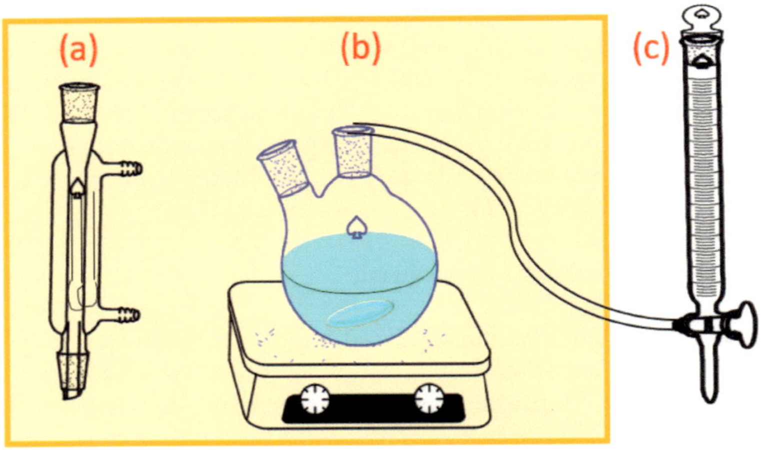

Figure 3.
Schematic representation of photocatalytic water splitting, (a) Light source fitted vertically in quartz tube having water circulation, (b) photoreactor, (c) a eudiometric tube.

are added. Lastly, produced H_2 in a reactor is collected in a eudiometric tube (c) for characterization on gas chromatography (GC). The eudiometer tube is with a saturated solution of sodium chloride (NaCl) to avoid the dissolution of evolved gases [16].

4.2 H_2S splitting

Irritating smell of H_2S gas is the reason for less research and development in this area as compared to water splitting. Earlier, Claus process is used for H_2S splitting which is mainly focused on sulfur and not on hydrogen. Photocatalytic H_2S splitting

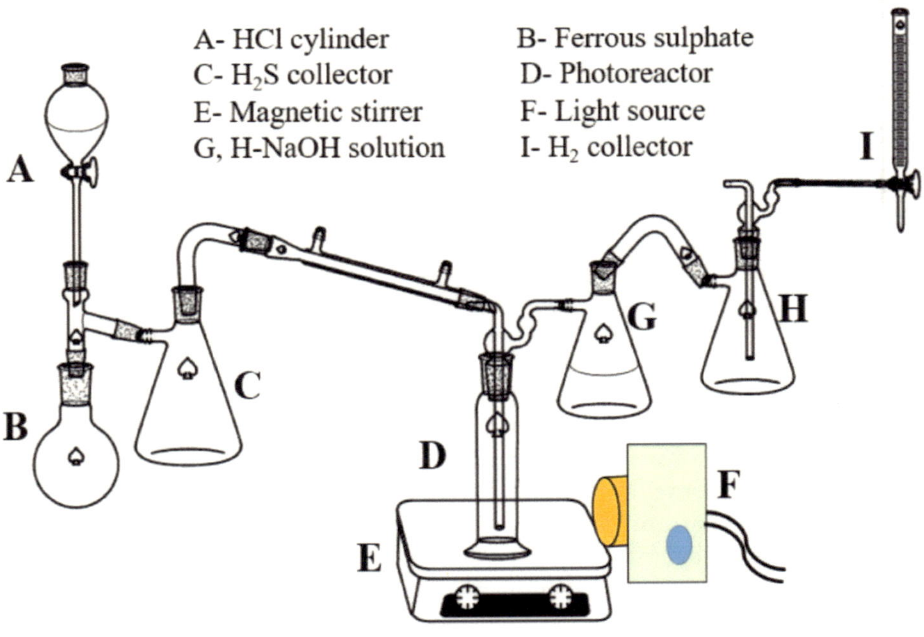

Figure 4.
Photocatalytic H_2S splitting reaction setup.

experiments for the generation of H_2 can be performed under light source or in sunlight depending upon the band gap of semiconductor photocatalyst [21, 22]. For light source (under lamp) an assembly is kept in a fuming hood to avoid bad smell and a schematic representation of a setup is shown in **Figure 4**. This setup can be divided in A to I parts as shown below.

For avoiding over heating of the photoreactor (D) throughout the photo-reaction water circulation is added, Photoreactors are of various capacity (100 to 1000 mL) can be used, the quantity of photocatalyst (0.05 to 1 gram) was decided in every photoreaction. A source of light (F) of intensity 450–600 W can be used as discussed in above water splitting set up. Hydrogen sulfide generated in (A-B) and collected extra amount in (C) is bubbled through the solution in the photoreactor under continuous stirring (E). Bubble rate differs as the capacity of photoreactor changes. The excess H_2S was trapped in NaOH solution (G-H). The amount of evolved H_2 was measured using graduated gas burette (I).

Large scale production of hydrogen using toxic H_2S is also possible like water splitting as discussed above in water splitting setup section.

5. Large scale hydrogen production and usage

Large scale production of hydrogen energy from water splitting is also possible by using this reaction setup. Appropriate photocatalyst with suitable band gap and light frequency source can leads to enhance hydrogen production. Quantity of photocatalyst and capacity of photoreactor promises the large-scale production of hydrogen energy. It can be achieved with proper engineering inside the lab as well as on roof of the lab under natural sunlight. Large scale production of hydrogen using toxic H_2S is also possible like water splitting as discussed above. Hydrogen is a clean energy which have applications in soil refining, methanol generation, steel production and ammonia production. Nowadays, it is also used for all type of transportation (alternative for Compressed Natural Gas, CNG), power generation, homes and fuel in jets and ships. Recently, the worlds first hydrogen-powered train rolled in Germany. After this many countries tested hydrogen fuel-cell passenger trains. Also, a UK-based car manufacturer company produced a two-seater hydrogen vehicle. The US, Europe and China are the top consumers of hydrogen mainly in refineries sector. Alas, all these emerging hydrogen applications are not economical than the other energy resources as hydrogen costs around USD 12–16/Kg.

6. Transportation and storage of hydrogen gas

Nowadays, the production of H_2 gas is not a new to a scientific community but the ways of transportation and storage of it is under continuous research. Transportation is mainly taking place either by road or by water routs in a cryogenic or compressed cylinders. As hydrogen is having low density the transportation is difficult and expensive so, for transporting H_2 from industry to working sites is achieved using pipelines [23]. In many cases gas is compressed and then filled in appropriate cylinders as per the quantity and requirement. Laboratories and research centers require hydrogen in small quantities and transportation of such cylinders can be takes place using small trolleys. On the other hand, the amount of hydrogen required for industrial purpose is in large quantities which are once received by transported kept in a gas bank.

Storage of hydrogen gas as a future fuel is important because one cannot produce it efficiently and in large amount at the domestic or small industry level.

For storage purpose research centers, industries and automobiles have well engineered tanks. The concept of adsorption is utilized for storage of gas, here a porous material with large surface areas like activated carbon are needed. In absorption metals are combined with hydrogen atoms to form a metal hydride. Palladium, zirconium and other transition metals are also reported for the large amount of gas absorption [24, 25].

Recently, photocatalytic solar hydrogen production from water splitting on a 100 m^2 scale panel, carbon films, closed systems, plates and sheets is demonstrated by K. Domen research group. A 100 m^2 array of panel reactors and other setup types effective for more than 90 days. Lately, chemical combination between hydrogen and boron or nitrogen is under investigation. During this combination hydrogen molecules are dissociated and forms hydrogen atoms which then goes in the empty spaces between the metals lattice structure [26].

7. Conclusions

Hydrogen produced from electrolytic, biological, steam reforming and photocatalytic methods can be utilized as a clean energy source. Still, plenty of research and development is needed to enhance the efficiency and to reduce the overall economy of the process. Though the various hydrogen production methods which are discussed here are better and less expensive, an effective method need to be developed for future energy crises. Photocatalytic hydrogen generation from water and hydrogen sulfide splitting using available setup is with opportunities to development for young researchers. Besides the transportation and storage of hydrogen gas it must be considered in a frame of the energy system globally.

Author details

Vikram Rama Uttam Pandit
The Poona Gujarati Kelwani Mandal's Haribhai V. Desai College, Pune, India

*Address all correspondence to: vikramupandit@gmail.com

References

[1] Tian B, Zheng X, Kempa T, Fang Y, Yu N, Yu G, et al. Coaxial silicon nanowires as solar cells and nanoelectronic power sources. Nature. 2007;**449**:885-889. DOI: 10.1038/nature06181

[2] Bastianoni S, Pulselli F, Tiezzi E. The problem of assigning responsibility for greenhouse gas emissions. Ecological Economics. 2004;**49**:253-257. DOI: 10.1016/j.ecolecon.2004.01.018

[3] Sagar A, Holdren J. Assessing the global energy innovation system: Some key issues. Energy Policy. 2002;**30**: 465-469. DOI: 10.1016/S0301-4215(01)00117-3

[4] Dove J. Student teacher understanding of the greenhouse effect, ozone layer depletion and acid rain. Environmental Education Research. 1996;**2**:89-100. DOI: 10.1080/1350462960020108

[5] Weatherhead E, Andersen S. The search for signs of recovery of the ozone layer. Nature. 2006;**441**:39-45. DOI: 10.1038/nature04746

[6] Papadimitriou V. Prospective primary teachers' understanding of climate change, greenhouse effect, and ozone layer depletion. Journal of Science Education &Technology. 2004;**13**:299-307. DOI: 10.1023/B:JOST.0000031268.72848.6d

[7] Schindler D. Effects of acid rain on freshwater ecosystems. Science. 1988;**239**:149-157. DOI: 10.1126/science.239.4836.149

[8] Riebesell U. Effects of CO_2 enrichment on marine phytoplankton. Journal of Oceanography. 2004;**60**:719-729. DOI: 10.1007/s10872-004-5764-z

[9] Tie S, Tan C. A review of energy sources and energy management system in electric vehicles. Renewable and Sustainable Energy Reviews. 2013;**20**:82-102. DOI: 10.1016/j.rser.2012.11.077

[10] Delmas R, Ascencio J, Legrand M. Polar ice evidence that atmospheric CO_2 20,000 yr BP was 50% of present. Nature. 1980;**284**:155-157. DOI: 10.1038/284155a0

[11] Esena M, Yuksel T. Experimental evaluation of using various renewable energy sources for heating a greenhouse. Energy and Buildings. 2013;**65**:340-351. DOI: 10.1016/j.enbuild.2013.06.018

[12] Saleska S, Harte J, Torn M. The effect of experimental ecosystem warming on CO_2 fluxes in a montane meadow. Global Change Biology. 1999;**5**:125-141. DOI: 10.1046/j.1365-2486.1999.00216.x

[13] Hussein A. Applications of nanotechnology in renewable energies-A comprehensive overview and understanding. Renewable and Sustainable Energy Reviews. 2015;**42**:460-467. DOI: 10.1016/j.rser.2014.10.027

[14] Kalamaras C, Efstathiou A. Hydrogen production technologies: Current state and future developments. Power Options for the Eastern Mediterranean Region. 2013;**2013**:1-9. DOI: 10.1155/2013/690627

[15] Kumar S, Himabindu V. Hydrogen production by PEM water electrolysis—A review. Materials Science for Energy Technologies. 2019;**2**:442-454. DOI: 10.1016/j.mset.2019.03.002

[16] Pandit V, Arbuj S, Hawaldar R, Kshirsagar P, Ambekar J, Mulik U, et al. Hierarchical CdS nanostructure by Lawesson's reagent and its enhanced photocatalytic hydrogen production.

RSC Advances. 2015;**5**:13715-13721. DOI: 10.1039/C4RA15138K

[17] Pandit V, Arbuj S, Pandit Y, Naik S, Rane S, Mulik U, et al. Solar light driven dye degradation using novel organo–inorganic (6,13-pentacenequinone/TiO_2) nanocomposite. RSC Advances. 2015;**5**:10326-10331. DOI: 10.1039/C4RA11920G

[18] Kumbhar D, Pandit V, Deshmukh S, Ambekar J, Arbuj S, Rane S. Synthesis of hierarchical zno nanostructure and its photocatalytic performance study. Journal of Nanoengineering and Nanomanufacturing. 2015;**5**:227-231. DOI: 10.1166/jnan.2015.1250

[19] Nevase K, Arbuj S, Pandit V, Ambekar J, Rane S. Synthesis, characterization and photocatalytic activity of tungsten oxide nanostructures. Journal of Nanoengineering and Nanomanufacturing. 2015;**5**:221-226. DOI: 10.1166/jnan.2015.1249

[20] Jawale V, Gugale G, Chaskar M, Pandit S, Pawar R, Suryawanshi S, et al. Two-and three-dimensional zinc oxide nanostructures and its photocatalytic dye degradation performance study. Journal of Materials Research. 2021;**36**:1573-1583. DOI: 10.1557/s43578-021-00174-w

[21] Pandit V, Arbuj S, Hawaldar R, Kshirsagar P, Mulik U, Gosavi S, et al. In situ preparation of a novel organo-inorganic 6,13-pentacenequinone–TiO_2 coupled semiconductor nanosystem: A new visible light active photocatalyst for hydrogen generation. Journal of Materials Chemistry A. 2015;**5**:4338-4344. DOI: 10.1039/C4TA06606E

[22] Pandit V, Arbuj S, Mulik U, Kale B. Novel functionality of organic 6,13-pentacenequinone as a photocatalyst for hydrogen production under solar light. Environmental Science & Technology. 2014;**48**:4178-4183. DOI: 10.1021/es405150p

[23] Mazloomi K, Gomes C. Hydrogen as an energy carrier: Prospects and challenges. Renewable and Sustainable Energy Reviews. 2012;**16**:3024-3033. DOI: 10.1016/j.rser.2012.02.028

[24] Schlapbach L, Züttel A. Hydrogen-storage materials for mobile applications. Nature. 2001;**414**:353-358. DOI: 10.1038/35104634

[25] Andersson J, Gronkvist S. Large-scale storage of hydrogen. International Journal of Hydrogen Energy. 2019;**44**:11901-11919. DOI: 10.1016/j.ijhydene.2019.03.063

[26] Nishiyama H, Yamada T, Nakabayashi M, Maehara Y, Yamaguchi M, Kuromiya Y, et al. Photocatalytic solar hydrogen production from water on a 100-m^2 scale. Nature. 2021;**598**:304-307. DOI: 10.1038/s41586-021-03907-3

Chapter 3

Quantitative of Mass Transfer in Liquid-Liquid Operations of Oil-Alcohol-Glycerin Systems

Benjamim H.L. Silva and Cesar A.M. Abreu

Abstract

The effects of mass transfer were quantified for the effective performance of mixtures between partially miscible phases, or for the promotion of their separations. To consolidate the analysis of heterogeneous liquid–liquid processes, variations in the composition of the liquid phases over the evolution of contact operations were considered, detailing the physical mechanisms involved in the mixtures of oil (soy, sunflower) and alcohol (methanol, ethanol), and in the separation between biodiesel and glycerin. Based on experimental evaluations, the average distribution coefficients for triglycerides (oil-alcohol) and glycerol (biodiesel-glycerin) were estimated at 1.31 and 1.46, and 3.42×10^{-2} and 4.06×10^{-2}, for soybean and sunflower, respectively, while their mass transfer coefficients, depending on their concentration ranges in the phase, varied in orders of magnitude from 10^{-2} s^{-1} to 10^{-4} s^{-1}. Including the values of the physical parameters, a heterogeneous model for the alkaline transesterification of soybean oil (methanol, ethanol, NaOH, 25°C, 40°C, 60°C, 600 rpm) was validated.

Keywords: Mass transfer, liquid–liquid, vegetable oil, biodiesel, glycerin

1. Introduction

The effects of mass transfer are highlighted in heterogeneous processes involving mixing, reaction and separation, affecting operations that seek to evaluate intimate contacts between two or more phases [1, 2].

In operations involving slightly miscible liquids, interactions occur between the phases through their contact interfaces, where the migration of components subjected to mass transfers with a predominance of diffusive effects occurs. The knowledge of these effects is associated with the dynamics of fluids in operations, highlighting actions for the purposes of mixing and separation.

In the last few decades, the search for independence from fossil fuels has led to significant developments in the field of renewable energies. Among them, biodiesel or green diesel has positioned itself prominently and its production is growing. In order to have this fuel more available, several processes converge to obtain it, among which the most widespread is the transesterification of vegetables oils.

From an operational point of view, this process has been improved and developed according to the needs of production and quality characteristics for energy purposes. Thus, in-depth knowledge of the phenomena involved in its operations involves detailing the interactions between the different partially miscible or immiscible liquid phases present, including oil, alcohol, biodiesel and glycerin. The different approaches to assess liquid–liquid interactions in the production of biodiesel identified as a heterogeneous process have highlighted the contributions of mass transfer, although associated with reaction effects [3–6]. However, this type an approach in which the mass transfer is not considered independently as a relevant step, prevents accurate descriptions of the effects that directly influence the performance of the process. By highlighting in detail the effects of mass transfer, a description of the supply of reagents (triglycerides) and release of the main co-product (glycerol) in the separation of biodiesel is provided [7].

The heterogeneous system formed in the transesterification operation occurs with the contact between the partially miscible polar and nonpolar phases: alcohol and triglycerides, and in its evolution includes the formation of the glycerin phase.

In batch operations, the effects of the mixture involve the dispersion of alcohol in the oil in the form of droplets, whose interfaces allow the dissolution of the triglycerides that are transferred internally [8–10]. On the other hand, the separation of glycerol involving the transfer of oil to glycerin is a determining step for obtaining high-purity biodiesel [11–13].

A more accurate knowledge of all the effects on the global kinetics of biodiesel production involving reaction and mass transfer can provide a reduction in operating time, decreasing them by intensifying the initial interactions between the phases and with the reactions, and in the separation of the final stages, where less interaction should take place.

Different investigations on the effects related to mass transfers between liquid phases in alkaline transesterification processes have indicated orders of magnitude of parameters representing these phenomena, although always combined with reaction effects, and for a fixed composition of each phase [14–16]. During biodiesel production, variations in composition result from the reactive process in which concentrations continue to evolve [5, 17].

Reflecting these effects, orders of magnitude of mass transfer parameters can be quantified according to the compositions of the interacting oil and alcohol phases and relative to the separation of glycerol from the mixture of esters [18–20].

The present work aims to consolidate the analysis of heterogeneous liquid–liquid processes using a detailed approach to quantify the interactions involved in the contacts between partially miscible phases. In this context, changes in the composition of the liquid phases throughout the evolution of operations were considered. Detailed knowledge of the physical mechanisms of the oil-alcohol and biodiesel-glycerol interaction processes was applied to describe the evolution of experimental concentrations. The mass transfer coefficients and the distribution coefficients of triglycerides (oil-alcohol) and glycerol (biodiesel-glycerin) were estimated based on experimental evaluations. These quantifications associated with the transesterification rates, allowed the orientation towards the best operating conditions and product quality in the production of biodiesel.

2. Evaluation of mass transfer and equilibrium between phases

Experiments carried out under non-reactive conditions provided measurements of triglyceride content in alcohol and residual glycerol in biodiesel. The values obtained in time evolution provided results described by a heterogeneous model. The adjustment of the model equations led to the quantification of the mass transfer and distribution equilibrium parameters.

2.1 Experimental

In assessing the effects of mass transfer and dissolution between the partially miscible phases of vegetable oil (O) and alcohol (A), and biodiesel (B) and glycerol (G), the mixtures and phase separations were carried out in an isothermal manner (60° C, 600 rpm). The oil-alcohol mixture (molar ratio A:O/6:1) was maintained for 50 min, when samples were collected every 2 min. Mixtures of glycerol and biodiesel (volumetric ratios G:B/ 0.50:0.50 - /0.40: 0.30) were maintained for 30 minutes, after which they were transferred to graduated hoppers where the phase separations occurred, while the corresponding times were counted. The levels of triglycerides accumulated in alcohol and residual glycerol in biodiesel were quantified by the corresponding chromatographic analytical methodologies (analytical method Hartman and Lago [21]; HPLC). The interfacial areas of contact between the phases were estimated based on the scattered droplet sizes measured by photographic technique (Canon EOSD60 camera, 28-300 mm lens, ImageJ® software). The specific interfacial area a ($a_{OL/A} = 6\varphi/d_{32}$) was calculated with 40–50 drops, diameter di, where d_{32} is the average Sauter diameter ($d_{32} = \Sigma d_i^2(\Sigma d_i^2)^{-1}$; i = 1, N) and φ is the volume of the dispersed phase.

2.1.1 Analyzes

Samples of 5.0×10^{-6} m^{-3} were taken from the reactor at times 2, 4, 6, 8, 10, 15, 20, 25, 30, 35, 40 and 50 minutes, neutralized with 2.0×10^{-6} m^3 aqueous hydrochloric acid (HCl, 15.0 mol m^{-3}), washed with 10.0×10^{-6} m^3 of distilled water saturated with sodium chloride and left in decanting funnels until complete phase separation (biodiesel, glycerin). The samples from the biodiesel phase were diluted in hexane and analyzed by gas chromatography (CG Master-Peak Simple II, SRI Instruments; Carbowax 20 M megabore column, 30 m, 0.54×10^{-3} m, FID detector), using methyl heptadecanoate (Sigma-Aldrich, purity ≥99.0%) as an internal standard. In the glycerin phase, glycerol analyzes were performed by high performance liquid chromatography (HPLC, Gilson chromatography, RI detector, Prevail Carbohydrate ES column, 0.25 μm, 0.25 mx 0.42×10^{-2} m) using an external standardization and the mobile phase acetonitrile / water (75:25 with flow of 1.0×10^{-6} m3 / min) at 25° C.

3. Results and discussion

3.1 Experimental results

The interactions between the liquid phases oil and alcohol, and biodiesel and glycerol were evaluated through measurements carried out in the respective phases (TG/alcohol, G/biodiesel) as a function of the time of operation. In relation to the two oils (soy, sunflower) used in the experiments, **Table 1** shows the composition (% by weight) in terms of its main triglycerides (TG).

For both oils, the unsaturated components (C18: 1, 18: 2) represented more than 80% of the composition, according to data already established [22].

Figure 1 shows the evolution of triglyceride concentrations (TG) in the alcohol phase (A) in each system (oil-alcohol) investigated, indicating the contents transferred over time.

It was observed that the transfer of triglycerides (TG) was faster in the soy / ethanol system, throughout the time domain, and that this also occurred in the sunflower/ethanol system, but only after 25 minutes of operation. The higher levels

Component	Soybean (% wt.)	Sunflower (% wt.)
C16:0	12.96	6.54
C18:0	4.23	3.23
C18:1	26.39	41.76
C18:2	50.41	48.09
C18:3	6.01	0.38

Table 1.
Composition of vegetable oils (% wt.).

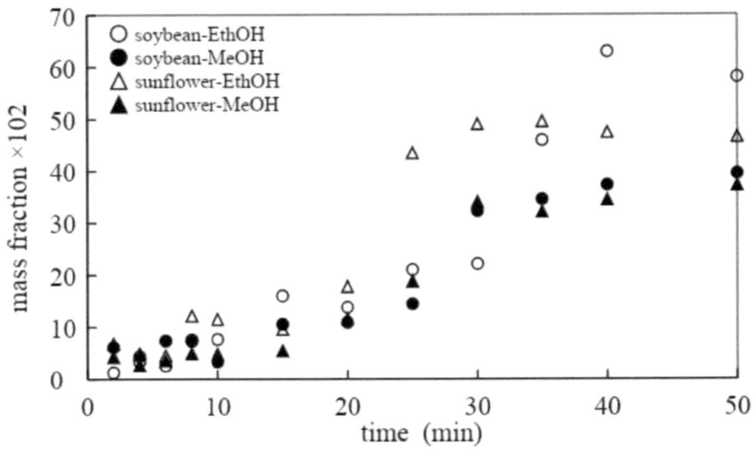

Figure 1.
Evolution of triglyceride concentrations (% wt.). Conditions: A:O/6:1, 60°C, 600 rpm.

obtained with ethanol at the end of the operation indicated that the distribution of TG between the oil and the alcoholic phase was more favorable for this alcohol. In fact, the TG of the two vegetable oils is more soluble in ethanol, of less polarity (Dielectric ct. 24.5) than in methanol (Dielectric ct.32.7).

3.2 Mass transfer and miscibility

Considering the phenomena involved in the mixing and miscibility of the oil in the alcohol, alcohol droplets are formed and dispersed in the oil, establishing interfaces through which the triglycerides are transferred, so that the solubilization and its accumulation in the phase occurs. When the biodiesel and glycerol phases are separated, the glycerol is transferred and released, decreasing its content in biodiesel and providing its decantation. To quantify the effects of the observed, the mass balance equations Eqs. (1) and (2) were formulated and applied to the related processes.

$$\frac{dC_{TG/A}}{dt} = K_{TG}\left(C_{TG/AI} - C_{TG/A}\right) \quad (1)$$

$$\frac{dC_G}{dt} = K_G(C_{GB} - C_G) \quad (2)$$

where $C_{TG/A}$, $C_{TG/Ao}$, C_G, C_{Go} are the concenrations of triglycerides and glycerol, respectively. In both cases, at the beginning of the contact between the phases,

$t = 0$, $C_{TG} = C_{TG/Ao}$, $C_G = C_{G0}$. K_{TG} *and* K_G *are* the respective mass transfer coefficients.

The solutions in Eqs. (1) and (2) (Runge Kutta 4th order; *software* gPROMS ModelBuilder®) describe the evolutionary behavior of the concentrations of triglycerides and glycerol in the alcohol and biodiesel phases, respectively. To do so, they must include the values of the volumetric mass transfer coefficients K_{TG} and K_G.

3.3 Predictions versus experimental

The evolution of the concentrations of triglycerides (TG) measured in each oil phase served to adjust the solution of Eq. (1) allowing to quantify $K_{TG/A}$ and α_{TG}. The values of specific interfacial area ($a_{TG/A}$) measured are included in the mass transfer coefficient. **Figure 2** shows for TG concentrations the parity between the predicted and experimental values, with indication of weak linear concordances (correlation coefficients, $R^2 \sim [0.82–0.91]$).

Although the parameter values were obtained from weak numerical adjustments, they were considered as initial orders of magnitude for the elaboration of more precise adjustments. The initial values of the mass transfer coefficients ($k_{TG/A}$) and the distribution parameters between the phases ($_{TG}$) and the specific interfacial areas measured are listed in **Table 2**.

Changes in the characteristics of the alcoholic phase (composition, density, viscosity) are considered during the supply of triglycerides via mixing. Thus, a

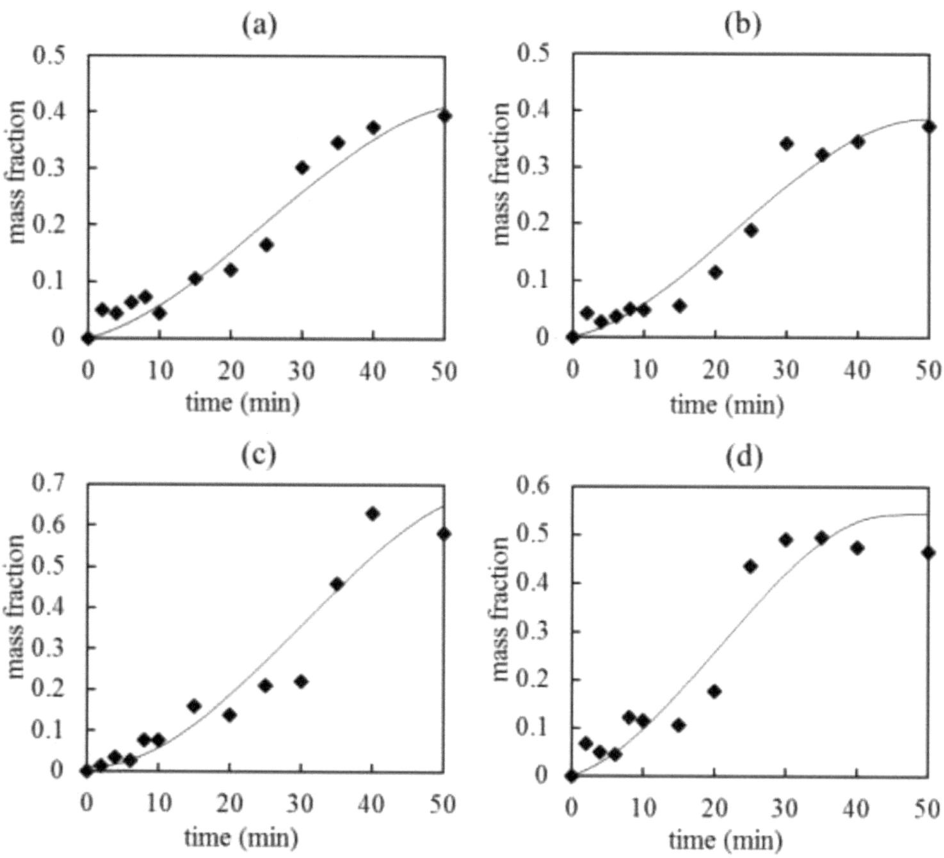

Figure 2.
Predicted versus experimental values of triglycerides mass fractions. Conditions: A:OL/6:1, C_{NaOH} = 0.50% wt., 600 rpm. a) Soybean/methanol, b) sunflower/methanol, c) soybean/ethanol, d) sunflower/ethanol.

Parameter	Soy/MeOH	Sunflower/MeOH	Soy/EthOH	Sunflower/EthOH
k_{TG} (ms^{-1})	3.60×10^{-7}	3.39×10^{-7}	5.22×10^{-7}	6.87×10^{-7}
α_{TG}	2.64	2.88	1.37	2.18
a (mm^{-1})	1.87	1.79	0.57	0.81

Table 2.
Mass transfer and equilibrium parameters of triglycerides. Oil/alcohol systems. Conditions: A:O = 6:1, 60°C, 600 rpm.

more realistic approach was applied to improve the predictions, admitting the effect of the concentration of triglycerides, focusing on the mass transfer coefficient expressed through the Darken equation (Eq. (3); [23, 24]).

$$K_{TG/A} = a_{OL/A}\lambda D_{TG/A}y_{TG/A} \tag{3}$$

where, $\lambda = \left(1 + y_{TG/A}d\ln\gamma_{TG}/dy_{TG/A}\right)$ is a non-ideality factor, and $D_{TG/ROH}$ is estimated by the Wilke-Chang correlation ([24], $D_{TG/A} = 7.4 \times 10^{-8}T[\psi M_A]^{1/2}$ $(\mu_A^{-1}V_{TG}^{-6})$). The parameters and properties introduced in the formulations are listed in **Tables 3** and **4**.

The values of the volumetric coefficients of mass transfer of triglycerides in the alcohol phases, as a function of the different concentrations are in **Table 5**.

3.3.1 Mass transfer of glycerol

For glycerol, the effects of mass transfer were evaluated by mixing it with the biodiesel phase. Then, the evolution of the residual content in biodiesel was monitored (**Figure 3**). **Figure 4** shows the evolution of the experimental and calculated concentrations by the solution of Eq. (2). From the biodiesel-glycerol mixtures, decreasing evolutions of the glycerol concentration were observed as the phase separation occurred. At the end of the experiments, no residual glycerol was detected in the biodiesel phase (measured by HPLC, droplet images), indicating that the phase separation was effective. Observing the behavior of the biodiesel

Parameter	Sobeany/MeOH	Sunflower/MeOH	Soybean/EthOH	Sunflower/EthOH
V_{Alcol} $(cm^3\ mol^{-1})$	40.45	40.45	58.43	57.16
M_B $(g\ mol^{-1})$	871.1	877.6	871.1	877.6
μ_B (cP)	16.7	16.4	16.7	16.4
α_{TG}	2.64	2.88	1.37	2.18
a (mm^{-1})	1.87	1.79	0.57	0.81

Table 3.
Parameters and properties [24, 25].

Parameter	Sobeany/MeOH	Sunflower/MeOH	Soybean/EthOH	Sunflower/EthOH
$D^0{}_{A/TG}$ $(m^2 s^{-1})$	6.52×10^{-10}	6.67×10^{-10}	4.65×10^{-10}	5.35×10^{-10}
λ $(m)^{-1}$	2.09×10^4	2.56×10^4	6.13×10^4	5.62×10^4

Table 4.
Estimated model parameters.

	Sobeany/MeOH	**Sunflower/MeOH**	**Soybean/EthOH**	**Sunflower/EthOH**
$C_{TG/A}$	$K_{TG/A}$	$K_{TG/A}$	$K_{TG/A}$	$K_{TG/A}$
(% *wt.*)	$(\times 10^4\, s^{-1})$	$(\times 10^4\, s^{-1})$	$(\times 10^4\, s^{-1})$	$(\times 10^4\, s^{-1})$
0.01	2.55	3.06	1.62	2.44
0.025	6.39	7.64	4.06	6.09
0.05	12.8	15.0	8.12	12.2
0.1	25.5	30.6	16.2	24.4
0.2	51.1	61.1	32.5	48.7
0.3	76.6	91.7	48.7	73.1
0.4	102	122	65.0	97.4
0.5	128	153	81.2	122
0.6	153	183	97.4	146
0.7	179	214	114	170

Table 5.
Volumetric coefficients of mass transfer triglycerides-alcohol. Concentration effect. Conditions: A:O = 6:1, 60°C, 600 rpm.

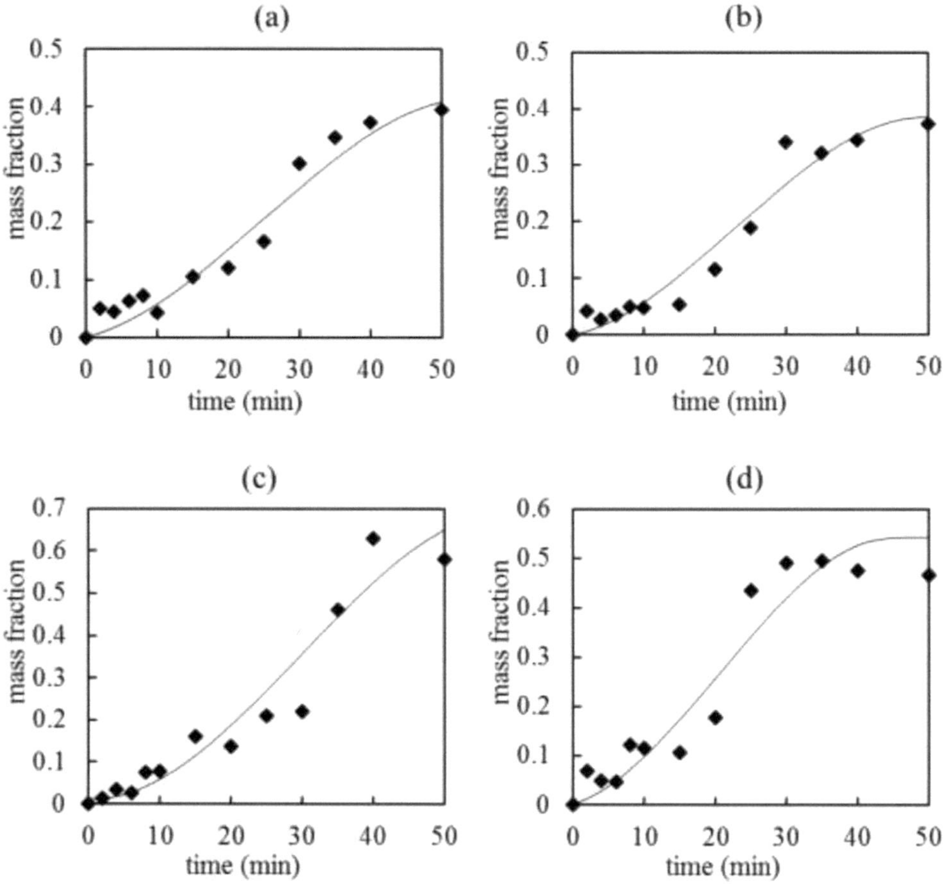

Figure 3.
Evolution of triglyceride mass fractions in the alcohol phase. Conditions: A:OL/6:1, C_{NaOH} = 0.50% wt., 600 rpm. a) Soybean/methanol, b) sunflower/methanol, c) soybean/ethanol, d) sunflower/ethanol.

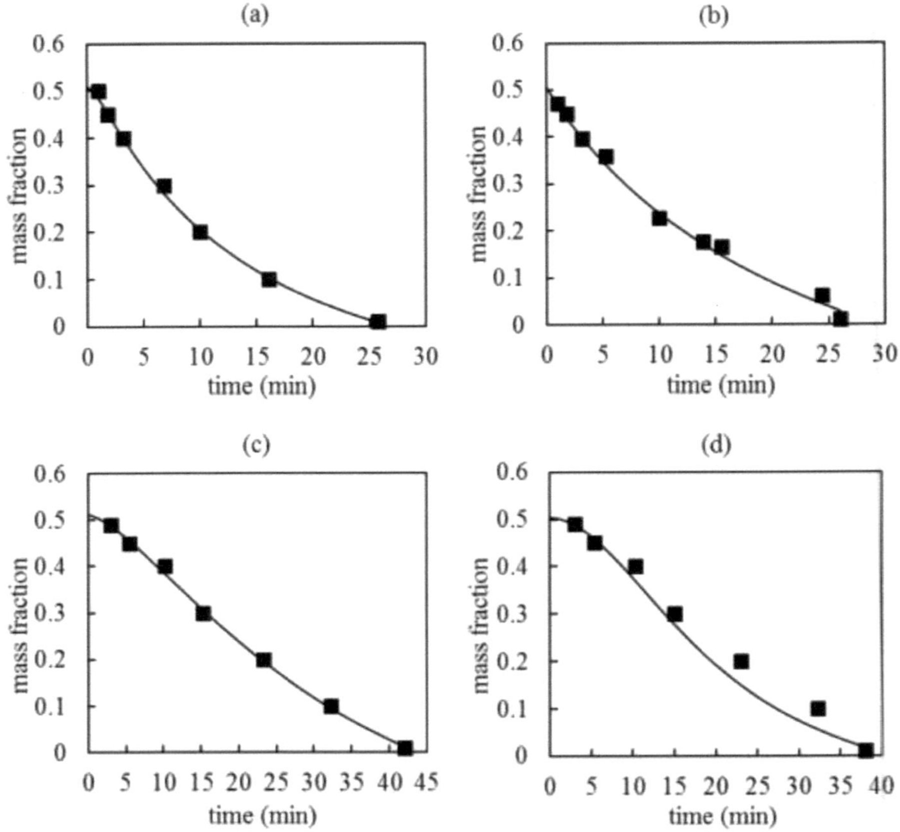

Figure 4.
Evolution of glycerol concentrations in the biodiesel phase. Conditions: R a:OL/6:1, C_{NaOH} = 0.50% wt., 600 rpm. a) Soy/methanol; b) sunflower/methanol; c) soy/ethanol; e d) sunflower/ethanol.

(ethyl)-glycerol mixture, it was found that the phase separation occurred more slowly than in the case of biodiesel (methyl)-glycerol.

The mass transfer coefficients of glycerol from biodiesel to glycerine (K_G) denote the transfer and also the speed with which the separation of the two phases (glycerine-biodiesel) occurred, where mean K_G values were 1.49×10^{-3} s^{-1} (soybean/methanol), 1.20×10^{-3} s^{-1} (sunflower/methanol), and 6.75×10^{-3} s^{-1} (soybean/methanol), 7.42×10^{-3} s^{-1} (sunflower/methanol), respectively. The evolutions represented for triglycerides and glycerol according to the model showed good levels of compatibility (R^2 = 0.9345–0.9806 linear parity, experimental vs. calculated; analysis of variance ANOVA F (95%) = 5.14; Snedecorv F: 5%).

In the context of kinetic effects for triglycerides and glycerol, orders of magnitude of mass transfer parameters (10^{-4} - 10^{-3} s^{-1}, [16, 18, 19]) were in the range of results obtained.

4. Application to the reactive process

From the point of view of using the quantified parameters via assessments of the effects of mass transfer in the mixture and in the separation of the phases involved in the biodiesel production processes, it is taken into account that its orders of magnitude are comparable to those obtained in previous studies [17, 26]. For purposes of application to the reactive process, the values of the mass transfer parameters, obtained in the physical evaluations, were used to simulate the alkaline transesterification kinetics of soybean oil (methanol, ethanol, NaOH). Thus, a heterogeneous model was formulated in terms of the mass balance equation as follows:

$$\frac{dC_{TG}}{dt} = K_{TG}(C_{TG0} - C_{TG}) - r_B \quad (4)$$

$$\frac{dC_G}{dt} = -K_G(C_{G0} - C_G) + r_B \quad (5)$$

where the biodiesel production rate (r_B) was introduced according the formulation by Silva et al. [5]:

$$r_B = \frac{dC_B}{dt} = k[(C_{TG0} + C_{A0}) - (C_G + C_B)] \quad (6)$$

As initial condition, at the start of the reaction operation, t = 0, $C_{TG} = C_{TG0}, C_A = C_{A0}, C_{B0} = C_{Go} = 0$. Using the values of the reaction rate constants estimated from methanolysis and ethanolysis operations (**Table 6**), conducted under kinetic-chemical regime [5], simulations of kinetic behaviors were performed at temperatures of 25°C, 40°C and 60°C.

Figure 5 shows the evolution of the concentrations of soybean alkyl esters, constituting the production of methyl and ethyl biodiesels. Their behaviors, already reported in different evaluations [14], show increases from the beginning of each operation to the levels reached in about 35 to 40 minutes of operation.

Temperature (°C)	Methanolise $k_{MeOH} \times 10^3$ ($m^3mol^{-1}\ min^{-1}$)	Ethanolise $k_{MeOH} \times 10^3$ ($m^3mol^{-1}\ min^{-1}$)
25	2.08	2.31
40	2.28	3.79
60	3.33	4.74

Table 6.
Reaction rate for alkaline transesterification of soybean (k). Conditions: A:OL/6:1, C_{NaOH} = 0.50% wt., 600 rpm [5].

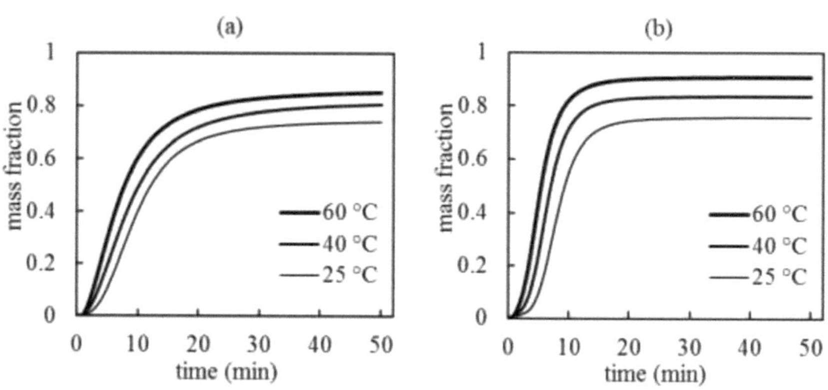

Figure 5.
Evolution of concentrations of biodiesel (alkyl esters) derived from soybean oil. Temperature effect. Conditions: A:OL/6:1, C_{NaOH} = 0.50% wt., 600 rpm. (a) Biodiesel (methyl esters) (b) biodiesel (ethyl esters).

5. Conclusions

An approach focused on the triglyceride mass transfer coefficients, based on their concentration in alcohol during the oil-alcohol mixing process, led to quantitative assessments, which are associated with those obtained for the glycerol mass transfer in biodiesel separation constituted the phenomenological base used in the predictions of behavior of the reactive processes of transesterification of vegetable oil.

A heterogeneous model was validated for the alkaline transesterification of soybean oil (methanol, ethanol, NaOH), including values of the physical parameters (distribution and mass transfer coefficients) determined experimentally, varying in the intervals [1.01–1.62] and [2.05–4.78] $\times 10^{-2}$, respectively, while the mass transfer coefficients were in the order of magnitude in the range 10^{-2} to 10^{-4} s^{-1} for triglycerides and [1.20–7.42] $\times 10^{-3}$ s^{-1} for glycerol. The kinetic behavior of biodiesel production was simulated using the specific reaction rates for each biodiesel produced by methanolysis and ethanolysis in the range of 25° C to 60° C.

Acknowledgements

Thanks to CNPq-National Council of Science and Technology of Brazil for the financial and logistical support dedicated to the execution of this research.

Author details

Benjamim H.L. Silva and Cesar A.M. Abreu*
Department of Chemical Engineering, Federal University of Pernambuco, Recife, Brazil

*Address all correspondence to: cesar@ufpe.br

References

[1] R.P. Chhabra and J.F. Richardson. Liquid mixing, Chapter 8 in Non-Newtonian Flow and Applied Rheology, Engineering Applications. ISBN 978-0-7506-8532-0, https://doi.org/10.1016/B978-0-7506-8532-0.X0001-7, Butterworth-Heinemann, Elsevier Ltda., 2008.

[2] Zeki Berk, Heat and Mass Transfer, Basic Principles, Chapter 3 in Food Process Engineering and Technology, ISBN 978-0-12-373660-4, https://doi.org/10.1016/B978-0-12-373660-4.00003-X, Academic Press, 2009

[3] Chanakaewsomboon, I., Tongurai, C., Photaworn, S. et al. Miscibility and mass transfer in biodiesel production observed by LCD digital microscope. Biomass Conv. Bioref. (2020). https://doi.org/10.1007/s13399-020-00728-8

[4] Thoai DN, Chanakaewsomboon I, Prasertsit K et al (2019) A novel inspection of mechanisms in conversion of refined palm oil to biodiesel with alkaline catalyst. Fuel 256:115831. h ttps://doi.org/10.1016/j.fuel.2019.115831

[5] Silva, S. P.; Sales, D. C. S.; Abreu, C. A. P.; Schuler, A. R. P.; Abreu, C. A. M., Kinetics of the biphasic liquid–liquid transesterification of vegetable oils into biodiesel. Reaction Kinetics, Mechanisms and Catalysis, V. 123, P. 529–542, 2018.

[6] Yusuff, A.S.; Ekwonu, M.C.; Ajala, I. I., A. A. Adeyi; Awofolaju, T. T.; Modelling And Simulation Of Transesterification Reaction In A Batch Reactor. Journal Of Bioprocessing And Chemical Engineering. 2014 V1i1. Doi: 10.15297/Jbce.V1i1.02.

[7] Atadashi, I. M.; Aroua, M. K.; Aziz A. A. Biodiesel Separation and Purification: A Review. Renewable Energy, V. 36, P. 437-443, 2011.

[8] Atapour, M.; Kariminia, H.; Moslehabadi, P. M. Optimization of biodiesel production by alkali-catalyzed transesterification of used frying oil. Process Safety and Environmental Protection, v. 92, p. 179–185, 2014.

[9] Berrios, M.; Skelton, R. L. Comparison Of Purification Methods For Biodiesel. Chemical Engineering Journal, V. 144, P. 459–465, 2008.

[10] Doell, R.; Konar, S. K.; Boocock, D. G. B. Kinetic Parameters of a Homogeneous Transmethylation of Soybean Oil. Journal of the American Oil Chemists' Society, v. 85, p. 271–276, 2008.

[11] Pinheiro, R. S.; Bessa, A. M. M.; Queiroz, B. A.; Duarte, A. M. S. F.; Sant'ana, H. B.; Santiago-Aguiar, R. S. Optimization of the methylic biodiesel purification process by intermediate of liquid–liquid equilibrium data for ternary systems containing methanol + water + (soybean, corn or brown shell of coconut) biodiesel. Fluid Phase Equilibria, v. 361, p. 30-36, 2014.

[12] Thoai DN, Chanakaewsomboon I, Prasertsit K et al (2019) A novel inspection of mechanisms in conversion of refined palm oil to biodiesel with alkaline catalyst. Fuel 256:115831. https://doi.org/10.1016/j.fuel.2019.115831

[13] Devjyoti Nath, Adisorn Aroonwilas and Amornvadee Veawab (January 25th 2017). Role of Mass-Transfer Interfacial Area in the Biodiesel Production Performance of Acid-Catalyzed Esterification, Frontiers in Bioenergy and Biofuels, Eduardo Jacob-Lopes and Leila Queiroz Zepka, IntechOpen, DOI: 10.5772/65657. Available from: https://www.intechopen.com/books/frontiers-in-bioenergy-and-biofuels/role-of-mass-transfer-interfacial-area-in-the-biodiesel-production-performance-of-acid-catalyzed-est.

[14] Thoai, D. N.; Tongurai, C.; Prasertsit, K.; Kumar, A. Review on Biodiesel Production by Two-Step Catalytic Conversion. Biocatalysis And Agricultural Biotechnology, V. 18, P. 10-23, 2019.

[15] Klofutar, B.; Golob, J.; Likozar, B; Klofutar, C.; Zˇagar, E.; Poljanšek, I. The transesterification of rapeseed and waste sunflower oils: Mass-transfer and kinetics in a laboratory batch reactor and in an industrial-scale reactor/separator setup. Bioresource Technology 101, 3333–3334, 2010.

[16] Frascari, D.; Zuccaro, M.; Paglianti, A.; Pinelli, D. Optimization of Mechanical Agitation and Evaluation of the Mass-Transfer Resistance in the Oil Transesterification Reaction for Biodiesel Production, Ind. Eng. Chem. Res. 2009, 48, 7540–7549

[17] Vineet K.Aniya, Radha K. Muktham, K. Alma, B. Satyavathi Modeling and simulation of batch kinetics of non-edible karanja oil for biodiesel production: A mass transfer study. Fuel 161, 137-145, 2015.

[18] Bhoi, R.; Singh, D.; Mahajani, S. Investigation of mass transfer limitations in simultaneous esterification and transesterification of triglycerides using a heterogeneous catalyst Reaction Chemistry & Engineering. The Royal Society of Chemistry 2017 DOI: 10.1039/c6re00218h.

[19] Castro Posada, J. H., & Ruíz Colorado, A. A. (2019). Kinetic model for FAME production using inmobilized Lipase in a solvent- free system. CT&F - Ciencia, Tecnología Y Futuro, 9(2), 99-108. https://doi.org/10.29047/01225383.129

[20] Lôbo, I. P.; Ferreira, S. L. C. Biodiesel: Parámetros de Qualidade e Métodos Analíticos, Química Nova, 32, No. 6, 2009, pp. 1596-1608. doi:10.1590/S0100-40422009000600Ο.

[21] Hartman, l.; Lago, B. C. A. Rapid preparation of fatty acids methyl esters from lipids. Laboratory Practice, v. 22, p. 475-477, 1973.

[22] Rinaldi, R.; Garcia, C.; Marciniuk, L. L.; Rossi, A. V.; Schuchardt, U.; Síntese de biodiesel: Uma proposta contextualizada de experiência para laboratório de química geral. Química Nova, v. 30, p. 1374-1380, 2007.

[23] Okino, T., Ending of Darken Equation and Intrinsic Diffusion Concept. Journal of Modern Physics, 2013, 4, 1495-1498. http://dx.doi.org/10.4236/jmp.2013.411180.

[24] Wilke, C. R., Chang, P., Correlation of Diffusion Coefficients in Dilute Solutions. AICHE J., V. 1, P. 264-270, 1655

[25] Perry, Robert H., Green, Don W. Perry's Chemical Engineers' Handbook 7th edition (1997). ISBN-13: 978-0071834087.

[26] Poljanšek, I; Likozar, B. (2011). Influence of Mass Transfer and Kinetics on Biodiesel Production Process, Mass Transfer in Multiphase Systems and its Applications, Prof. Mohamed El-Amin (Ed.), ISBN: 978-953-307-215-9, InTech, Available from: http://www.intechopen.com/books/mass-transfer-in-multiphase systems-and-its-applications/influence-of-mass-transfer-and-kinetics-on-biodiesel-production-process

Chapter 4

Exploitation of Excess Low-Temperature Heat Sources from Cogeneration Gas Engines

Darko Goričanec and Danijela Urbancl

Abstract

The chapter presents an innovative technical solution for the use of low-temperature excess heat from the combined heat and power (CHP) of gas engines using gas or liquid fuel for district heating, building heating or industry. The primary fuel efficiency of CHP gas engines for heat production can be significantly increased by using the low-temperature excess heat of the exhaust gasses and the cooling system of the CHP gas engine, which are released into the environment thereby also reducing CO_2 emissions. District heating hot water systems generally work with higher temperatures of the heating water, which is transported to the heat consumer via the supply line, and the cooled heating water is returned to the CHP gas engine via the return line. In order to make use of the excess low-temperature heat of the exhaust gasses and the cooling system of the CHP gas engine, a condenser must be installed in the exhaust pipe in which the water vapor contained in the exhaust gasses condenses and a mixture of water and glycol is heated, which later leads to the evaporator of the high-temperature heat pump (HTHP). The cooled heating water is returned from the heat consumer via the district heating return pipe to a condenser of one or more HTHPs connected in series, where it is reheated and then sent to a CHP gas engine, where it is reheated to the final temperature. The Aspen plus software package is used to run a computer simulation of one or more HTHPs connected in series and parallel to the district heating system and to demonstrate the economics of using the excess heat from the exhaust gasses and the cooling system of the CHP gas engine.

Keywords: Rational use of energy, Low-temperature energy sources, CHP gas engine, High-temperature heat pump, District heating, Economic analysis

1. Introduction

More efficient energy consumption, and thus reduced consumption of non-renewable energy sources, can significantly reduce energy costs, mitigate climate change, improve the quality of life and reduce the EU's dependence on imported oil and gas. To achieve these goals, energy efficiency must be improved throughout the energy chain, from production to final consumption. EU action therefore focuses on sectors where savings can be greatest, such as energy consumption for heating and cooling buildings. In 2007, the EU set three main targets: a 20% reduction in greenhouse gas emissions (compared to 1990 levels), 20% of energy consumption from renewable sources in the EU and a 20% improvement in energy efficiency.

The 20% energy efficiency target was adopted with the adoption of Energy Efficiency Directive 2012/27/EU in 2012 [1].

The development of energy consumption since 2014 shows that the EU's energy consumption targets for 2020 have not been met. The crisis of COVID has severely affected the economy, reducing energy consumption in 2020. However, unless the European economy becomes more energy efficient, the subsequent economic recovery will lead to a resurgence of energy consumption. EU Member States have set up a working group to discuss with stakeholders the reasons for the increase in energy consumption in 2014 and 2017 and possible measures to address the problem.

The new edition of the Energy Efficiency Directive (EU) 2018/2002 entered into force in December 2018. The directive contains several new elements and some updates to previous directives. The EU's main goal is to achieve 32.5% energy efficiency by 2030 (compared to forecasts of expected energy consumption in 2030), with a clause on a possible upgrade by 2023. In accordance with the Energy Union and Climate Action Regulation (EU) 2018/1999, each Member State must draw up an integrated national energy and climate change plan (NECP) for the period 2021–2030, covering a 10-year period from 2021 to 2030 and describing how it intends to contribute to the objectives for 2030 in terms of energy efficiency, use of renewable energy sources and greenhouse gas emissions [1].

Moreover, increasing the efficiency of new or existing heat generation plants is one of the priorities in line with the EU's commitments to reduce GHG (greenhouse gases) emissions and achieve several environmental goals [1].

With the development of techniques and the growing demand for energy, more emphasis is now being placed on the use of renewable energy sources, in line with EU directives and the adoption of new legislation. The focus is on the rational use of energy and energy self-sufficiency of commercial and public buildings with all types of energy (electricity, heating and cooling).

There are only a few studies dealing with the coupling of heat pumps and CHP (combined heat and power) engines. An experimental study was conducted to increase the heating capacity of an electric heat pump using heat recovered from the generator of a gas engine [2]. Mancarella presented an approach for energy and CO2 emission modeling of CHP systems coupled with electric heat pumps [3], Blarke and Dotzauer [4] developed a novel CHP concept with a compression heat pump and cold storage using exhaust heat. Similar concepts were presented in [5], where Blarke compared an electric boiler and heat pumps with respect to decentralized CHP in West Denmark. Capunder et al. [6] presented an optimization model to evaluate the techno-economic and environmental characteristics of different multi-generation options.

The objectives for the use of surplus low temperature energy sources in CHP gas engines are:

- to implement a new l solution for the use of surplus low-temperature heat sources generated by the operation of CHP gas engines,
- to enable the smooth operation of new or existing CHP gas engines in heat generation for district and high temperature heating systems,
- to generate a greater amount of usable heat with the same consumption of primary fuel for the needs of high temperature heating,
- to maximize the overall fuel efficiency of CHP gas engines,
- selecting the technologically and economically optimal number of high-temperature heat pumps.

The purpose of using redundant low- temperature sources of CHP gas engines is:

- the need for a more efficient use of energy in heat production for district or high temperature heating systems, as today's energy efficiency is mainly limited to the insulation of pipes in district heating and the efficient use of energy by the end user,

- the need to use the excess low-temperature heat from CHP plants to increase the efficiency of the primary fuel,

- surplus low-temperature energy sources from CHP gas engines are used by high-temperature heat pumps to heat the return pipe water,

- the need to reduce CO_2 emissions to the environment while making major economic savings,

- the requirement to comply with EU recommendations and requirements concerning energy efficiency and environmental protection.

2. Operating the conventional CHP device

The CHP plant enables the simultaneous production of heat and electricity within one unit. The unit converts the chemical energy of the fuel by means of a steam turbine, gas turbine or internal combustion engine into mechanical energy, which is driven by a generator via a shaft, which converts the invested mechanical energy into electrical energy. A byproduct of this process is also the useful heat for high-temperature heating and the low-temperature waste heat, which is released into the environment via the cooling system and the exhaust system [6, 7].

The need for operation of the CHP plant is influenced by the season, the outdoor temperature and the specific needs of the end users. The mode of operation changes accordingly and adapts to the current demand for useful heat and electricity. In relation to the amount of usable heat and electricity, a proportional share of excess low-temperature heat is also produced.

3. Exploitation of excess low-temperature heat sources from CHP gas engines

When operating CHP gas engines, many low-temperature heat sources are generated, which are released into the environment via the cooling and exhaust system of the gas engine. These low- temperature heat sources are too low to be used directly to heat the return water of the district heating system. To be able to use this low-temperature heat source of the CHP gas engine, it is necessary to raise the temperature to the required temperature level using a high-temperature heat pump or several high-temperature heat pumps connected in series.

The principle of using surplus low-temperature heat sources of the CHP gas engine with built-in high-temperature heat pump is shown in **Figure 1**, where they are shown and marked: drive unit for combined heat and power (CHP), gas engine (ICE) with internal combustion, generator (G) for electricity generation, heat exchangers (HE1 - HE3), high-temperature heat pump (HP), piping system (P1 - P19) for heat distribution, heat consumer (HC), valves (V1 - V3), dampers

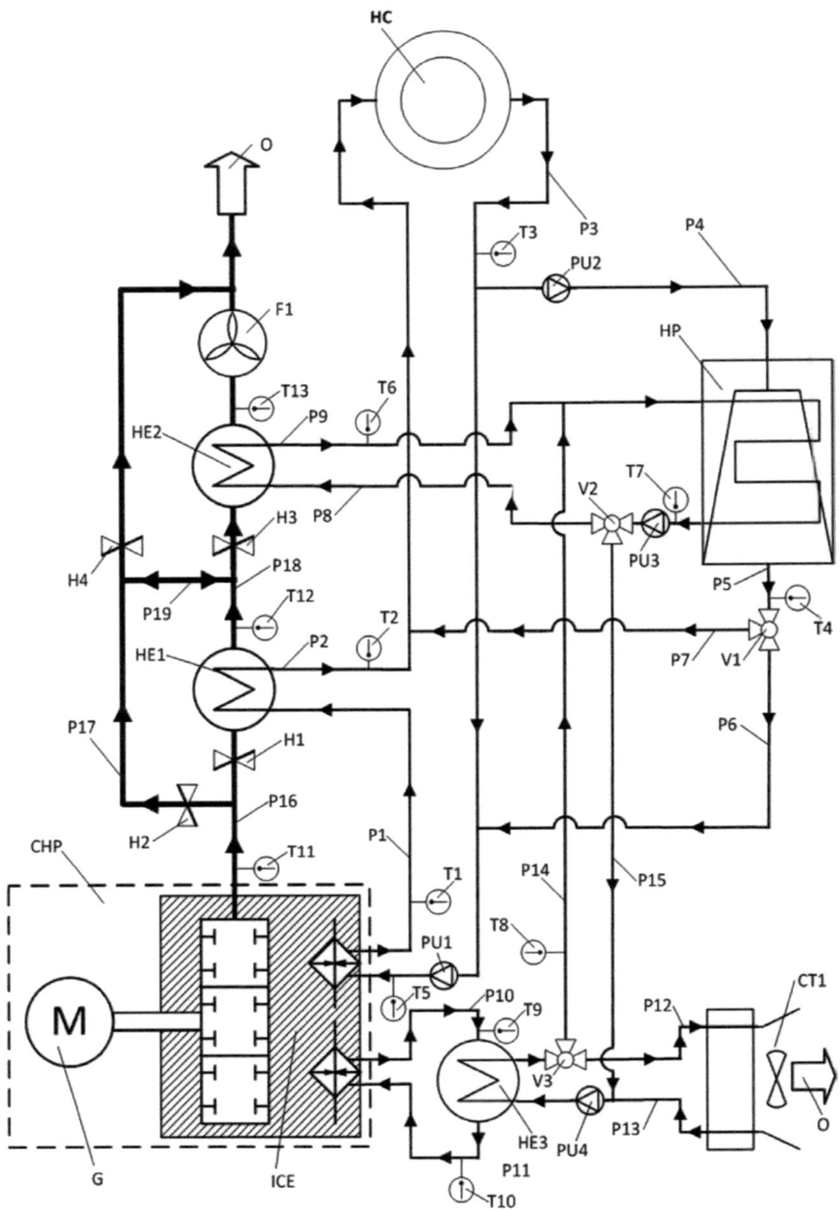

Figure 1.
The principle of exploiting the low-temperature sources of the CHP plant using the high-temperature heat pump [7].

(H1 - H4), fan (F1), pumps (PU1 - PU4), backup cooling system (CT1), external environment (O) and temperature sensors (T1 - T13) [7].

The drive unit of the combined heat and power plant (CHP) is directly connected to the high temperature district heating system of the end user (HC) with integrated cooling system or heat exchanger with piping (P1 and P3). The primary medium or heat transfer medium in a high-temperature district heating system is water, which transfers the thermal energy of the combustion engine cooling system to the end consumer (HC) of the high-temperature district heating system.

The operation of a combustion gas engine (ICE) produces hot exhaust gasses which are discharged through the pipe (P16) into two heat exchangers (HE1 and

HE2) connected in series. The first of the heat exchangers (HE1) on the exhaust system is designed to use the high-temperature heat of the exhaust gasses, which it transfers directly to the primary medium of the high-temperature heating system. After the high-temperature heat is dissipated in the heat exchanger (HE1), the exhaust gasses continue their path through the pipe (P18) to the heat exchanger (HE2). Due to the low temperature of the exhaust gasses in the pipe (P18) and the heating of the primary medium in the pipe (P8) (HE2), the heat of the primary medium is not suitable for further direct use in a high temperature heating system. To use this low-temperature source, a high-temperature heat pump (HP) is therefore used, into which the condenser directs the return flow of the district heating system. The heat exchanger (HE2) heats the secondary medium (water or a mixture of water and glycol) by further cooling the exhaust gasses and condensing the water contained in the exhaust gasses. This medium circulates in a closed circuit between the heat exchanger (HE2) and the evaporator of the high-temperature heat pump (HP). In the evaporator of the high-temperature heat pump (HP) the secondary medium evaporates the refrigerant of the heat pump (HP).

In a similar way, the low-temperature heat source intercooler 2nd stage and gas engine lubricating oil (ICE) is used via a pipe connection (P10 and P11) to the heat exchanger (HE3).

Due to the use of this low-temperature heat source, the evaporator of the high-temperature heat pump (HP) is connected to a secondary heat exchanger (HE3), which contributes part of the low-temperature heat to the evaporation of the working medium in the evaporator of the high-temperature heat pump (HP). One or more heat exchangers can be integrated in the series or parallel connection of CHP gas engines and high temperature heat pumps to use the excess low temperature heat sources which are now released into the environment.

To illustrate the process of using low-temperature heat sources from CHP gas engines, data on the operation of the commercially available CHP plant were obtained [8]. In **Table 3** nominal power of the 3.3 MW CHP plant given. The estimated operating time of the CHP plant depends on the heat consumer's demand. In the presented example are about 4000 h/year. The operation mode of the district heating pipe network or HC heat consumers is 90/60°C in winter and 90/55°C in summer.

3.1 Excess low-temperature heat from the CHP gas engine

The excess low temperature heat of the CHP gas engine is released unused to the environment in several ways. The most common are:

- heat from exhaust gases with a temperature of 120°C, which is released to the environment through the exhaust system of the CHP gas engine. During the operation of the CHP gas engine, hot exhaust gas with a temperature of approximately 365°C is fed into the heat exchanger HE1 (**Figure 1**), where it is cooled to a temperature of 120°C and then released to the environment. Exhaust gas at a temperature of 120°C that is discharged to the environment represents a significant low-temperature energy potential.

- heat of the external cooling system (CT) of the CHP gas engine. The cooling system is used to cool the compressed air during the 2nd stage intercooler. In this process, the 2nd stage intercooler transfers the compressed air heat to a water-glycol mixture, which is heated to 45°C and fed into an air cooling system (CT), where it is cooled to 40°C. The heat flow of the cooling system (CT), which is dissipated to the environment as low-temperature waste heat, is 197 kW.

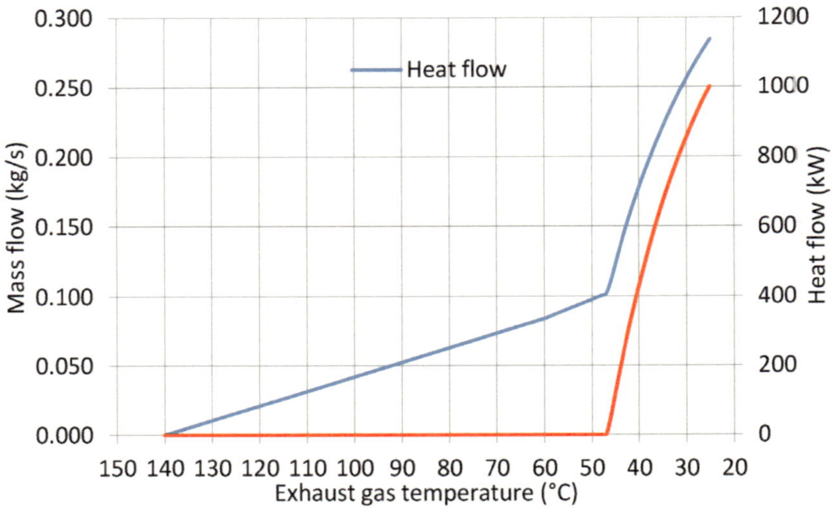

Figure 2.
The obtained low-temperature heat flow with an additional cooling of the flue gasses from the CHP gas engine [9].

In order to use the low temperature heat of the exhaust gas with a temperature of 120°C, it is necessary to install the condenser heat exchanger HE2 (**Figure 1**) in the exhaust system of the gas engine, where the exhaust gases should be cooled down to a temperature of 25°C. For this purpose, a computer simulation of the cooling of the exhaust gases was carried out using the Aspen plus software, the results of which are shown in **Figure 2** [9].

The heat flux obtained with additional cooling of the flue gasses from the CHP unit is presented on **Figure 2**. the specifications of which are shown in **Table 3**, where the flue gasses are first cooled from a temperature of 120°C to a temperature of 46°C, at which the water from the flue gasses starts to condense in the condenser heat exchanger HE2 (**Figure 1**). **Figure 2** shows the mass flow of condensed water from the flue gas as a function of the temperatures of the flue gasses.

The diagrams in **Figure 2** show that the CHP exhaust gas are first cooled from a temperature of 120°C to a temperature of 46°C, yielding approximately 400 kW of heat. However, further cooling of the exhaust gases causes condensation of water vapor, which is present in the exhaust gases. Most of the heat, about 700 kW, is obtained by cooling the exhaust gases from 46–25°C, producing about 0.250 kg/s of condensate.

By further cooling the exhaust gases of the CHP gas engine to about 5°C, an additional 600 kW of heat could be obtained. By exploiting the heat released from the surface of the CHP gas engine and heating the air in the room where the CHP device is installed, an additional 202 kW of heat could be obtained. This means that by further cooling the exhaust gases from 25–5°C and utilizing the heat released from the external surfaces of the CHP unit, approximately 800 kW of low-temperature heat could be extracted, which could be utilized by a high-temperature heat pump and thus further increase the primary fuel efficiency of the CHP gas engine to about 117% relative to the LHV of natural gas. The temperature to which the flue gases would be cooled depends on the economics of operation of high-temperature heat pumps, because as the temperature of the low-temperature source decreases, the average COP of the heat pumps decreases rapidly. By lowering the evaporation temperature of the refrigerant in the evaporator of the high-temperature heat pump, the pressure ratio of the compressor increases, so more power is required for the electric motor drive of the compressor, which results in a lower COP.

3.2 High-temperature heat pump

High-temperature heat pumps have a high added value and contribute a lot to energy dependence reduction. They can be used in all industries where waste heat flows of different fluids are generated. They allow an economically and ecologically efficient use of low-temperature resources to improve specific energy use in processes, increase efficiency and consequently reduce CO_2 emissions through the reduced consumption of fossil fuels for heat generation. The high-temperature heat pump has created the possibility of using heat from renewable or non-renewable low-temperature energy sources to meet the needs of technological processes or high-temperature heating systems.

Heat pumps have been around for many years, about as long as refrigeration units. The rapid development of heat pumps was triggered by the first oil crisis. People then began an intensive search for a replacement for fossil fuels and corresponding technological solutions. Laws related to pollution became stricter, people became aware of pollution and its effects, and energy prices increased. Heat pumps became popular due to their energy efficiency and environmental friendliness. In most of the cases, heat pumps were used for cooling purposes, while they were used for heating buildings only in case of low temperature heating up to 60°C. With the high temperature heat pump, low temperature heat sources up to 55°C can be used so that the heat potential is used to produce hot water up to 85°C [10]. This heat can be used for heating buildings or in industrial processes, and simultaneously cold water can be produced (down to 10°C), for air conditioning needs [11, 12].

The single-stage high temperature heat pump operation is based on the deprivation of heat from a low-temperature fluid (water) to get it to a higher temperature level.

For high-temperature heat pumps different low-temperature heat sources can be used:

- geothermal water with temperature around 55°C,
- the flue gasses heat,
- low-temperature energy sources from industry,
- waste heat of cooling systems, such as cold-storage chambers, meat processing industry,
- sea water heat, heat of lakes, groundwater, rivers, etc.

High-temperature heat pump efficiency is determined by a heating number, which is the ratio between the heat flow generated in the condenser for heating requirements and the electricity consumed to drive the compressor. The evaporator heat flow indicates how much heat was generated from the low temperature energy source and the condenser heat flow indicates how much heat was generated for heating purposes. The determination of the power required to drive the compressor allows the determination of the power consumption for the compression of the refrigerant, which is a substance with special physical properties [13]. The use of the high temperature heat pump allows:

- Use of low temperature heat sources in an area where the infrastructure for high temperature heating already exists,

- improve energy efficiency,
- replacement of the obsolete technology with the new one, which offers much higher efficiency and more environmentally friendly energy production,
- Reducing the emission factor (CO2/kWh), thereby reducing greenhouse gas emissions, and
- rease in heat production rates and a quick return on investment.

The operating characteristics of the 500-kW high-temperature heat pump as a function of the speed of a compressor, the required temperature of the output water, and the water temperature of a low-temperature source are given in **Figures 3** and **4** [10]. The results are given for different operating conditions using the working fluid R717 (NH3) and the commercial 50-bar piston compressor for the hot water temperatures from 65 to 85°C. Other refrigerants were found to be less suitable due to lower enthalpy difference between vapor and liquid

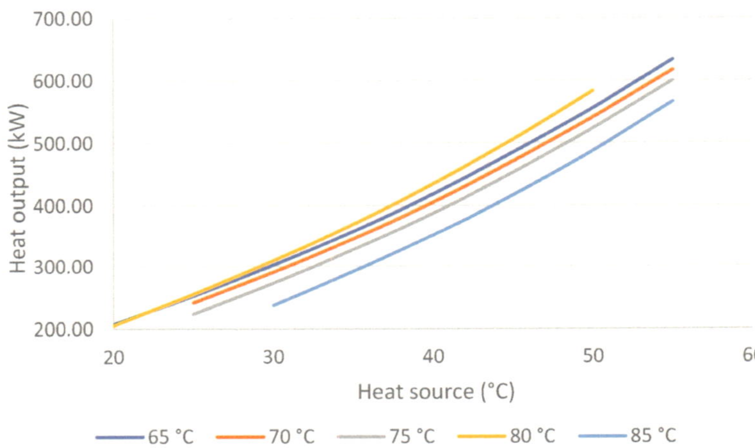

Figure 3.
The heat output in dependence of source inflow temperature for different temperatures of hot water.

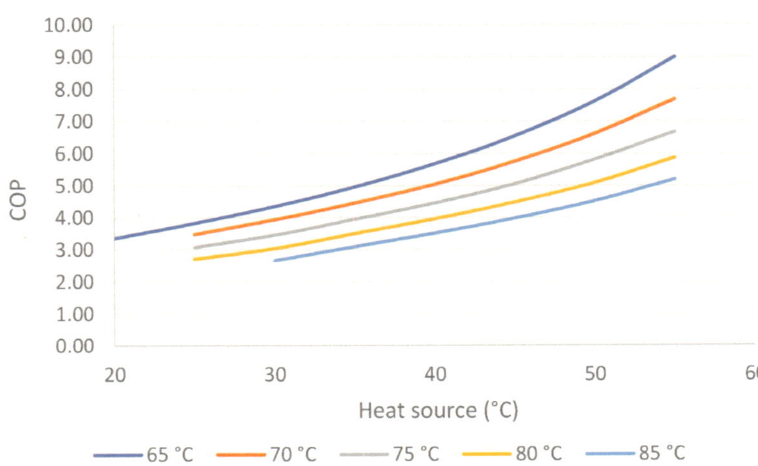

Figure 4.
The COP in dependence of source inflow temperature for different temperatures of hot water.

phases, lower heat flux, and lower COP. The capacity of the compressor in the HTHP can be controlled in steps (970; 1,450; and 1,600 rpm) or by a stepless control of the electric motor driving the compressor.

4. Process simulation with AspenPlus software package

In order to be able to use the energy of the exhaust gasses at a temperature of 120°C, a retrofit in the form of an additional heat exchanger (HE2) in the exhaust gas system is required, which would cool the exhaust gasses to 25°C. In this way, 1,100 kW of heat could be recovered. The second low temperature heat source is the cooling system 2nd stage intercooler (HE3) where 197 kW of heat could be recovered. The temperature of both low-temperature heat sources is too low to be used directly for high-temperature heating, but can still be used by using the HTHP, which raises the temperature level to a level suitable for high-temperature heating. A high temperature heat pump (or several of them) is integrated into the CHP system, as shown in **Figure 1**, to use the low temperature heat exchangers HE2 in HE3. The heat flow recovered from the high temperature heat pump is used to reheat the water return to a temperature of 70°C. In case the return water temperature is too high, the heated water is sent directly to the supply line through valve V1 - **Figure 1**. The total low-temperature heat flow obtained by the CHP unit with a nominal capacity of 3.3 MW with heat exchangers HE2 and HE3 is approximately 1,297 kW. To utilize the 1,297 kW from the low-temperature heat source by heating the return water from the high-temperature heating system from 60–70°C, which is the maximum water temperature allowed to enter the CHP unit, would require four high-temperature heat pumps with a rated capacity of 500 kW. The low-temperature energy source of HE2 and HE3 is utilized with the circulation of a glycol-water mixture, to which four high-temperature heat pumps are connected in sequence.

To utilize the excess low-temperature heat from the CHP gas engine, whose operating data are given in **Table 3**, a computer simulation of four series-connected high-temperature heat pumps with a rated capacity of 500 kW was performed using the Aspen Plus software package. The operating data of a 500 kW high temperature heat pump at a compressor speed of 1450 rpm are given in **Table 3**. Heat generation with a 500 kW high temperature heat pump can be modified with a frequency controlled electric motor drive of a reciprocating compressor with a maximum permissible speed of 1600 rpm.

Figure 5 schematically shows the serial connection of four high-temperature heat pumps with some results of computer simulation. Four series-connected high-temperature heat pumps consist of four compressors (COMP1, COMP2, COMP3, COMP4), four refrigerant evaporators (EVAP1, EVAP2, EVAP3, EVAP4), four condensers (COND1, COND2, COND3, EXP3) and four expansion valves (VALVE1, VALVE2, VALVE3, VALVE4). The low-temperature heat source of four series-connected high-temperature heat pumps is water or a mixture of water and glycol heated in a heat exchanger (HE2 and HE3), shown in **Figure 1**.

Water or a mixture of water and glycol, first heated slightly to 50°C in a heat exchanger HE2 and HE3 and fed through line P9 and P14 shown in **Figure 1** and **Figure 5** successively through all four evaporators (EVAP1, EVAP2, EVAP3, EVAP4). Chilled water or a mixture of water and glycol at about 23°C leaving the EVAP 4 evaporator is returned to the heat exchanger HE2 and HE3 via line P8 and P15. In each of the four evaporators of high-temperature heat pumps, the evaporation pressure of the refrigerant (ammonia) is different because it depends on the temperature and the available heat flow obtained in each individual evaporator. Refrigerant vapors leaving the evaporators are compressed by compressors

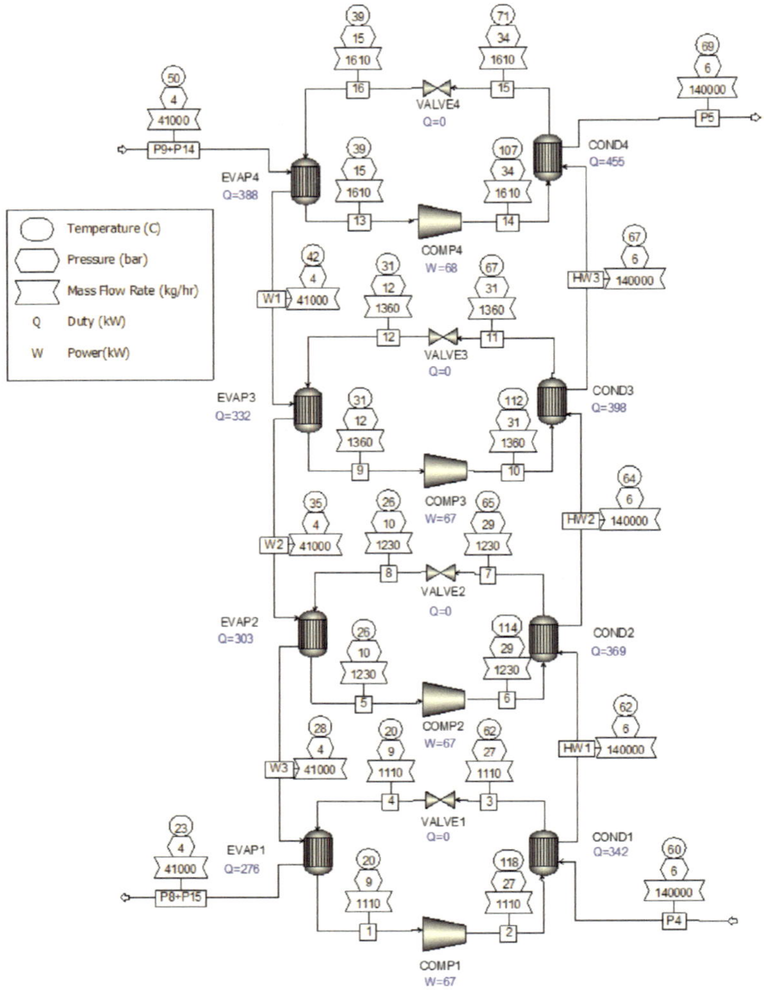

Figure 5.
The scene of four series-connected high-temperature heat pumps with the results of computer simulation using the Aspen plus software package.

(COMP1, COMP2, COMP3, COMP4) to the pressure required to condense the refrigerant in series-connected condensers (COND1, COND2, COND3, COND4) when heating the district heating return water.

The return water of the district heating system is fed to the first condenser via pipe P4, as shown in **Figures 1** and **5**, and leads sequentially to each individual condenser of the high-temperature heat pump. In each condenser, the heating water heats up a little until the desired temperature is reached in the last condenser. The water of the district heating system heated in this way is then fed to the CHP gas engine via pipe P5, in which it is heated to the final temperature of the district heating system.

To compare the operating characteristics of the system of four series-connected high-temperature heat pumps for the exploitation of low-temperature CHP gas engine heat sources, a computer simulation of one high-temperature heat pump with the same operating characteristics as four parallel-connected 500 kW high-temperature heat pumps is made.

Figure 6 shows a diagram of one high-temperature heat pump consisting of a single compressor (COMP), a refrigerant evaporator (EVAP), a condenser (COND) and an expansion valve (VALVE). The low temperature heat source of a high

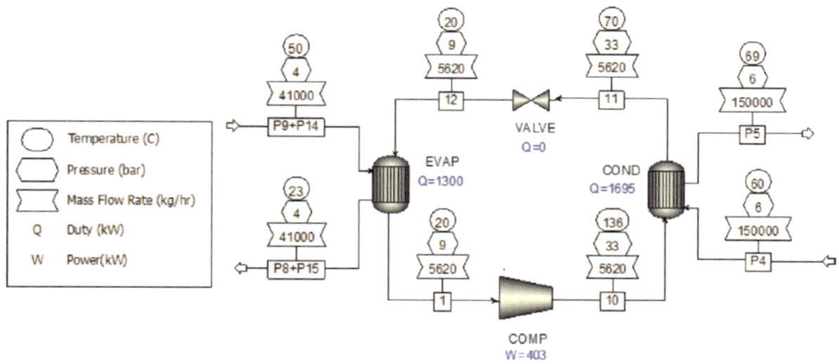

Figure 6.
Schematic diagram of a high-temperature heat pump with the results of a computer simulation using the Aspen plus software package.

temperature heat pump is water, or a mixture of water and glycol heated in the heat exchanger HE2 and HE3 shown in **Figure 1**. The water or mixture of water and glycol is heated to about 50°C and through line P9 and P14 shown in **Figures 1** and **6**, is fed to a high temperature heat pump evaporator (EVAP), where it is cooled to about 23°C and returned to the heat exchanger HE2 and HE3.

Refrigerant vapors leaving the evaporator (EVAP) are compressed in the compressor (COMP) to the pressure required to condense the refrigerant in the condenser (COND) when heating the return water of the district heating system. In the condenser, the return water of the district heating system is heated to the desired temperature and then returned to the CHP gas engine via pipe P5, where it is heated to the final temperature.

A summary of the results of the computer simulation of the four high-temperature heat pumps connected in series in **Figure 5** is shown in **Table 1**. The computer simulation results presented in **Table 1** show that the average COP of all four high temperature heat pumps is 5.81. To drive the frequency-controlled electric motor drives of the compressors in all four high-temperature heat pumps, 269 kW of electricity would be required. By lowering the temperature of a low-temperature heat source, the COP and the output of the series-connected high-temperature heat pumps also decrease. The COP of high-temperature heat pumps also decreases as the district heating return heating water increases. The calculation was made by heating the heating return water from a temperature of 60°C to a temperature of 70°C in series-connected condensers of four high-temperature heat pumps.

A summary of the results of the computer simulation of a high temperature heat pump in **Figure 6** is shown in **Table 2**. The results of the computer simulation in

HTHP sequence number	Required power compressor (kW)	Compressor pressure ratio (Pa)	Evaporator heat flux (kW)	Condenser heat flux (kW)	COP
1	67	3.00	276	342	5.10
2	67	2.90	303	369	5.51
3	67	2.58	332	398	5.94
4	68	2.27	388	455	6.69
In total	269		1,299	1,564	5.81

Table 1.
Summary of the results of the computer simulation of the four series connection of high-temperature heat pumps from ***Figure** 5.*

HTHP sequence number	Required power compressor (kW)	Compressor pressure ratio (Pa)	Evaporator heat flux (kW)	Condenser heat flux (kW)	COP
1	403	3.67	1,300	1,695	4.21

Table 2.
Summary of the results of the computer simulation of the single high-temperature heat pump from ***Figure 6****.*

Table 2 show that the average COP value of a high temperature heat pump is 4.21, which is much less than four serial connected high temperature heat pumps. The reason for this is that it is necessary to overcome the greater temperature difference between the outlet temperature of the low temperature heat source and the desired temperature of the heating water return of the heating system with a high temperature heat pump. To drive the frequency-controlled electric motor drive of a

	CHP	CHP + HTHP	
Natural gas LHV	9.5	9.5	kWh/Nm^3
Energy input (LHV)	7,351	7,351	kW
Mechanical output	3,428	3,428	kW
Electrical output	3,349	3,349	kW_e
Recoverable thermal output:			
• Intercooler 1st stage	883	883	kW_{th}
• Lube oil	290	290	kW_{th}
• Jacket water	463	463	kW_{th}
Exhaust gas cooled to 120°C	1,399	1,399	kW_{th}
Total directly recoverable thermal output (90°C)	3,035	3,035	kW_{th}
Heat to be dissipated:			
• Intercooler 2nd stage	197		kW_{th}
• Surface heat	202	202	kW_{th}
Exploited low temperature available heat:			
• Additional flue gas cooling to 25°C		1,100	kW_{th}
• Intercooler 2nd stage		197	kW_{th}
Total exploited low temperature thermal output		1,297	kW_{th}
Heat generated with HTHP (70°C)		1,564	kW_{th}
Absorbed power HTHP		269	kW_e
Total exploited thermal output		4,332	kW_{th}
Total thermal output CHP + HTHP		4,599	$kW_{th\ total}$
Total energy output generated	6,384	7,679	kW_{total}
Net electrical output	3,349	3,080	kW_e
Electrical efficiency	45.6	41.9	%
Thermal efficiency	41.3	62.5	%
Total efficiency CHP + HTHP	86.8	104.4	%

Table 3.
Technical data for CHP device [6] rated power 3.3 MW and technical data for CHP device rated power 3.3 MW with series-installed high-temperature heat pumps [7].

compressor of a high-temperature heat pump, a higher electrical power (403 kW) is therefore required, since a higher-pressure ratio between the evaporator pressure in the evaporator and the condensing pressure in the condenser must be created.

The energy consumption efficiency of natural gas and the utilization of excess low-temperature heat of a CHP gas engine with four series-connected high-temperature heat pumps are given in **Table 3**.

Electricity is needed to drive the frequency-controlled electric motors of the high-pressure compressors of high-temperature heat pumps. The electricity can be drawn from the grid or electricity from a gas engine with combined heat and power generation can be used. In **Table 3**, we have used 269 kW of electricity generated by a CHP unit to drive four high-temperature heat pumps connected in series, so that the electrical efficiency of the CHP fell from 45.6% to 41.9%. The total heat produced by the CHP and the four CHP units to use the excess low temperature heat from the CHP gas engine increased from 41.3% to 62.5%. The overall energy efficiency of the primary fuel of the CHP was increased by 17.6%, from 86.8% to 104.4%.

5. Economics of excess heat recovery of CHP gas engines

The economics of the exploitation of low-temperature heat sources from a selected CHP gas engine with the four HTHP, was calculated in MS Excel. For the HTHP drive, instead of the electricity produced by the CHP device, we can use electricity from the electricity grid.

The price of electricity, when we take it from the grid, is made up of production costs, transport costs, distribution costs, nominal power, various taxes, etc. and averages 95.2 €/MWh. The price of electricity produced in combined heat and power generation and sold on the electricity market averages 43.5 € /MWh. For the

Data on heat production		
Average COP four HTHP	5.81	
Average operation power four HTHP	78.2%	
Rated power of HTHP	4 x 500	kW_{th}
Average heat flow four HTHP	1,564	kW_{th}
Working hours per year	4,000	h/a
Yearly production of heat with four HTHP	6,256	MWh/a
Electricity consumption of four HTHP	269	kW_e
Specific electricity consumption of HTHP	0.172	kWh_e/kWh_{th}
Energy price		
Price of heat	0.0480	EUR/kWh
Price of electricity	0.0435 / 0.0952	EUR/kWh
Price of produced heat with HTHP	0.0075 / 0.0164	EUR/kWh
Economic data		
Investment into HTHP system	720,000	EUR
Discount rate	2.0	%

Table 4.
Basic operating data of the high-temperature heat pump (HTHP), energy prices and other economic data.

economic calculation of the surplus heat recovery of CHP gas engines with HTHP, electricity prices vary from country to country.

Two calculations have been made:

- with electricity to power the compressor in the high-temperature heat pump taken from the grid (95.2 €/MWh) and
- with electricity produced by the CHP unit (43.5 €/MWh).

Basic technical and economic data are given in **Table 4**. The average price of heat was defined as the production price of heat produced from natural gas with CHP device and gas boilers district heating system.

The economic calculation of the exploitation of low-temperature heat sources from a CHP plant with the HTHP using electricity from a CHP plant (43.5€/MWh) is shown as a diagram in **Figure 7**.

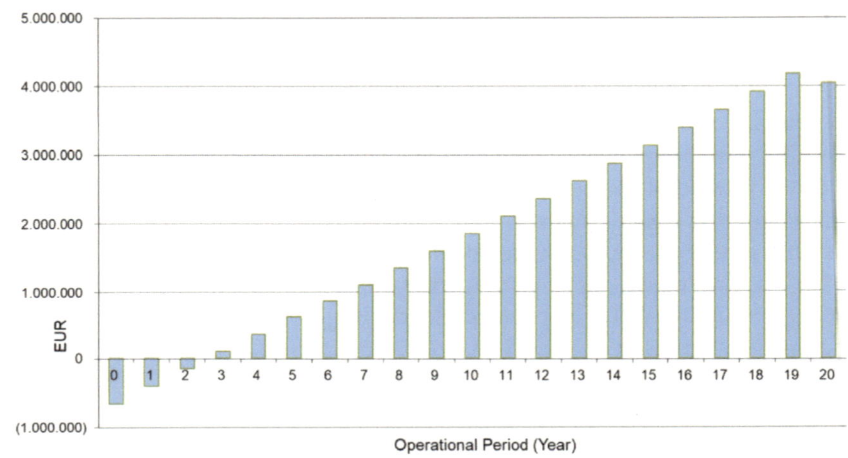

Figure 7.
Cumulated discounted cash flow at the price of the electricity at 43.5 €/MWh.

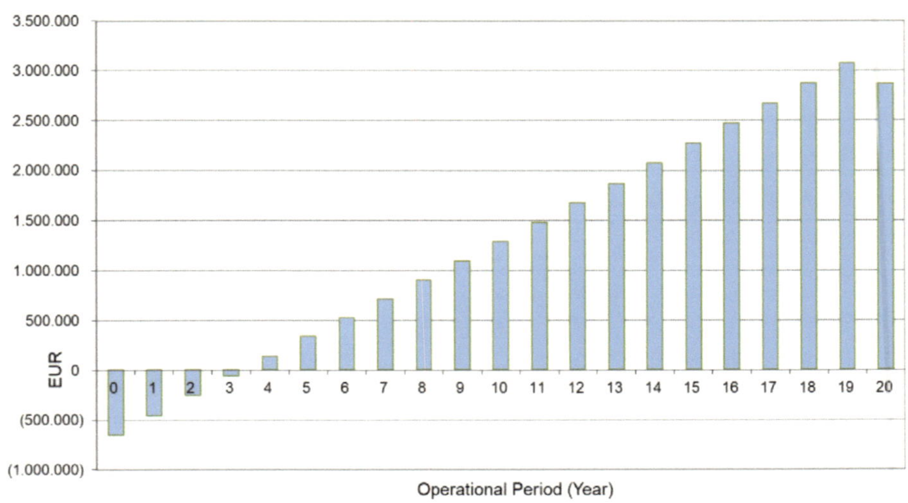

Figure 8.
Cumulated discounted cash flow at the price of the electricity at 95.2 €/MWh.

The diagram in **Figure 7** shows the internal rate of return IRR = 39.4% and payback of the investment exploiting excess heat from CHP device with four series-connected high-temperature heat pumps is approximately 2.5 years.

The economic calculation of the exploitation of low temperature heat sources from the CHP device with four series-connected high-temperature heat pumps with the electricity taken from the grid (95.2 €/MWh) is shown as a diagram in **Figure 8**.

The diagram in **Figure 8** shows the internal rate of return IRR = 30.5% and payback of the investment exploiting excess heat from CHP device with four series-connected high-temperature heat pumps is approximately 3.3 years.

6. Conclusions

An innovative technical solution for increasing the efficiency of the primary fuel and thus reducing CO_2 emissions by retrofitting existing or new CHP gas engines with high-temperature heat pumps is presented. High-temperature heat pumps use the excess heat of the exhaust gasses and the heat of the cooling system of the CHP gas engine and heat the return water of the district heating system.

The described principle of using the excess heat of CHP gas engines can also be used for the use of low-temperature heat sources, including hot water boilers and other types of CHP equipment.

In order to use the excess heat of the exhaust gasses and the cooling system of the CHP gas engine, it is necessary to install a heat exchanger in the exhaust system, where they are further cooled by cooling and condensation of the water contained in the exhaust gasses. The heat generated in this way is too low to be used directly for high-temperature heating, but it can also be used to heat the return water of the district heating system by using high-temperature heat pumps.

To illustrate how the innovative technology of using low-temperature sources of CHP gas engines works, a computer simulation of four series-connected high-temperature heat pumps was carried out using the Aspen plus software package. With a system of series-connected high-temperature heat pumps, the overall efficiency of natural gas can be increased by 17.6% (from 86.8% to 104.4% in terms of LHV of natural gas) by using low-temperature heat sources of CHP gas engine. An increase of 17.6% in the primary energy efficiency of the natural gas CHP gas engine is achieved when the exhaust gasses from the gas engine are cooled to a temperature of 25°C.

The energetic efficiency of the primary fuel of the CHP gas engine can be significantly increased to approx. 117% in relation to the LHV of natural gas by additional cooling of the exhaust gasses (up to approx. 5°C) and utilization of the heat emitted from the surface of the CHP gas engine to the environment. The temperature to which the exhaust gasses would be cooled depends on the economic efficiency of the operation of high-temperature heat pumps, because when the temperature of a low-temperature source drops, the average COP of heat pumps drops rapidly.

The presented technical solution for increasing the primary fuel efficiency of the CHP gas engine by using the excess heat of the exhaust gasses and the cooling system of the gas engine is also very economical with a very short return on investment.

During the computer simulation of the presented technical solutions for increasing the efficiency of the primary fuel of the CHP gas engine using high temperature heat pumps, the operating parameters of the manufacturer of the CHP gas engine and the high temperature heat pump were taken into account.

Author details

Darko Goričanec and Danijela Urbancl*
Faculty of Chemistry and Chemical Engineering, University of Maribor, Maribor, Slovenia

*Address all correspondence to: danijela.urbancl@um.si

References

[1] https://ec.europa.eu/energy/topics/energy-efficiency/targets-directive-and-rules/eu-targets-energy-efficiency_ro [Accessed 11/11/2020]

[2] Kim CM, Lee J, Chang SD, Hwang Y. An experimental study of the enhanced heating capacity of an electric heat pump (EHP) using the heat recovered from a gas engine generator. Int J Refrig 2009;32:1702-10

[3] Mancarella P. Cogeneration systems with electric heat pumps: energy-shifting properties and equivalent plant modelling. Energy Convers Manag 2009;50:1991-9

[4] Blarke MB, Dotzauer E. Intermittency-friendly and high-efficiency cogeneration: operational optimisation of cogeneration with compression heat pump, flue gas heat recovery, and intermediate cold storage. Energy 2011;36:6867-78

[5] Blarke MB. Towards an intermittency-friendly energy system: comparing electric boilers and heat pumps in distributed cogeneration. Appl Energy 2012;91:349-65

[6] Capuder T, Mancarella P. Techno-economic and environmental modelling and optimization of flexible distributed multi-generation options. Energy 2014;71:516-33

[7] Patent WO2016051246A1: Method and apparatus for cogeneration power plant waste heat source utilization by incorporated water source high temperature heat pump

[8] https://www.innio.com/de [Accessed 11/11/2020]

[9] Caf A., Urbancl D., Trop P., Goricanec D. Exploitation of low-temperature energy sources from cogeneration gas engines. Energy. 1 Avg. 2016, vol. 108, str. 86-92

[10] http://www.mayekawa.eu/en/products/mycom/standard-units/heat-pump-unit/plus-heat [Accessed 11/11/2020]

[11] Kapustenko PO, Ulyev LM, Boldyryev SA, Garev AO. Integration of a heat pump into the heat supply system of a cheese production plant. Energy 2008;33:882-9

[12] Korfitsen E, Kristensen APR. Ammonia high pressure heat pumps in food refrigeration applications. Int J Refrig 1998;21:212-8

[13] Kulcar B, Goricanec D, Krope J. Economy of exploiting heat from lowtemperature geothermal sources using a heat pump. Energy Build 2008;40:323-9

Chapter 5

Perspective Chapter: Device, Electronic, Technology for a M.E. M.S. Which Allow the Extraction of Vacuum Energy Conform to Emmy Noether Theorem

Patrick Sangouard

Abstract

This theoretical work corresponds to the hope of extracting, without contradicting EMMY NOETHER's theorem, an energy present throughout the universe: that of the spatial quantum vacuum! This article shows that it should be theoretically possible to maintain a continuous periodic vibration of a piezoelectric structure, which generates current peaks during a fraction of the vibration period. Electronics without any power supply then transform these alternating current signals into a usable direct voltage. To manufacture these different structures, we also present an original microtechnology for producing the regulation and transformation electronics, as well as that necessary for controlling the very weak interfaces between the Casimir electrodes and that of the return electrodes! These vibrations are obtained by controlling automatically and at appropriate instants the action of the attractive Casimir force by a repulsive Coulomb force applied to return electrodes. The Casimir force appearing between the two electrodes of a reflector deforms a piezoelectric bridge, inducing a displacement of the barycenter of the ionic electric charges of the bridge. This internal piezoelectric field attracts opposing moving charges, from the mass, on either side of the piezoelectric bridge used by the Coulomb force used to generate an opposing Coulomb force.

Keywords: Casimir, coulomb, vacuum quantum energy extraction, piezoelectric, MEMS, NEMS

1. Introduction

1.1 Obtaining an electric current from vacuum?

We know that the quantum vacuum, the energy vacuum, the absolutely nothing, does not exist!

This statement has been proven multiple times and noted in particular by:

- Lamb's shift (1947) of atomic emission frequencies.

IntechOpen

- By the force of Van der Waals which plays a very important physicochemical role and had an interpretation quantum 1930 [London] when two atoms are coupled to the same fluctuations in vacuum.
- By Hawking's radiation theory, predicted in 1974 and observed on September 7, 2016.
- By the experimental verification (1958) of the existence of a force equated by Casimir in 1948. This so-called Casimir force was measured for the first time in 1997.

1.2 Brief presentation of Casimir's force

The vacuum energy is the zero-point energy of all fields (tensorial and scalar) in space, which for the standard model includes the electromagnetic field, gauge fields, fermionic fields, as well than the Higgs field. In quantum field theory, this vacuum energy defined as zero, is the ground state of fields! In cosmology, vacuum energy is a possible explanation for Einstein's cosmological constant. It has been observed and shown theoretically that this so-called zero-point energy, is non-zero for a simple quantum harmonic oscillator, since its minimum energy is equal to $E = h\,\nu/2$ with ν the natural frequency of the oscillator, and h the Planck's constant.

Originally [1], the Casimir effect is derived from statistical fluctuations in total vacuum energy and is the attraction (in general) between two plates separated by a vacuum. In this approach, this Casimir energy is the part E_{CA} of the vacuum energy which is a function of the z_S separation of the Casimir plates, with:

$$E_{CA} = S\left(\frac{\pi^2\hbar c}{740 z_S^3}\right)$$

This Casimir energy is proportional to the reduced Planck constant h, to the speed of light c and to the surface S of the reflectors (in the limit where the edge effects of the plates are negligible, which then imposes large dimensions of the reflectors compared to that of the separation of the plates).

The force of Casimir F_{CA} between the two reflectors is then the derivative compared to z_s of this energy thus: This Casimir force,

$$F_{CA} = \frac{d(E_{CA})}{dz} = S\left(\frac{\pi^2\hbar\, c}{240\, z_S^4}\right) \tag{1}$$

proportional to the surface, defines a pressure F_{CA}/S which depends only on the distance $z_s{}^4$ between the reflecting plates. This local approach greatly facilitates the formulations of Casimir's forces [1, 2]. The force of Casimir is then attractive and can be understood like a local pressure namely, the so-called virtual radiation pressure and exerted by vacuum fluctuations on the mirrors. These homogeneous, isotropic, and constant fluctuations of the vacuum, manifested by radiation and virtual particles, are modified by the presence of reflecting mirrors. These particles, however real, are called virtual because their lifetimes are noticeably short (for an electron of the order of 6.10^{-22} seconds) before recombining to return to a vacuum!

The presence of the reflective plates excludes wavelengths longer than the distance z_s between the plates. They thus induce a pressure difference of the virtual particles generated by the vacuum between the internal and external space of the 2 plates. This difference results in a force that pushes the plates together.

The Casimir force between two perfectly conductive and smooth plates, without conductive charge, at zero temperature is written as the difference of the radiation pressures calculated outside and inside the cavity and is then written (**Figure 1**):

$$F_{ca} = S\left(\frac{\pi^2 \hbar c}{240\ z_S^4}\right)$$

The famous physicist Evgueni Mikhaïlovitch Lifchits gave a general formula, which supplements that of Casimir because it considers the effect of temperature [3]. Indeed, when the temperature is no longer zero, the radiation of the black body must then be considered and the Casimir force at temperature T then becomes that of Lifchits [4].

$$F_{ca} = S\left(\frac{\pi^2 \hbar c}{240\ z_S^4} + \frac{\pi^2}{45}\frac{(kT)^4}{(\hbar c)^3} - \frac{kT\pi}{z_S^3}\exp\left(-\frac{\pi\hbar c}{kTz_S}\right)\right) \tag{2}$$

With k the Boltzmann constant and T the temperature. This modification is important for large distances, typically L ≥ 3 mm at ambient temperature.

We will use Eq. (1) to calculate and simulate the structure defined in the following diagram (**Figure 2**) because the intervals between electrodes are much smaller than μm and the effect of temperature is negligible.

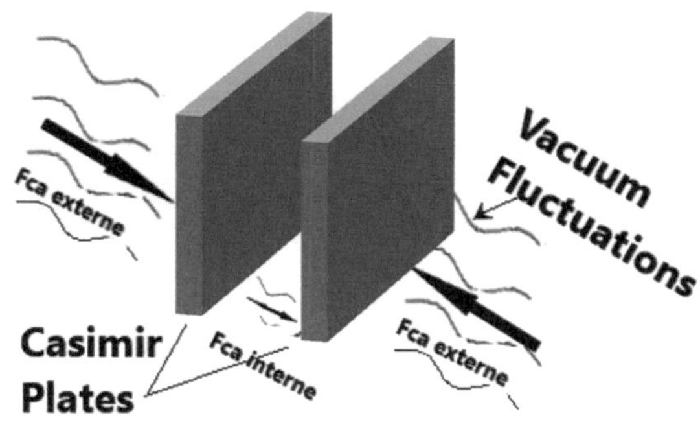

Figure 1.
Casimir effect.

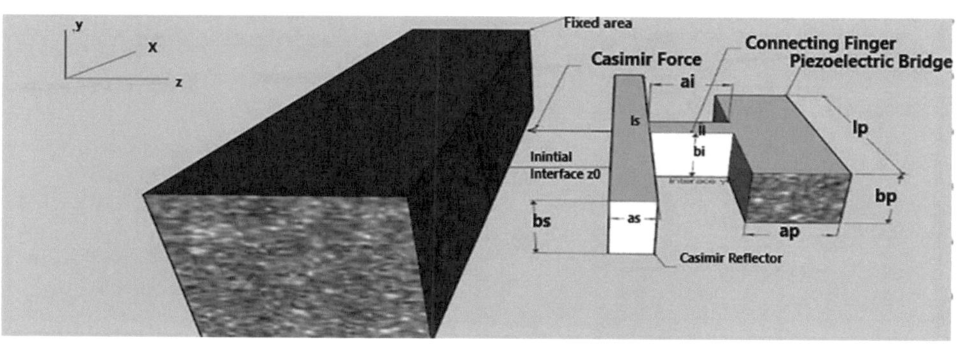

Figure 2.
General representation of the structure.

1.3 Extract energy from the vacuum?

The term *vacuum energy* is sometimes used by some scientists claiming that it is possible to extract energy—that is, mechanical work, heat … ., from the vacuum and dispose thus, ideally, a gigantic and virtually inexhaustible source of energy. Of course, these different hypotheses arouse great skepticism among many scientific researchers because they call into question a principle demonstrated mathematically by the theorem of the mathematician Emmy Noether in 1915, which involves the conservation of energy (like all invariances). This theorem is accepted in physics and has never been faulted until now!

In fact, the problem is less to extract energy from the vacuum than to extract it without spending more energy that we cannot hope to recover! This principle of Noether's theorem, still observed at the macroscopic scale, suggests that extracting energy from a vacuum would require at least as much energy, even probably more, than the process of its recovery would provide.

Thus, a cyclic system, on the model of a piston engine going from a position n°1 to n°2, then from n°2 to n°1, the existence of the Casimir force in $1/zs^4$, therefore greater in position (2) than in (1), would then imply spending more energy to return to (1), which would necessarily require an added energy!! This problem, like that of perpetual motion, then implies that this hope of extracting energy from a vacuum seems impossible and cannot be done with at least zero energy balance! But this is forgetting that an energy is not limited to a force but is, for example, the product of a force (intensity variable) by a displacement (position variable) (see **Figure 3**).

Indeed, imagine that the piston is a piezoelectric bridge, and that the deformation of this bridge is caused by the Casimir force. The deformation of this piezoelectric bridge induces fixed electric charges of opposite sign on each of its faces 1 and 2. Imagine that opposite mobile charge moves from the mass on each side of piezoelectric area. So, if it is possible to propagate, at the right moment, a part of the mobile charges of face 1 for example, on an electrically insulated surface opposite to face 2, then a Coulomb force opposite to the attractive Casimir force would practice. If this Coulomb force is greater than the Casimir force, for example by a factor of at least 2, then the total force $F_t = F_{CA} - F_{CO}$, applied to the center of this piezoelectric bridge deforms it in the other direction, decreases then cancels out its deformation thus the electric charges on the two faces of the piezoelectric bridge. The disappearance of electric charges suppresses the Coulomb force (see **Figures 4** and **5**). The system would then return

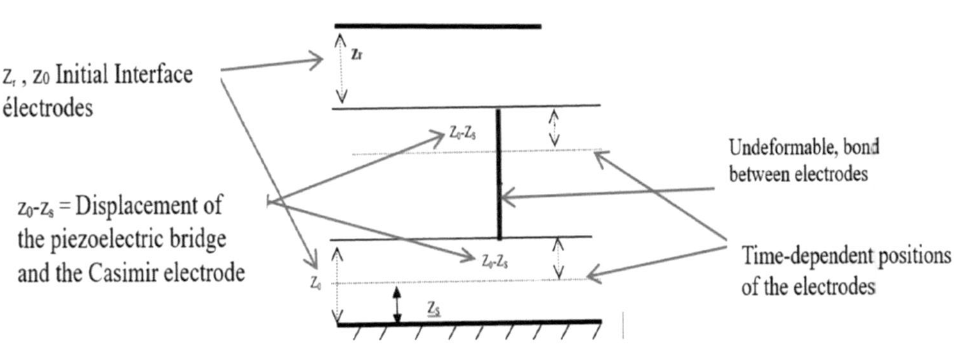

Figure 3.
Nomenclature of the different positions for the moving Casimir electrode.

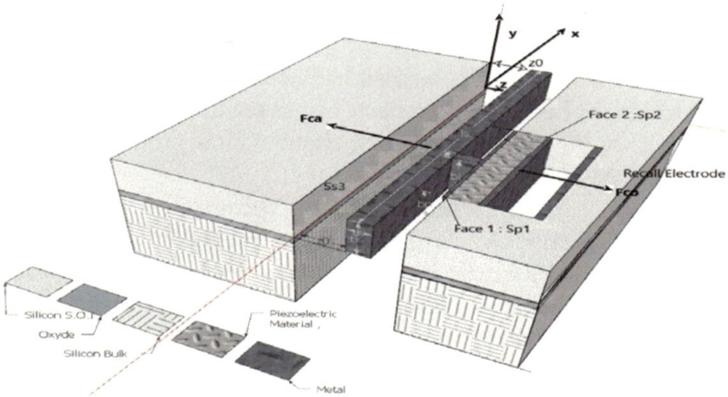

Figure 4.
Different view of the device without electronics.

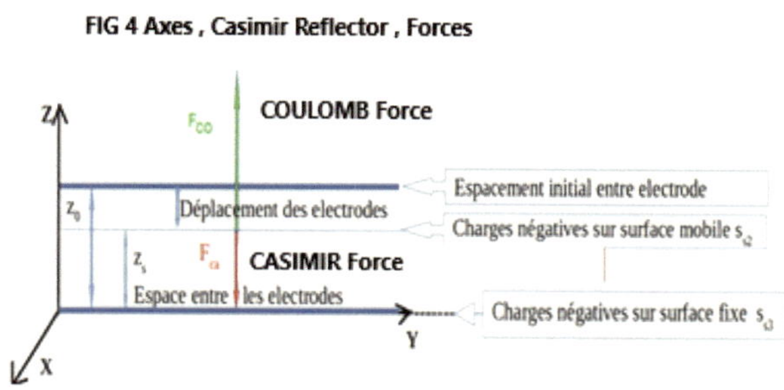

Figure 5.
Axes, forces, Casimir's electrodes.

to its original position and physical characteristics. Everything would start again, causing vibrations of the piezoelectric bridge of the Casimir reflector device without any external energy input!

Then, on the assumption that all the transient states of the system do not require any input of external energy and are only consequences of a primary cause which is the energy of the vacuum, the principle of Emmy Noether should not be contradicts!

The fixed electric charges on the two metallized faces of the piezoelectric bridge are of opposite signs and attracts from the mass of the mobile charges of opposite signs (**Figure 3**).

Let us imagine that the whole of the return electrode is in two parts of equal areas but separated by a switch circuit consisting of MOSN and MOSP enriched, in parallel and of threshold voltage $Vt_{NE} = -Vt_{PE}$ (**Figure 3**). The first part of this metallic return electrode and surface S_{p1} consists of one of the faces of the piezoelectric bridge and carries mobile electric charges $Q_m = -Q_F$.

The second part of this metal electrode is earthed via another switch circuit made up of depleted MOSN and MOSP, in series and of the same threshold voltage as the enriched MOSN and MOSP: $Vt_{ND} = Vt_{PE} > 0$ and $Vt_{PD} = Vt_{NE} < 0$ (see **Figure 3**).

These switch circuits (circuit 1 = MOSN and MOSP enriched in parallel, circuit 2 = MOSND and MOSPD in depletion and in series), are designed so that they open and close in opposition.

When circuit 1 opens or closes, at the same time switch circuit 2 closes or opens, thus isolating this return electrode from the ground (see **Figure 3**). This opposite behavior of the switch circuits can be seen in **Figure 3**.

So, when circuit 1 is open, the two parts of the return electrodes are grounded through circuits 2, on the other hand when circuit 1 closes the two parts of the return electrodes are isolated!

The electric field inside a perfect conductor being zero, the mobile charges attracted to the first part of the return electrode are redistributed on the two parts of this electrode in the ratio of the homogenization surfaces, that is to say ½ and are opposite the other electric charge face n°2 of the bridge! One then develops, between these isolated electrodes an attractive Coulomb force which is in the opposite direction to the Casimir force and can be greater than it (see **Figures 3–5**).

Now, we know that in the case of a deformation perpendicular to the polarization of a piezoelectric layer and caused by an F_{CA} force, the fixed charges Q_F induced by the deformation of this piezoelectric layer are proportional to the Casimir force F_{CA} and are therefore in $1/z_s^4$, with [5, 6].

$$Q_F = (d_{31}.F_{CA}.l_P)/a_p, \quad (3)$$

With d_{31} = piezoelectric coefficient (CN^{-1}),l_p,a_p respectively length and thickness (m) of the piezoelectric bridge (**Figure 5**). These fixed electric charges on the two metallized faces of the piezoelectric bridge have opposite signs and attract mobile charges of opposite signs from the mass (**Figure 5**). Thus, when it is effective, the Coulomb return force F_{CO} is in $1/z_s^{10}$ because on the one hand in $(Q_F/2)^2$ (therefore in $1/zs^8$) but also in $1/(z_r + z_0 - z_s)^2$ because depending of the distances $z_r + z_0 - z_s$ between return electrode n°1 and face n°2 of the piezoelectric bridge. With z_r the initial distance between the opposite face of the piezoelectric bridge and the return electrode, zs = distance between Casimir electrodes, time dependent, and z_0 = initial distance between Casimir electrodes (see **Figures 3–6**) and **Figure 2** bis. We will choose in the following MATLAB simulations (unless otherwise specified), the same interface between return electrode z_r as that attributed to the initial interface z_0 between Casimir reflectors. The distances over which the free Casimir electrode moves, correlated with

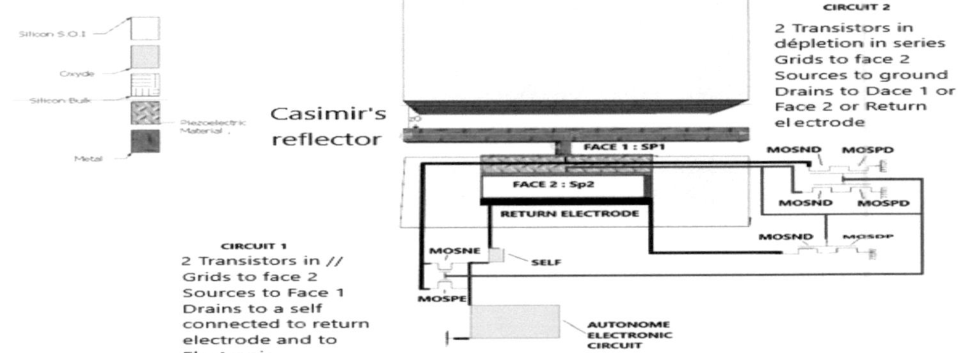

Figure 6.
General configuration of the device: MOS grid connections (face 2 of the piezoelectric bridge: Red), source connections (face 1 of the piezoelectric bridge: Green.

the deformation of the piezoelectric bridge are very small and less than 100 A°. The variations z_0,z_e,z_s are therefore <100 A° and are very small compared to the dimensions of the piezoelectric bridge and that of the Casimir electrodes. Although the rigorous calculation is possible, for the sake of simplification we will first consider that the Coulomb return electrodes remain strictly parallel (**Figures 3–5**).

Considering that the Coulomb force is zero when the piezoelectric bridge has no deformation, we thus obtain an attractive Coulomb force of direction opposite to that of Casimir with:

$$F_{CO} = \frac{Q_F^2}{4\pi\varepsilon_0\varepsilon_r}\left(\left(\frac{1}{z_r+z_0-z_S}\right)^2\right) = \left(S_S\frac{\pi^2\hbar c\, d_{31}l_P}{240\ 2^* a_P}\left(\frac{1}{z_s^4}-\frac{1}{z_0^4}\right)\right)^2\left(\frac{1}{4\pi\varepsilon_0\varepsilon_r}\right)\left(\frac{1}{z_r+z_0-z_S}\right)^2 \quad (4)$$

We note that $F_{CO} = 0$ when the bridge has no deflection (zs = z0), so no electrical charges! With z_r = interface between the face 2 and the return electrode.

This Coulomb Force in $1/z_s^{10}$ can therefore become greater than that of Casimir which is in $1/z_s^4$.

Applied to the piezoelectric bridge, it reduces its deformation, and therefore the induced charges. So, Coulomb's force diminishes and then vanishes when the bridge goes to the starting position n°1 since $zs = z_0$.

For these Coulomb return electrodes generate only a Casimir force that is negligible compared to that of the reflector, it will be necessary to choose interfaces z_r greater than $2z_0$. For example, if $z_r = 2.z_0$ and the surface of Casimir reflector $S_{s2} = 5^* S_p$ then the Casimir force between the return electrodes will be about 100 times weaker than that linked to the reflector, which is negligible (**Figure 3**).

On the other hand, as F_{CO} depends on the charge accumulated on the piezoelectric bridge in $(1/zs)^8$, the electric voltage that the MOS switches will have to withstand, before being triggered increases with the interface z_r, since the position where $F_{CO} = F_{CA}$ decreases with z_r (see **Figure 7**).

This leads to an increase in the threshold voltages of the different MOS with the increase interface z_r.

This electrostatic attraction of the piezoelectric bridge is possible because the fixed electric charges generated by the deformation of the piezoelectric bridge attract mobile electric charges Q_m of opposite signs from the mass.

Note that if switch circuit n°1 is open, we have seen that circuit n°2 is closed and connected to ground, so the second part of the return electrode is to ground.

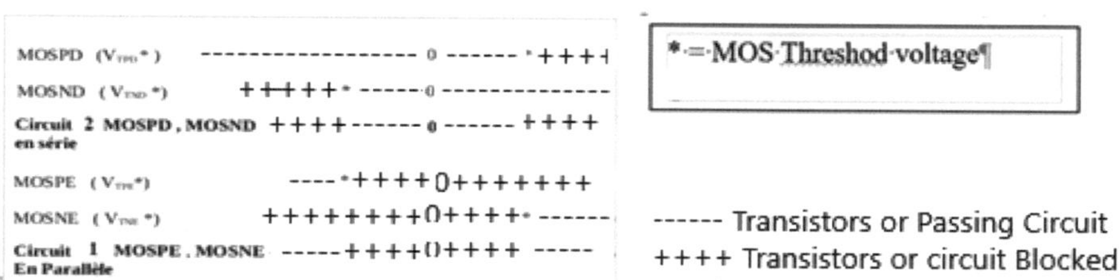

Figure 7.
Distribution of the threshold voltages of enriched and depleted N and P MOS switches.

Conversely, when circuit n°1 is closed then circuit n°2 is open, isolating the second part of the return electrode.

On the other hand, the mobile loads of face 2 by triggering, at the appropriate moment and depending on the threshold voltages, the automatic closing of circuit 1 and the opening of circuit 2, the charges of face 1 are distributed uniformly over the surfaces of the face n°1 of the piezoelectric bridge and the return electrode.

They create an attractive force F_{CO} of Coulomb opposed to that of Casimir which can be superior to him in modulus. The total $F_{CA} - F_{CO}$ force then becomes repulsive and, applied to the piezoelectric bridge decreases and cancels out its deformation, which consequently automatically removes the electric charges on it and initiates the reopening of circuit 1 and the closing of circuit 2.

Thus, the repulsive force of Coulomb disappears, and the force of Casimir becomes preponderant again which allows this cycle to start again!

It seems the spatial and temporal omnipresence of the attractive Casimir force, with the spontaneous appearance and at the appropriate moment of the Coulomb force described above, then generate vibrations of the mobile Casimir reflector plate!

We will calculate the frequency of these vibrations with MATLAB.

Note that during the movement of the piezoelectric bridge from (1) to (2), only the force of Casimir FCA is exerted, because the circuit 2 is conducting and connects the return electrode to the mass suppressing the action of the force Coulomb. Note also that the fixed electrode of the Casimir reflector is constantly earthed (see **Figure 3**).

During the short homogenization time of the mobile charges on the fixed return electrode, an alternating current peak I_a is recovered which generates an alternating voltage peak Ua through its crossing of an integrated inductor (therefore without any additional energy).

This weak and ephemeral but always present electric power $U_a.I_a$ is at the frequency of vibration of the structure. It then activates suitable electronics that must transform—without any external power source—this alternating voltage U_a into a direct voltage U_c which can be used (see Chapter 5).

This electronics was designed when I was working at ESIEE and on abandoned sensors. It works very well in SPICE simulation (see Part V).

If all the components of this project are successful (principle of extracting energy from the vacuum + device generating current peaks at the vibration frequency of the system and converted in peak of voltage by a coil + transformation electronics + technology for realization the device selected), all without any additional energy, the principle of Noether should be validated and the vacuum could then be considered as a simple medium, with which it is possible to exchange energy!

2. Description of the principle used to "extract" energy from the vacuum

As a preamble, we hope and suppose that the events which induce the attractive force of Casimir are exerted in a universal, isotropic, perpetual, and immediate way, if the conditions of separation between reflecting Casimir plates are suitable.

Let therefore be a Casimir reflector device consisting of:

1. a metallized and mobile parallelepiped electrode, of surface $S_{s1} = S_{s2}$ on its 2 lateral faces.

2. a fixed metallized surface S_{s3} separated by a distance z_0 (**Figures 4–6**).

In order for the movement of the movable plate of this Casimir reflector to create electric charges that can be used to induce an attractive Coulomb force, it is necessary that the movement of this movable plate naturally induces a deformation of a structure creating electric charges. A piezoelectric device is therefore required, rigidly connected to the mobile Casimir electrode, so that its induced naturally deformation leads to the appearance of electric charges! And this without any other energy being involved (**Figures 3–5**).

Of course, it is also necessary that the movement of this Casimir reflector mobile plate S_{s2} can be stopped at a chosen and predefined value before the bonding of the surface S_{s2} on the fixed surface S_{s3} takes place! Otherwise, we just definitively collapse the two reflector plates, and no energy extraction is possible! In addition, it is necessary that the mobile system returns to its initial position (or slightly exceeds it).

As said previously, we can then imagine that the attractive Casimir force exerted between the facing surfaces S_{S2} and S_{S3} and which moves the mobile reflector plate S_{S2}, induces a deformation of a parallelepiped piezoelectric bridge of surface S_{P1} and S_{P2}, rigidly linked, by a metal finger, to this reflecting mirror (**Figures 3–5**).

We observe, in **Figure 5**, that the surfaces $S_{p1} = S_{p2} = b_p{}^* \, l_p$, green or red metallic of the insulating piezoelectric bridge, are connected for:

1. for S_{p1} on the face n°1, through the metal finger (green) to the mobile plate of the Casimir surface reflector $S_{s2} = b_s{}^* \, l_s = S_{s1}$ which forms one of the electrodes of the Casimir reflector. Thus, the metallic surfaces S_{p1} and the metallic parallelepiped $S_{s1} = S_{s2}$ are equipotential

2. for S_{p2} on the face n°2, at the grids of the switch circuits n°1 and n°2.

The deformations caused by the attractive force of Casimir, then produce fixed electric charges for example $Q_{fn1} = -Q_{fp2}$, on the faces S_{p1} and S_{p2} of the insulating piezoelectric bridge. These fixed charges in turn attract, from the immediate environment (mass or earth) to which they are connected by circuits 2, mobile electric charges Q_{mp1} and Q_{mn2}, respectively. These charges are distributed over the metallized surfaces deposited on the insulating piezoelectric bridge, therefore on S_{p2} and the gates of the transistors of circuits 1 and 2 as well as on S_{p1}, the metal block, the sources of the transistors of circuit 1 and the two coupling capacitors of the electronic transformation circuit (see **Figures 3** and **8**).

Let S_{MOS} be the surface of the gates of the MOS of switch circuits 1 and 2. The mobile charges, for example positive Q_{mp2}, located on the surface S_{p2} of face 2 going to the gate of a MOSNE enriched transistor in the ratio $Q_{mp2MOS} = Q_{mp2}{}^* \, S_{MOS}/S_{p2}$, then produce a positive voltage $V_G = Q_{mp2\,MOS}/C_{OX}$ on the gate of the MOS transistors, with C_{ox} the gate capacitance of the MOS transistors.

Depending on the sign of these mobile charges on the gates of the MOSNE or MOSPE transistors, they can turn one of them ON, if they are sufficient to induce a voltage V_G greater than their threshold voltage V_{TE}, positive for the MOSNE transistor and negative on the MOSPE transistor in parallel.

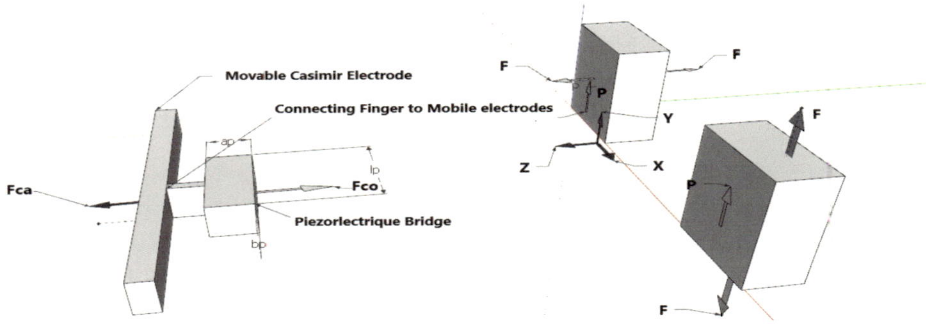

Figure 8.
Polarization and applied force and load on a piezoelectric block.

The nature of these charges depends on the initial polarization of the deposited piezoelectric parallelepiped and on the direction of the deformation imposed by the Casimir force. The sign of these mobile charges on the surfaces Sp1 and Sp2 depending on the real polarization obtained during the realization of the piezoelectric material of this bridge, it is the reverse which occurs if the mobile charges are negative on Sp2, hence the parallel setting of switches! (see **Figure 3**).

As long as this voltage on face 2 of the bridge, for example positive, is less than the threshold voltage V_{TNE} of this MOSNE transistor, the latter remains blocked! Consequently, the mobile charges Qmn1 = −Qmp2 located on the other face Sp1 of the deformed piezoelectric device (connected by a metal block to Ss2) and connected to the sources of the MOSNE and MOSPE remain on these surfaces and do not propagate on the surface of the return electrode. The MOS switches N and P depleted in series from the switch circuit 2 are then on and connect this return electrode to ground (**Figure 3**).

On the other hand, if this voltage V_G becomes greater than the threshold voltage V_{TNE} of the enriched MOSNE, it becomes conducting, and circuit 2 is then blocked, so the mobile charges Q_{mn1}, located on S_{p1} and the metal block can cross the MOSNE to homogenize the charge density on all the return electrodes. These electric charges pass through the inductance L_{IN} in series (**Figure 3**).

When one of the two MOSNE or MOSPE enriched transistors of circuit 1 turns on, then the depleted MOS N and P switches of circuit 2 are blocked. The return electrode, no longer connected to ground, therefore does not discharge these mobile electrical charges and is isolated (**Figure 3**).

Let $S_{p1} = S_{p2} = l_p{}^* \ b_p$ be the surface area of the faces of the piezoelectric bridge, S_{bloc} = the surface of the metal block of the Casimir reflector (**Figure 5**). Let $S_r = S_{p2} = S_{p1}$ be the surface of the return electrode facing the metallized face S_{p2} of the piezoelectric bridge.

The mobiles charges for example negative $Q_{mn1} = -Q_{mp2}$ which was initially distributed on the metallic surfaces S_{p1} are distributed, after the closing of the MOSNE switch, on the surfaces $S_{p1} + S_r$. They induce between the faces S_{p2} and S_r, electric charges of opposite sign, an attractive force of Coulomb, parallel and opposite to the attractive force of Casimir.

These same electrical charges opposite on the surfaces Sp2 and Sr. become after distribution **as Sp1 = Sr.** This charge Q_{mn1f} remains on the return electrode because circuit 2 is blocked (**Figures 3** and 5). If the threshold voltages of the transistors are positioned according to $V_{TND} \cong V_{TPE} < 0 < V_{TNE} \cong V_{TPD}$, then we have the following configurations depending on the value of the voltage V_G.

As a result, when circuit 1 is blocked, there is no Coulomb electrostatic attraction between Sp2 and Sr. because the metal return electrode Sr. is grounded and therefore free of charges! However, when circuit 1 is on, circuit 2 is then blocked, so as the metallic return electrode Sr. is isolated, the charges of opposite sign and present on the electrodes S_{p2} and S_r induce an attractive Coulomb electrostatic force!

This attractive Coulomb force as a first approximation is written:

$$F_{CO} = \left(\left(S_S \frac{\pi^2 \hbar c}{240} \frac{d_{31} l_P}{2^* a_P} \left(\frac{1}{z_s^4} - \frac{1}{z_0^4} \right) \right)^2 \left(\frac{1}{4\pi\varepsilon_0 \varepsilon_r} \right) \left(\frac{1}{z_r + z_0 - z_S} \right)^2 \right) \quad (5)$$

This attractive force triggered by an input of mobile electric charges of opposite sign on the surfaces S_{p2} and S_r is exerted only when the Casimir force between the two reflectors reaches a defined value, dependent on the threshold voltage of the MOSNE transistor.

It will therefore be necessary to adjust the threshold voltage of all these MOSs adequately. Initially, when the piezoelectric beam is not deformed, the electric charges on the faces S_{p1},S_{p2},S_{s2} and S_{s3} of the Casimir reflector are zero! The face S_{s2} of the Casimir sole plate, distant from z_0 from the face S_{s3} is then attracted against the fixed face S_{s3}, only by the force of Casimir. This force is communicated via the connecting finger at the center of the face S_{p1} of the piezoelectric bridge and then deforms it.

We will admit in the remainder of this presentation that we are in the case of mode 31 and that, when the piezoelectric bridge undergoes a deformation then the fixed electric charges of the piezoelectric bridge attract negative mobile charges on the electrode Sp1 from the mass and positive mobiles on the other S_{p2} electrode.

If, the threshold voltage of the MOSNE is adjusted so that the Coulomb force is triggered only when $F_{CO} = p\ F_{CA}$ with p proportionality factor 2, then the total repulsion force Ft variable in time and applied to the piezoelectric bridge becomes (**Figures 9** and **10**).

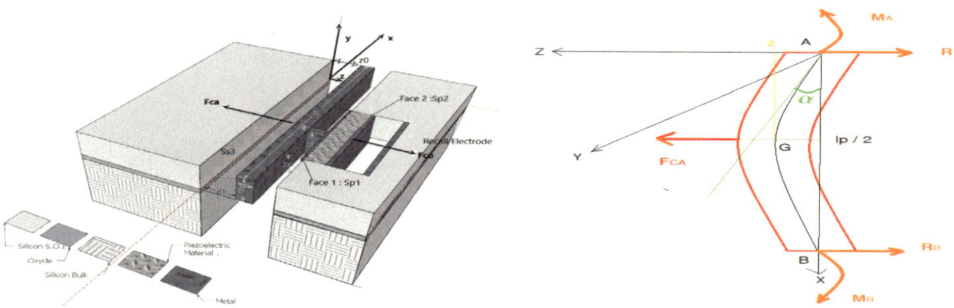

Figure 9.
Piezoelectric bridge cutting reactions and bending moment, deflection.

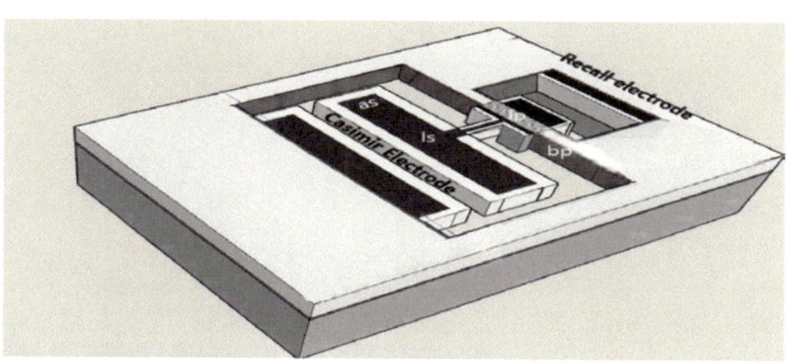

Figure 10.
Final structure with the metal oxides surrounding the metal electrodes.

$$\overrightarrow{F_T} = \overrightarrow{F_{CA}} - \overrightarrow{F_{CO}} = (1-p)\overrightarrow{F_{CA}} \Rightarrow \overrightarrow{F_T} < 0$$

Becoming repulsive, this force Ft (dependent on time) induces a deformation of the piezoelectric bridge in the other opposite direction, and the piezoelectric bridge returns or slightly exceeds (because of inertia) its neutral position, without initial deformation, therefore towards its position without any electrical charge.

The variation in time of these mobile charges follows, as a first approximation, a law of distribution of the charges on a short-circuited capacitor. Indeed, the fixed electrode S_r initially at zero potential since at ground, is now isolated by switch circuit 2 which is open and isolates it from ground! This temporal variation of the charges is given by the well-known exponential form of discharge of a capacitor according to the formula:

$$Q_{mn} = Q_{mn2}\ Exp\left(-\frac{t}{R_m C_S}\right) \tag{6}$$

This variation in mobile charges stops when these electrical charges Q_{mn} are uniformly distributed over the two electrodes S_{p2} and Sr. and are equal to $Q_{mn2}/2 = -Q_{mn1}/2$ on the two electrodes. Therefore, at time $t_e = R_m C_s \ln(2)$ (Eq. (5)), t_e being the time to reach equilibrium, with R_m the ohmic resistance of the metal track L_{in} of the inductance, C_s the capacitance formed by the electrodes S_{p2} and S_r and the input capacitances of the electronics (**Figure 11**).

This homogenization of electric charges within a metallic conductor:

1. occurs when the gate voltage of the MOSs constituting circuits 1 and 2 exceeds their threshold voltage.

2. induces an attraction of the piezoelectric structure in the direction opposite to that of Casimir.

3. decreases the deformation of the piezoelectric bridge and brings the gate voltage back below the threshold voltage.

4. The transistor MOS only turns off after the charges are homogenized during the short time t_e.

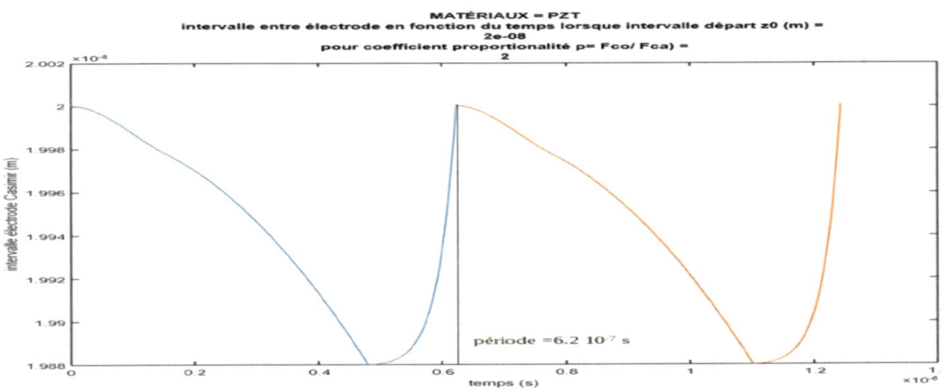

Figure 11.
Interval between casimir electrodes as a function of time for a proportionality coefficient FCO / FCA = 2.

We therefore obtain a current peak during this homogenization with a duration t_e of the order of a nanosecond! This current peak I_{IN} circulating for the duration of time t_e is:

$$IIN = d\frac{Qmn}{d}t'' \; I_{IN} = -\frac{Q_{mn2}}{R_m C_s}\left(Exp\left(-\frac{t}{R_m C_s}\right)\right) \tag{7}$$

We therefore obtain a current peak during this homogenization with a duration t_e of the order of a nanosecond! A current peak is obtained at time t = 0. With $Q_{mn2}/2$ the charge which is distributed uniformly over the two electrodes S_{p2} and Sr. The time t is counted from the closing of one of the transistors of circuit 1 and the opening of the switches of circuit 2. This current peak I_{IN} crossing an inductance L_{IN} during the time t_e, induces a voltage U_{IN} at the terminals of this inductance L_{IN} as a function of time according to the usual formula:

$$\begin{aligned} U_{IN} &= L_{IN}\frac{d(I_{IN})}{dt} = L_{IN}\frac{Q_{mn2}}{R_m C_s}(Exp\left(-\frac{t}{R_m C_s}\right) \\ &= L_{IN}\frac{Q_{mn2}ln(2)}{R_m C_s t_e}(Exp\left(-\frac{t}{R_m C_s}\right) = \mathrm{L_{IN}I_{IN}}\ln\frac{2}{t}e \end{aligned} \tag{8}$$

There is therefore a voltage peak across the inductance and the electronics appearing without power supply at time t = 0! As the deformations of the piezo-electric bridge cancel each other out during its "rise", the mobile charges on the surfaces S_{p1} as well as S_{p2} also cancel each other out! As a result, the gate voltage on circuit 1 and 2 MOSs drops below the threshold voltages and circuit 1 blocks. Circuit n°2 turns on again and connects face 2 of return electrode to ground, so the electrical charges on the bridge and the Sr. electrode cancel each other out! (see **Figure 5**).

The force of Casimir F_{CA}, still present, again attracts the metallic surface S_{S2} against S_{S3} and the events described above are repeated. Casimir's force deforms this bridge again and it seems that all starts all over again! The consequence is that the structure made up of the piezoelectric bridge, the connecting finger, the metal block forming the mobile Casimir electrode starts to vibrate, with a frequency dependent:

1. Of the Casimir restoring force, and of the Coulomb return electrode Force therefore:

    a. of the starting z_0 and z_r separation interface

    b. geometric dimensions of the different electrodes,

2. Properties of the piezoelectric bridge,

3. The choice of threshold voltages of the different MOS transistors

4. The choice of conductive metal!

As we will see, this frequency is lower than that of the first resonant frequency of the mobile structure if the initial interface z_0 is not weak enough (< 150 A°) to induce a sufficient Casimir force (see Chapters V and X)! An AC voltage peak U_{IN} is therefore automatically recovered at the terminals of the solenoid L_{IN}. This AC voltage peak can then be rectified to a DC voltage of a few volts, by suitable electronics operating without power supply (see amplification electronics without VI power supply).

Before moving on to theoretical calculations and mathematical simulations of the structure we wish to emphasize that the alternating signal U_{IN} is obtained without the input of any external energy!

In conclusions it seems (except errors) that all the electro-physical phenomena leading to a vibration of the structure and to the production of a voltage modulation are only the consequence of a first phenomenon which is at the origin of the Force of Casimir induced by fluctuations in vacuum energy.

They occur naturally and automatically without the input of any external energy except that of a vacuum ... without contradicting Noether's theorem!

3. Calculation of the current generated by the Casimir structure

If the initial separation interface z_0 is greater than 150 A°, the forces present are too weak to induce a vibration frequency of the device corresponding to its first resonant frequency (see Chapter V). We sought the numerical solutions of the differential equations obtained and unfortunately insoluble analytically when the device does not vibrate at its first resonant frequency!

3.1 Calculation of the frequency of vibration of the Casimir structure

Let us calculate the evolution in time of the force of Casimir which is applied between the two electrodes separated by an initial distance z_0. Apply the theorem of angular momentum to this vibrating structure. The angular momentum of the device is

$$\overrightarrow{\sigma^S_{A_{x,y,z}}(structure)} = \overline{\overline{I^S_{A_{x,y,z}}}}\ \overrightarrow{\Omega^S_A} \tag{9}$$

With the angular momentum vector of the structure, the inertia matrix of the structure with respect to the reference (A, x, y, z) and the rotation vector of the piezoelectric bridge with respect to the axis Ay with α the low angle of rotation along the y axis of the piezoelectric bridge.

$$\text{We have because } z << l_p\ sin\,(\alpha) = \sin\,(2z/l) \approx \frac{2\ z}{l_P} \Rightarrow \overrightarrow{\Omega^S_A}$$

$$= \begin{pmatrix} 0 \\ d\alpha/dt \\ 0 \end{pmatrix}\quad with\, d\alpha/dt \approx \frac{2\ dz}{l_P\ dt}$$

Let (Gp, x, y, z), (Gi, x, y, z), (Gs, x, y, z) be the barycentric points respectively of the piezoelectric bridge, of the metal connecting finger and of the metal block constituting the sole mobile of the Casimir reflector. We have (**Figure 5**):

$$\overrightarrow{AG_{P,x,y,z}} = \frac{1}{2}\begin{pmatrix} l_P \\ b_P \\ a_P \end{pmatrix}\quad \overrightarrow{AG_{I,x,y,z}} = \frac{1}{2}\begin{pmatrix} l_P & +l_i \\ b_P & +b_i \\ a_P & +a_i \end{pmatrix}\quad \overrightarrow{AG_{S,x,y,z}} = \frac{1}{2}\begin{pmatrix} l_P & +l_i+l_s \\ b_P & +b_i+b_s \\ a_P & +a_i+a_s \end{pmatrix}.$$

The inertia matrix of the bridge, in the frame of reference (Gp, x, y, z) is

$$\overline{\overline{I^P_{GP}}} = \frac{m_P}{12}\begin{pmatrix} a_P^2+b_p^2 & 0 & 0 \\ 0 & l_P^2+b_p^2 & 0 \\ 0 & 0 & l_P^2+a_p^2 \end{pmatrix} \tag{10}$$

The inertia matrix of the finger is, in the frame of reference (Gi, x, y, z) is:

$$\overline{\overline{I^I_{GI}}} = \frac{m_i}{12}\begin{pmatrix} a_i^2 + b_i^2 & 0 & 0 \\ 0 & l_i^2 + a_i^2 & 0 \\ 0 & 0 & l_i^2 + b_i^2 \end{pmatrix} \tag{11}$$

The inertia matrix of the reflector is, in the frame of reference (G_S, x, y, z) is:

$$\overline{\overline{I^C_{GS}}} = \frac{m_S}{12}\begin{pmatrix} a_S^2 + b_S^2 & 0 & 0 \\ 0 & l_S^2 + a_S^2 & 0 \\ 0 & 0 & l_S^2 + b_S^2 \end{pmatrix} \tag{12}$$

The total inertia of the structure becomes in the reference (A, x, y, z), $I^S{}_{A,x,y\ z} = I^P{}_{A,x,y,z} + I^I{}_{A,x,y,z} + I^C{}_{A,x,y,z}$ with A at the edge of the embedded piezoelectric bridge and $I^P{}_{A,x,y,z}, I^I{}_{A,x,y,z}, I^C{}_{A,x,y,z}$ is the inertia matrix obtained taking Huygens' theorem applied to this structure. The angular momentum theorem applied to the whole structure gives

$$\frac{d\left(\overrightarrow{\sigma^S_{A,x,y,z}}\right)}{dt} = \overline{\overline{I^S_{A,x,y,z}}}\,\frac{d\left(\overrightarrow{\Omega^S_A}\right)}{dt} \Rightarrow \overline{\overline{I^S_{A,x,y,z}}}\,\frac{2}{l_P}\begin{pmatrix} 0 \\ d^2z/dt^2 \\ 0 \end{pmatrix}$$

$$= \overrightarrow{\sum_A \text{Moments on the structure}} = \overrightarrow{M_A} + \overrightarrow{M_B} + \overrightarrow{F_{CA}} \wedge \begin{pmatrix} 2/l_P \\ 0 \\ 0 \end{pmatrix}$$

$$\text{with } \overrightarrow{F_{CA}} = \begin{pmatrix} 0 \\ 0 \\ F_{CA} \end{pmatrix} \tag{13}$$

We know (see X), according to the axis of Az that the moments: $M_{AY} = M_{BY} = -F_{CA}\ lp/8$, Therefore: Σ Moments on the structure relative to the axe $Ay = 1/4 * lp * F_{CA}$.

Any calculation done

$$\overline{\overline{I^S_y}}\,\frac{2}{l_P}\,\frac{d^2z}{dt^2} = \frac{l_P}{4}F_{CA} = \frac{l_P}{4}S_S\frac{\pi^2\hbar c}{240}\,\frac{1}{z^4} \tag{14}$$

with I_{Sy} the inertia of the structure relatively to the axe Ay. See Eq. (14) below.

$$\overline{\overline{I^S_y}} = \rho_P a_{P b_P l_P}\left(\frac{l_P^2 + a_P^2}{12} + \frac{l_P^2 + a_P^2}{4}\right) + \rho_I a_{i b_i l_i}\left(\frac{l_i^2 + a_i^2}{12} + \frac{(l_P + l_i)^2 + (a_P + a_i)^2}{4}\right)$$
$$+ \rho_S a_{S b_S l_S}\left(\frac{l_S^2 + a_S^2}{12} + \frac{(l_P + l_i + l_S)^2 + (a_P + a_i + a_S)^2}{4}\right) \tag{15}$$

with ρ_P, ρ_I, ρ_S, respectively the densities of the piezoelectric bridge, the intermediate finger and the mobile electrode of the Casimir reflector.

We then obtain the differential equation which makes it possible to calculate the interval between the two electrodes of the Casimir reflector as a function of time during the "descent" phase when the Coulomb forces are not present.

$$\frac{d^2z}{dt^2} = \frac{l_P^2}{8\ .I_Y^S} S_S \frac{\pi^2\hbar c}{240} \frac{1}{z^4} = \frac{B}{z^4} \quad with\ B = \frac{l_P^2}{8I_Y^S} S_S \frac{\pi^2\hbar c}{240} \tag{16}$$

Coulomb forces do not intervene yet because the MOS switches in parallel of circuit 1—before the inductance Lin—are open and the MOS switches in series of circuit 2—after the inductance L_{in}—being closed, the return Coulomb electrode is to earth. The fixed Casimir electrode is always to earth (see **Figures 3** and 5). Coulomb forces will intervene when the gate voltage $V_G = Q_{mp2\ MOS}/C_{OX}$ on the MOSs of circuit n°1 exceeds the threshold voltage of one of them and when circuit n°2 of the depleted N and P MOSs in series will be open (**Figures 3** and 5)! Then the switches of the circuit of the parallel MOS transistors will close.

The switches of the series MOS circuit will open and the charge Q_{mn1} initially present exclusively on the electrode of the bridge and of the metallic block will be distributed uniformly over the second part of coulomb electrodes according to:

$$Q_{mn1f} = Q_{mn1} \frac{S_{P1}}{S_{P1} + S_R} \approx \frac{Q_{mn1}}{2}.$$

Because $S_r = S_{p1}$, Just at the moment of closing circuit n°1 and opening circuit n°2 (**Figure 5**) we have F_{CO} = -p F_{CA} with p a coefficient of proportionality ≥2 defined by the threshold voltages of the MOS interrupters. The total force F_T exerted in the middle of the piezoelectric bridge just at the start of the charge transfer becomes $F_T = F_{CA} - F_{CO} = F_{CA} - p^*\ F_{CA} = F_{CA}\ (1\text{-}p)$.

The "descent" time of the free Casimir electrode will therefore stop when F_{CO} = -p F_{CA}.

However, we know that: $F_{ca} = S\left(\frac{\pi^2\hbar c}{240 z_S^4}\right)$

1. The Casimir force is variable in time and its equation is Eq. (1):

2. The mobile charge on the Casimir electrodes variable also in time (Eq. (3)) is:

$$Qmn2 \approx \frac{Qmn}{2} = \frac{d_{31}F_{CA}\,l_P}{2a_P}$$

3. The Coulomb force (4), variable over time, acting in opposition to the Casimir force (Eq. (4)):

$$F_{CO} = pF_{CA} \Rightarrow \left(S_S \frac{\pi^2\hbar c}{240} \frac{d_{31}l_P}{2^*\,a_P}\left(\frac{1}{z_s^4} - \frac{1}{z_0^4}\right)\right)^2 \left(\frac{1}{4\pi\varepsilon_0\varepsilon_r}\right)\left(\left(\frac{1}{z_r + z_0 - z_S}\right)^2\right) = pS_S \frac{\pi^2\hbar c}{240} \frac{1}{z_s^4}$$

The differential Eq. (15) unfortunately does not have a literal solution and we programmed on MATLAB the solution of this differential equation "descent" and calculated the duration of this “descent” of free Casimir electrode. The duration of the “descent” depending on the desired value of the coefficient of proportionality p, which is regulated by the values of the threshold voltages of the MOS transistors and defined during the manufacture of the device. The “descent” of the free Casimir electrode stops when the inter electrode interface z_s is such that:

$$z_s^4\left(\left(\frac{1}{z_r+z_0-z_S}\right)^2\left(\frac{1}{z_S^4}-\frac{1}{z_0^4}\right)\right)=p\frac{3840\pi\varepsilon_0\varepsilon_r}{\pi^2\hbar c S_S}\left(\frac{a_P}{d_{31}l_P}\right)^2 \tag{17}$$

This programmable equation gives the time t_d of the “descent” of the structure submitted to the Casimir force and:

a. depend on the coefficient of proportionality p,

$$F_T=(1-p)F_{CA}\Rightarrow(1-p)S_S\frac{\pi^2\hbar c}{240}\frac{1}{z_{sm}^4}<0 \quad if p>1$$

b. is calculable and will stop when the inter-electrode interface z_s has a value z_{sm} satisfying Eq. (16). At the instant of the appearance of the Coulomb force, with z_{sm} the value solving 16, the total force is therefore:

1. The total force, variable over time and exerted at the center of the piezoelectric bridge, becomes (Eq. (17)):

$$\begin{aligned}F_T&=F_{CA}-F_{CO}\\&=S_S\left(\frac{\pi^2\hbar\ c}{240}\right)\left(\frac{1}{z_s^4}-S_S\frac{\pi^2\ \hbar\ c}{240}\left(\frac{d_{3\ 1}\ l_P}{a_P}\right)^2\right)\left(\frac{1}{16\ \pi\ \varepsilon_0\ \varepsilon_r}\right)\left(\frac{1}{z_s^4}-\frac{1}{z_s^4}\right)^2\left(\frac{1}{(z_r+z_0-z_S)^2}\right)\end{aligned} \tag{18}$$

The piezoelectric bridge subjected to this force then rises towards its neutral position. The Casimir inter electrode interval increases causing the Casimir force to decrease! As the deformations of the piezoelectric bridge decrease, the electric charge present on the piezoelectric face’s decreases, which consequently leads to a drop in the Coulomb Force. The F_T force therefore rapidly approaches the starting F_{CA} force, during the “ascent” of the Casimir electrodes.

Let us calculate the duration of this “rise” of the mobile electrode of the Casimir reflector triggered when $F_{CO}=p^*\ F_{CA}$.

To know the time taken by the structure to “go back” to its neutral position, we must solve the following differential equation (Eq. (18)):

$$\begin{aligned}\frac{d^2z}{dt^2}&=\frac{l_P^2}{8.I_Y^S}(F_{CA}-F_{CO})\\&=\frac{l_P^2}{8.I_Y^S}\left(lsb_S\frac{\pi^2\hbar c}{240}\frac{1}{z_s^4}-lsb_S\frac{\pi^2\hbar c}{240}\frac{d_{31}l_P}{2^*a_P}\frac{1}{4\pi\varepsilon_0\varepsilon_r}\left(\frac{1}{z_s^4}-\frac{1}{z_0^4}\right)^2\left(\frac{1}{z_r+z_0-z_S}\right)^2\right)\end{aligned}$$

By posing $A1=lsb_S\left(\frac{\pi^2\hbar c}{240}\right)$ the differential Eq. (17) concerning the “ascent” of the bridge is written:

$$\frac{d^2z}{dt^2} = \frac{l_P^2}{8.\ I_Y^S}(F_{CA} - F_{CO})$$
$$= \frac{l_P^2}{8.\ I_Y^S} A1\left(\frac{1}{z_s^4} - A1\left(\frac{d_{3\ 1} l_P}{a_P}\right)^2\right)\left(\frac{1}{16\ \pi\ \varepsilon_0\ \varepsilon_r}\right)\left(\frac{1}{z_s^4} - \frac{1}{z_0^4}\right)^2\left(\frac{1}{(z_r + z_0 - z_S)^2}\right) \tag{19}$$

This differential Eq. (18) has no analytical solution and can only be solved numerically. We programmed it on MATLAB with the inter-electrode distance z_s belonging to the interval $[z_{sm}, z_0]$. The properties and dimensions of the different materials used in this simulation are as follows (**Table 1**).

The metal used for the Casimir reflector block is Aluminum with a density of 2.7 g cm^{-3}.

In these MATLAB calculations we considered that the metal of the electrodes and metal block was oxidized over a thickness allowing to have an interface between Casimir electrodes of 200 A° (see Chapter 5) which modifies the mass and the inertia of the vibrating structure. It turns out that the choice of aluminum as the metal deposited on these electrodes is preferable given:

1. The ratios between the thickness of the metal oxide obtained and that of the metal attacked by the growth of this oxide during its thermal oxidation (see Chapter V).

	PZT	**AlN**	**LiNbO3**	**PMN-PT: (1-x) Pb(Mg1/3–Nb1/3)O3-xPbTiO3**
Young Modulus $(\mathbf{kg^*\ m^*\ s^{-2}})/\mathbf{m^2}$	Ep = 8.9*10^{10}	Ep = 32*10^{10}	Ep = 2.45*10^{9}	Ep = 150*10^{9}
Volumic mass ($\mathbf{kg\ m^{-3}}$)	dp = 7600	dp = 3255	dp = 4700	dp = 7920
Piezoelectric coefficient d31 of the beam $(\mathbf{C}/(\mathbf{kg^*\ m^*\ s^{-2}}))$	d31 = 200*10^{-12}	d31 = 2.400*10^{-12}	d31 = 6*10^{-12}	d31 = 1450*10^{-12}
Length piezoelectric beam lp (m)	50 10^{-6}	50 10^{-6}	50 10^{-6}	50 10^{-6}
Width piezoelectric beam bp (m)	150 10^{-6}	150 10^{-6}	150 10^{-6}	150 10^{-5}
Thickness piezoelectric ap (m)	10 10^{-6}	10 10^{-6}	10 10^{-6}	10 10^{-6}
Connecting finger length li (m)	10 10^{-6}	10 10^{-6}	10 10^{-6}	10 10^{-6}
Width finger connection bi (m)	150 10^{-6}	150 10^{-6}	150 10^{-6}	150 10^{-6}
Thickness finger connection ai (m)	10 10^{-6}	10 10^{-6}	10 10^{-6}	10 10^{-6}
Mobile Casimir electrode block length ls (m)	500 10^{-6}	500 10^{-6}	500 10^{-6}	500 10^{-6}
Mobile Casimir electrode block width bs (m)	150 10^{-6}	150 10^{-6}	150 10^{-6}	150 10^{-6}
Casimir mobile electrode block thickness as (m)	10 10^{-6}	10 10^{-6}	10 10^{-6}	10 10^{-6}

Table 1.
Table of characteristics used for MATLAB and ANSYS simulations.

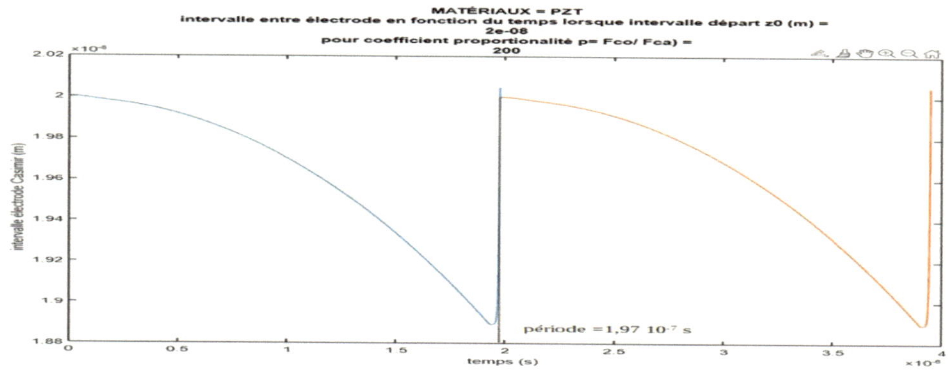

Figure 12.
Vibrations of the structure for a coefficient of proportionality p = FCO / FCA = 200: PZT.

2. As its density is weak, we chose aluminum for the purpose to increasing and optimizing the vibration frequency of the structure by minimizing the inertia of the Casimir's reflector and parallelepiped block transferring the Casimir force to the piezoelectric bridge. The mass M of the vibrating structure is then:

$$M = d_{pm}{}^{*}(a_s{}^{*}b_s{}^{*}l_s + a_i{}^{*}b_i{}^{*}l_i) + d_{om}{}^{*}2^{*}z_{of}{}^{*}\left(a_{so}{}^{*}b_{so} + b_{so}{}^{*}l_{so} + a_{so}{}^{*}l_{so)}\right) + d_p{}^{*}\left(a_p{}^{*}b_p{}^{*}l_p\right);$$

With d_{pm} the density of the metal, a_s,b_s,l_s the geometries of the final metal part of the Casimir electrode sole, d_{om} the density of the metal oxide, a_{so},b_{so},l_{so} the geometries of the oxidized parts around the 6 faces of the metal block, dp the density of the piezoelectric parallelepiped (see **Figure 12**):

3.2 Calculation of the current peak

Let us estimate the duration of the current peak linked to the circulation and homogenization on the return electrodes of the mobile charges. Let R_m be the ohmic resistance of the metals used for the surface electrodes S_{p1} + the L_{IN} solenoid + the S_r electrode (see **Figure 3**) and Cs the capacitance formed by the gap between the return electrodes Sp2 and S_r. Then the current peak circulating during the transition of the mobile loads between Sp1 to S_R via circuit n°1 and the L_{IN} solenoid is, as we have seen:

$$I_{IN} = \frac{Q_{mn2}}{R_m C_S} Exp\left(-\frac{t}{R_m C_S}\right)$$

The time t being counted from the closing of the MOSNE switch. The duration of this current peak is estimated at $t_e = R_m.C_s$. Log (2), when the charges on each electrode will be $Q_{mn2}/2$. This current peak is present even if the switch transistors may close some time after, because the mobile charges have already propagated.

- $R_m \cong r_m{}^{*}\, l_m/S_m$, with ρ_m the resistivity of the metallic conductor in the circuit between electrodes (the solenoid + the electrodes themselves), l_m its total length of this conductor, S_m its section

- $Cs = \varepsilon_0.\varepsilon_{om}.l_p.b_p/z_r$, the inter-electrode return capacitance, with ε_0 the permittivity of vacuum, v the relative permittivity of the metal oxide, l_p and b_p the geometries of the return electrode. A calculation of the duration of the homogenization of the electric charges and therefore of the duration of the current peak (based on an estimate to propagate in a L_{IN} coil of about 10^{-5} Henri) gives $t_e \cong 10^{-9}$s. This current peak, passing through a L_{IN} solenoid develops a voltage peak $U_{IN} = L_{IN} I_{INP}/t_e = L_{IN}{}^{*}\ Q_{mn2}/(2.te^{*}\ R_m.C_s)$ which will be exploited by integrated electronics without any power supply described in Chapter IV.

We present below the results of the MATLAB simulations carried out by numerically calculating the differential Eqs. (15) and (18). These numerical calculations give the vibration frequency of the structure which, as we will see, vibrates at a frequency lower than its first resonant frequency (IV).

This vibration frequency depends on the characteristics of the structure (Nature of the piezoelectric material, nature of the metallic conductors, initial interface z_0 and z_r between Casimir electrodes and return electrodes, geometric dimensions of the Casimir reflectors, coefficient of proportionality $p = F_{CO}/F_{CA}$ ...)) (See IV and Annex).

4. Simulation of devices with different piezoelectric bridge

We will see that this device vibrates at a frequency lower than its first resonant frequency and that its vibration frequency depends on the characteristics of the structure (Nature of the piezoelectric material, nature of the metallic conductors, starting interface z_0 *and* z_r *between Casimir electrodes and return Coulomb electrodes, dimensions of the Casimir reflectors, coefficient of proportionality* $p = F_{CO}/F_{CA}$ *...). Except precision the interface* z_r *between the Coulomb's electrode is chosen the same that those of Casimir reflector* z_0

4.1 Piezoelectric materials = PZT (lead zirconia titanium)

4.1.1 Interface between Casimir electrodes as a function of time for different trigger values of MOS transistors

For a starting interface between Casimir electrode of $z_0 = 200^{*}\ 10^{-10}$ (m) and a coefficient of proportionality $p = F_{CO}/F_{CA} = 2$, we obtain the following evolution in time of the Casimir interface:

We notice a phase of rise of the Casimir electrode faster than that of descent. The period of vibrations is 6.18 10^{-7} s therefore with a vibration frequency of 1.613 10^6 Hertz, while the first resonant frequency of the same structure is 6.54 10^6 hertz.

The moving electrode drops to $z_s = 198.8$ Angstroms from the fixed electrode S_{S3}. The current peak for this coefficient of proportionality $p = 2$ is 2.58 10^{-8} A. This current is obtained by adjusting the threshold voltage of the enriched and depleted MOS transistors to a value Vt = 0.6553 V for a length L = width = W = 4 10^{-6} m and with a grid oxide thickness SiO2 = tox of 250 10^{-10} m (see **Figure 13**)! Let us simply change the coefficient $p = F_{CO}/F_{CA}$ of proportionality to p = 200, then we get (see **Figure 14**):

We notice for the ratio $p = F_{CA}/F_{CO}$ = 200 (**Figure 14**), a phase of "rise" of the Casimir electrode also much faster than that of "descent" but also more dynamic than for the ratio of previous p = 2 The vibration frequency of the device of 5.07 10^5 hertz, while the first resonant frequency of the structure is still 6.54 10^6 hertz!

The moving electrode is now approaching to $z_s = 188.9$A° of the fixed electrode S_{S3}, so the vibration amplitude of the structure is 200–188.9 = 11.1 Angstroms!

This current is obtained by adjusting the threshold voltage of the enriched and depleted MOS transistors to a value Vt = 6.89 V for the same geometries as above

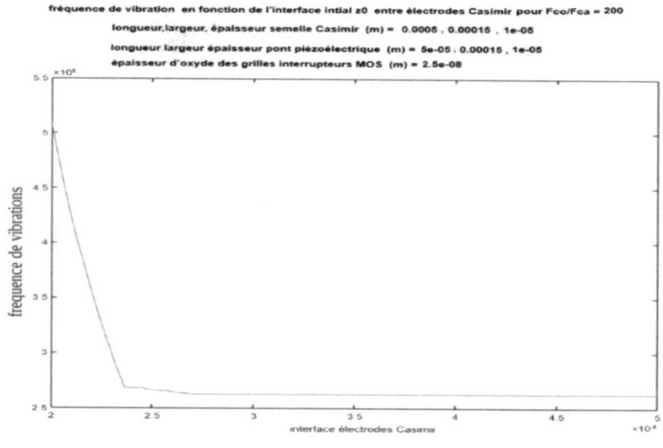

Figure 13.
Structure vibration frequency = f (starting interface zo): F_{CO} / F_{CA} chosen = 200: PZT.

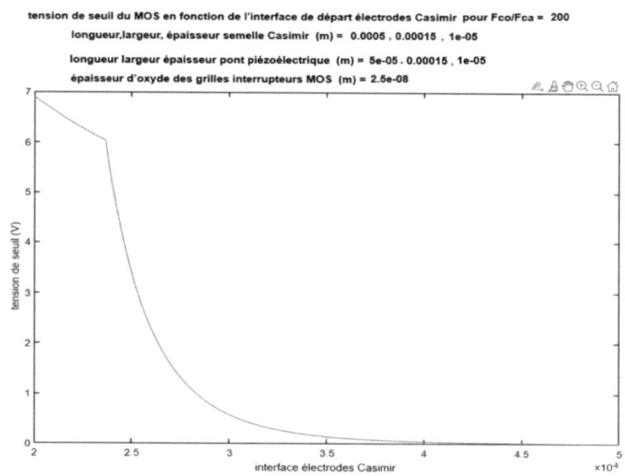

Figure 14.
MOS threshold voltage = f (starting interface zo): F_{CO} / F_{CA} chosen = 200: PZT.

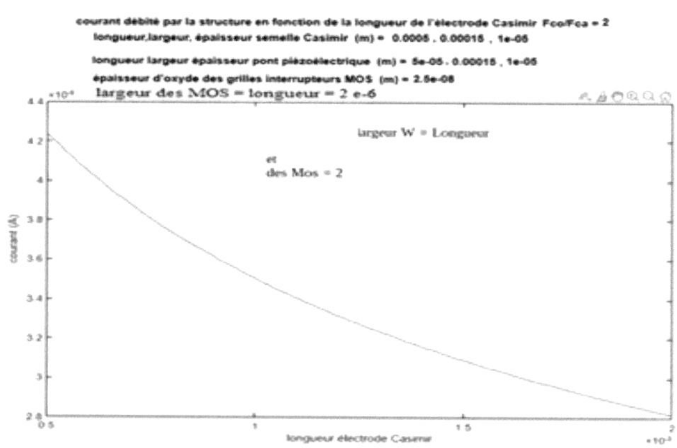

Figure 15.
Maximum current = f (length of the Casimir electrode ls), starting interface = 200 A °, selected coefficient of proportionality p = F_{CO} / F_{CA} = 2.

(see **Figure 15**). We must therefore adjust the threshold voltages to precisely adjust the ratio $p = F_{CO}/F_{CA}$ for which the Coulomb force is triggered. This is a point that can be easily obtained technologically (see the technological part of this report)! In conclusion, as the vibration frequency of the structure depends, among other things, on the coefficient of proportionality p and therefore on the current that one wishes to obtain, *the structure does not vibrate at its first resonant frequency*.

4.1.2 Variation of the starting interface z_0 between Casimir electrodes: PZT

We notice (**Figure 16**) that the vibration frequency of the structure drops as the initial space between the Casimir electrodes increases, which is related to a decrease in the Casimir Force and therefore makes sense. The vibration frequency depends on the chosen F_{CO}/F_{CA} ratio. This frequency drops and stabilizes around 2.6 MHz as the electrode interface increases by a ratio of 200. It is much lower than the first resonant frequency of the structure which is 6.85 Megahertz (for this structure). The vibration frequency approaches that of first resonance if the starting z_0 interface is less than 200 Angstroms.

We chose an initial interface of 200 A° for reasons of technological feasibility (see VI)!

We also notice (**Figure 17**) that the threshold voltage of the Enriched and Depleted MOS transistors increases with a decrease in the starting interface

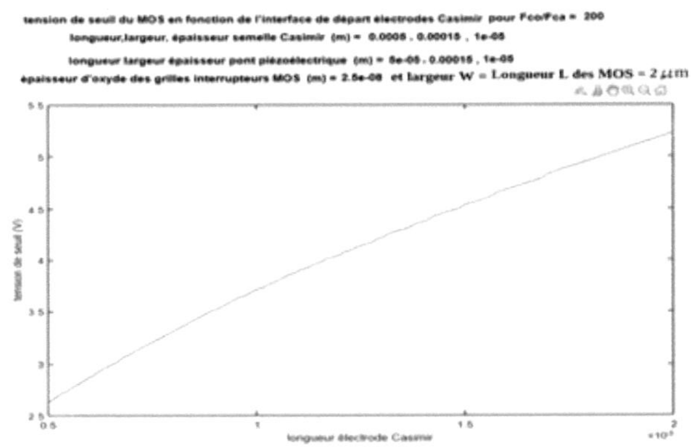

Figure 16.
Threshold Voltage = f (length of the casimir electrode ls), starting interface = 200 A °, selected coefficient of proportionality p = F_{CO} / F_{CA} = 2.

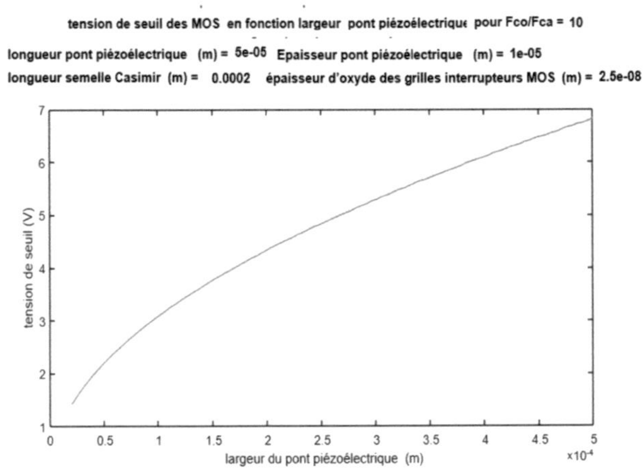

Figure 17.
Threshold voltage of the MOS = f (width of the casimir electrode bp), starting interface = 200 A °, selected coefficient of proportionality = p = F_{CO} / F_{CA} = 10.

between Casimir electrodes. This seems logical, since the Casimir force increasing, the deflection of the bridge and therefore the charges generated on its faces do the same. It is therefore necessary that the threshold voltage of the MOS transistors be greater so that the voltage V_G on the gates of the MOS does not trigger them!!

4.1.3 Current and threshold voltage function of the length ls of the Casimir electrode: PZT

We obtain (**Figure 18**) a small decrease in current with the increase in the length of this electrode. However, a significant increase in the threshold voltage (**Figure 19**), which is understandable since the inertia of the structure increases.

We obtain (**Figure 18**) a small decrease in current with the increase in the length of this electrode and a significant increase in the threshold voltage (**Figure 19**), which is understandable since the inertia of the structure increases.

4.1.4 Variation of the width bp of the piezoelectric bridge: PZT

We now vary the width b_p of the piezoelectric bridge. We obtain an increase in the threshold voltage of the MOS by increasing the width b_p of the piezoelectric

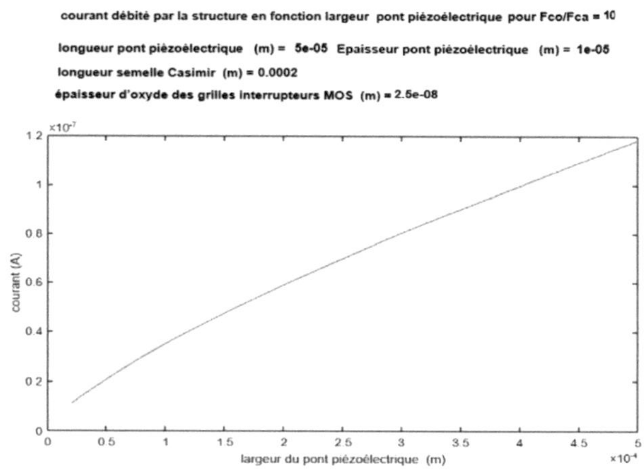

Figure 18.
Maximum current = f (width of the casimir electrode bp), starting interface = 200 A °, selected coefficient of proportionality = p = F_{CO} / F_{CA} = 10.

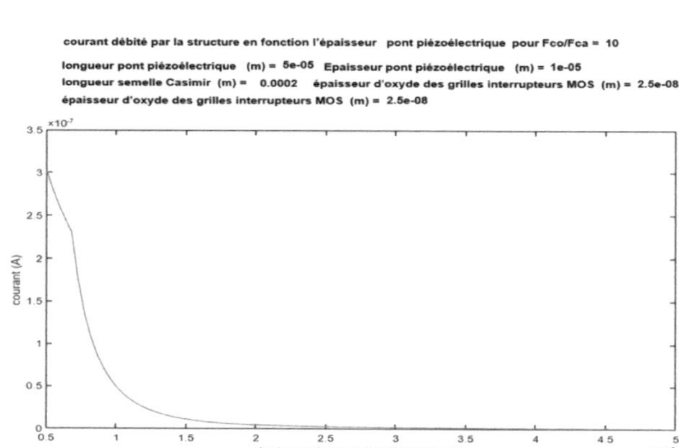

Figure 19.
Current of the MOS = f (Thickness of piezoelectric film ap), start Interface = 200 A ° with a choice F_{CO} / F_{CA} =10.

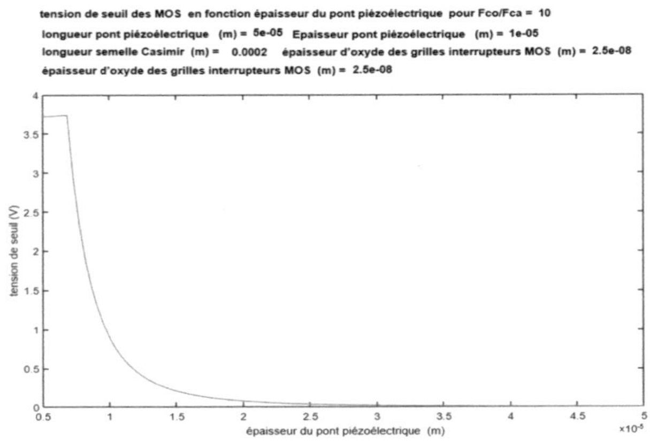

Figure 20.
Threshold of the MOS = f (Thickness of piezoelectric film ap), start Interface = 200 A ° with a choice F_{CO} / F_{CA} =10.

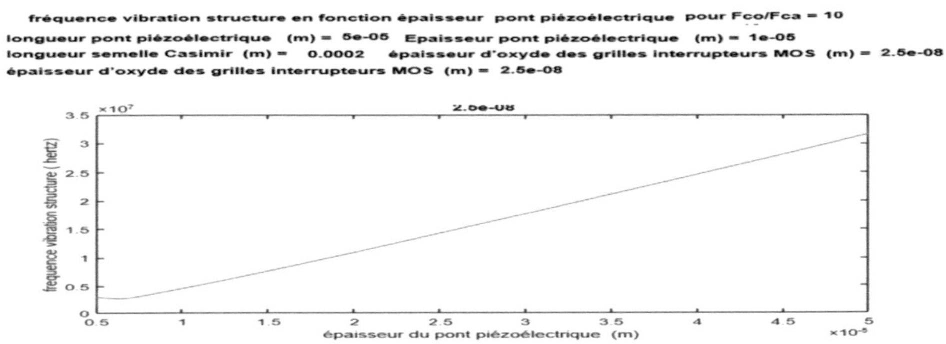

Figure 21.
Structure vibration frequency as a function of the thickness ap of the piezoelectric bridge, starting interface zo to 200 A °, Ratio p = F_{CO} / F_{CA} = 10.

bridge. However, the current delivered by the structure varies little with the width of the piezoelectric bridge (**Figures 20** and **21**). These considerations give that:

For reasons of technological convenience, it will be preferable to choose a thickness of around 20 μm!

4.1.5 Variation of the thickness ap of the piezoelectric bridge: PZT

If we increase the thickness a_p of the piezoelectric bridge, we obtain a decrease in the current (**Figure 22**) and of the threshold voltage of the MOS (**Figure 23**), but an increase in the vibration frequency (**Figure 24**).

The vibration frequency increases linearly with the thickness of the piezoelectric bridge (**Figure 25**).

4.1.6 Variation of the proportionality ratio $\mathbf{p} = \mathbf{F_{CO}}/\mathbf{F_{CA}}$*: PZT*

In a non-intuitive way, the current simply increases linearly by a factor of 40 (**Figure 26**) if we increase the proportionality ratio $p = F_{CO}/F_{CA}$ by a factor of 500. On the other hand, the threshold voltage of the MOS switches increases by a factor 8 for the same variation of the interface (**Figure 27**).

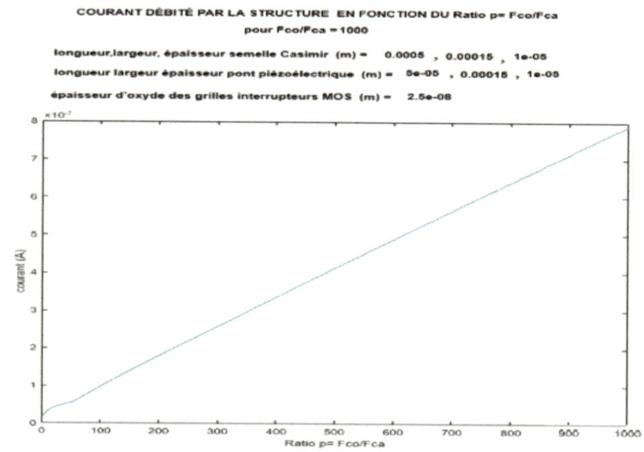

Figure 22.
Current of the MOS = f (ratio = F_{CO} / F_{CA}), start Interface = 200 A ° piezoelectric material = PZT.

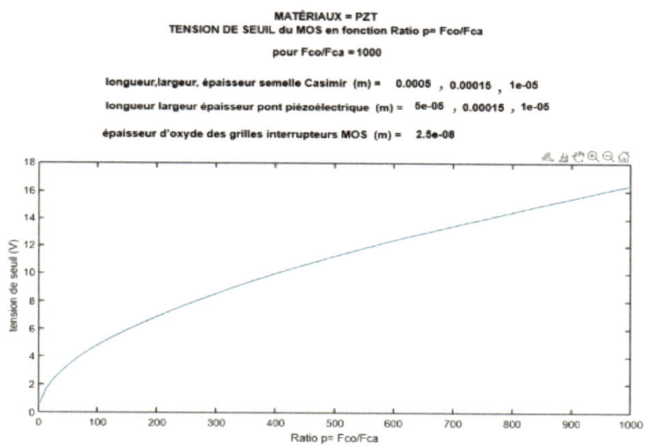

Figure 23.
Threshold voltage of the MOS = f (ratio = F_{CO} / F_{CA}), start Interface = 200 A ° piezoelectric material = PZT.

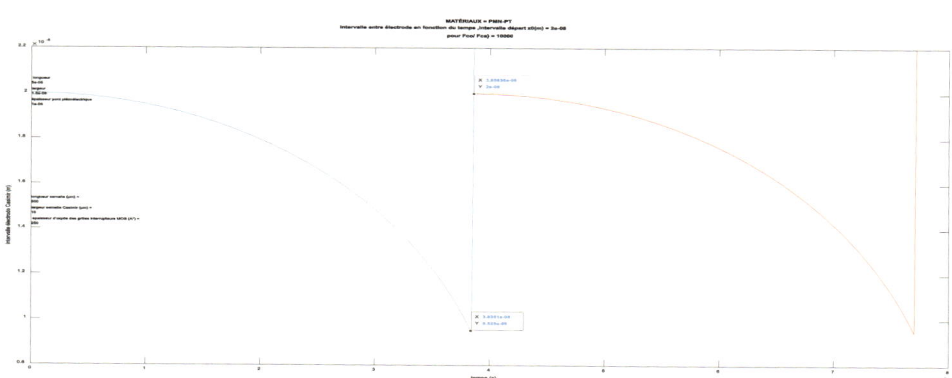

Figure 24.
Plot of the evolution of the Casimir inter-electrode interval as a function of time over two periods and an F_{CO} / F_{CA} ratio = 1000: Casimir inter-electrode interface = 200 A°.

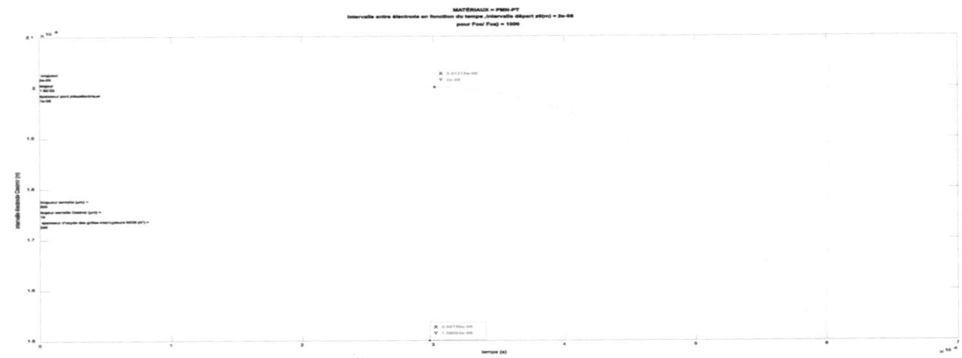

Figure 25.
Plot of the evolution of the Casimir inter-electrode interval as a function of time over two periods and an F_{CO} / F_{CA} Ratio = 1000: Casimir inter-electrode interface = 200 A °.

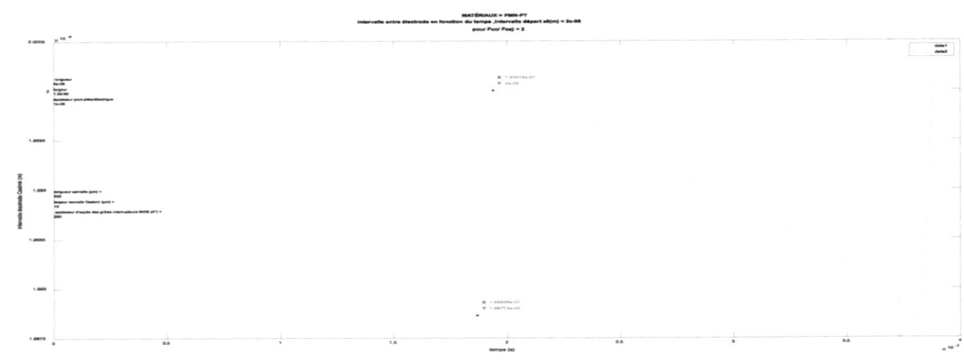

Figure 26.
Plot of the evolution of the Casimir inter-electrode interval as a function of time over two periods and a ratio F_{CO} / F_{CA} = 2. Casimir inter-electrode interface = 200 A °.

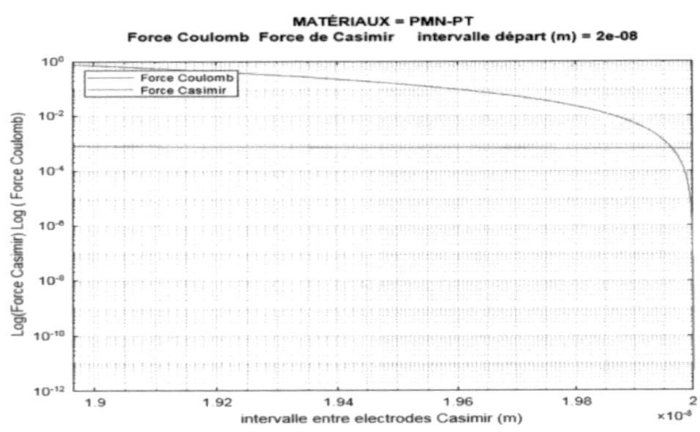

Figure 27.
Materials = PMN-PT: Coulomb and Casimir force as a function of the inter-electrode interface. Start interface = 200 A °.

The MOS N or P switch transistors enriched in parallel have the following geometries: Width W = 4 mm and length L = 4 mm

4.2 Use of other piezoelectric materials

In the presentation above we used PZT but, in order to increase the density of electric charges at the terminals of the piezoelectric bridge, piezoelectric material

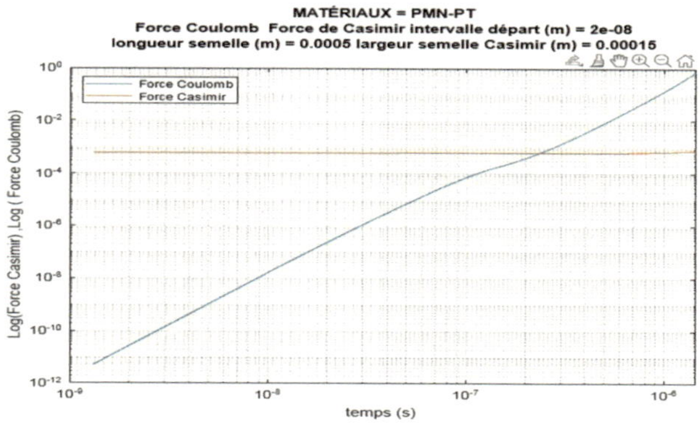

Figure 28.
Materials = PMN-PT: Coulomb and Casimir force as a function of time. Start interface = 200 A °.

PMN-PT can be used which can be deposited by RF-magnetron sputtering and of composition, for example: PMN-PT = (1-x) Pb (Mg1/3 – Nb1/3) O3-xPbTiO3; $d_{31} = 1450^* \ 10^{-12}C/(kg^* \ m^* \ s^{-2})$ and a Young's modulus of $Ep = 150^* \ 10^9$ (**Figure 28**). We will also simulate the results obtained with AlN (aluminum nitride) $(d_{31} = 2.4^* 10^{-12}\text{et } Ep = Ep = 32^* 10^{10})$, another piezoelectric material or AlN widely used in microelectronics because it is easily removable and lead-free!

4.2.1 Piezoelectric material = PMN-PT

With the MATLAB simulation of a structure using PMN-Pt we obtain the evolution over time of the Casimir and Coulomb forces as well as the F_{CO}/F_{CA} ratio of **Figures 7, 11, 13, 24, 29–39** below. For a ratio of 1000, the maximum current delivered by the vibrating structure, the threshold voltage of the MOSE and MOSD and the vibration frequency of the structure are respectively: 1.2 10^{-4}A, Vt = 3.2 V and 957,000 Hertz

4.2.1.1 Evolution of the Casimir interface as a function of time during two periods: PMN-PT

The F_{CO}/F_{CA} ratio = 10,000 induces a period of 3.85 10^{-6} s and a rise time of 21.3 10^{-9} s with a deflection of the bridge of 105 A°. The structure vibrates at

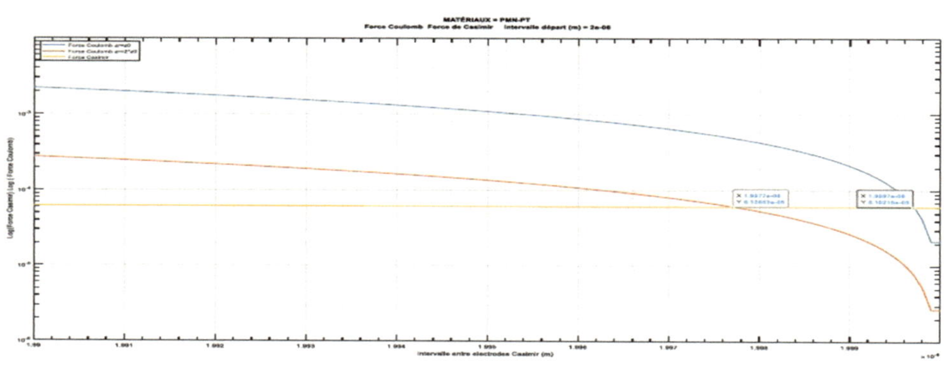

Figure 29.
Materials = PMN-PT: Coulomb force for zr = 200 A ° (Blue)and zr = 400 A ° (Red) and Casimir force (Yellow , zo =200 A °) as a function of the inter-electrode interface Starting interface = 200 A °.

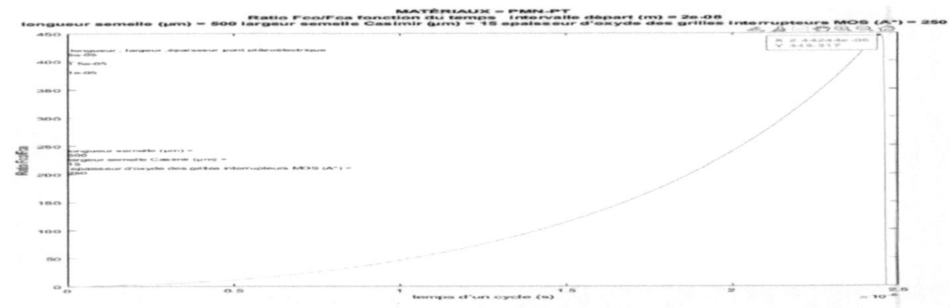

Figure 30.
Materials = PMN-PT: Ratio p = F_{CO} / F_{CA} as a function of time. During a period of vibration. Start interface = 200 A °, Maximum ratio chosen = 450 A °.

259.7 kHz. At the rise sequence, the structure exceeds the initial 200 A° by 20 A° due to inertia (**Figure 29**).

The ratio F_{CO}/F_{CA} = 1000 induces a period of 2.96 10^{-6} s and a rise time of 44.5 10^{-9} s with a deflection of the bridge of 50 A° The structure vibrates at 337.8 kHz.

For p = 1000 (**Figure 24**), we notice a vibration amplitude of 50 A° with a period of 2.96 10^{-6} s, with faster rise of the mobile electrode producing a slight rebound of 5A, because of the inertia of the structure.

For the ratio $F_{CO}/F_{CA} = 2$ (**Figure 30**) a vibration amplitude of just 0.27 A° and a period of 1.86$10^{-7}$ s is obtained This low deformation of the PMN-PT piezoelectric bridge is mainly due to the extremely high piezoelectric coefficient d_{31} of 1450 (pC/N) of PMN-PT compared to 120 (pC/N) for PZT (**Figure 28**). It is also observed that weak overshoot of the initial interface (200 A°) for the mobile electrode increases with the ratio (F_{CO}/F_{CA}).

4.2.1.2 Evolution of the forces of Casimir and Coulomb: PMN-PT

We obtain the evolution of the Casimir and Coulomb forces as a function of the inter-electrode interface (**Figure 13**) and over time (**Figure 31**) as well as the

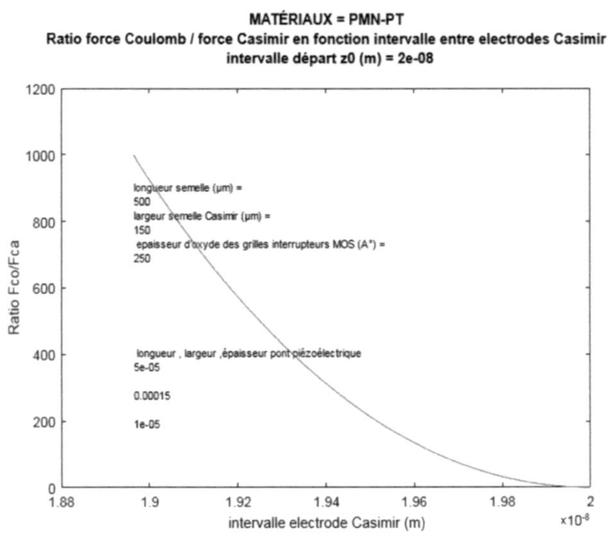

Figure 31.
Materials = PMN-PT: Coulomb Force / Casimir Force ratio as a function of the Casimir inter-electrode interface. Start interface = 200 A °.

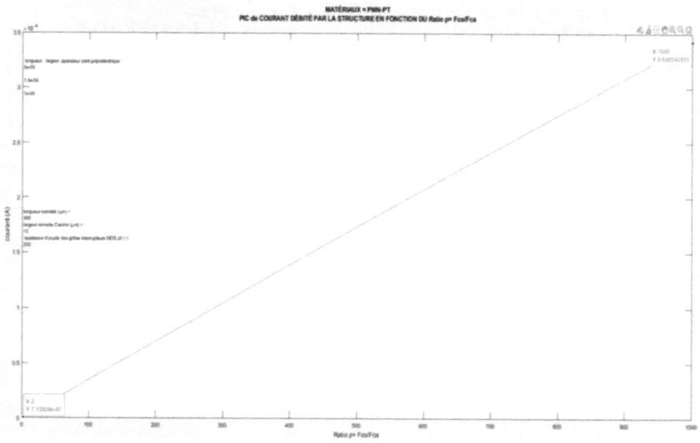

Figure 32.
Materials = PMN-PT: peak current delivered by the structure as a function of the F_{CO} / F_{CA} ratio. Start interface = 200 A °.

F_{CO}/F_{CA} ratio as a function of time for an entire period (**Figure 7**). For an interval between Casimir electrode z_0 = 200 Angstroms, we observe (**Figure 7**) that the Coulomb return force becomes more important than the Casimir force, if we induce a deflection of the piezoelectric bridge of 20 A° more for an interval z_r = 400 Angstroms between return electrodes than for z_r = 200 that.

The attraction of the electrodes by the Casimir force induces a deformation of the piezoelectric bridge, therefore electric charges, which can be used in the Coulomb force. The break circuits triggered at time t = 2.44 10^{-6} s suddenly induce a rise of the mobile electrode, therefore a sudden decrease in electric charges. We observe the gradual evolution towards the chosen ratio of 450 and then the sudden drop in this ratio as the electrodes regain their initial position (**Figure 32**).

4.2.1.3 Ratio as a function of Casimir interval and current peak as a function of the ratio: PMN-PT

We observe (**Figure 33**) that for PMN-PT a deflection of 10 A° and a length of the piezoelectric bridge of 150 µm of the mobile Casimir electrode is sufficient to have an Fco/Fca ratio = 1000. A Ratio of 2 gives a peak current of 7 10^{-7} A, while a ratio of 1000 produces a peak current of about 3.5 10^{-4} A (**Figure 34**) for the same period of homogenization of the charges of about 10^{-9} s!

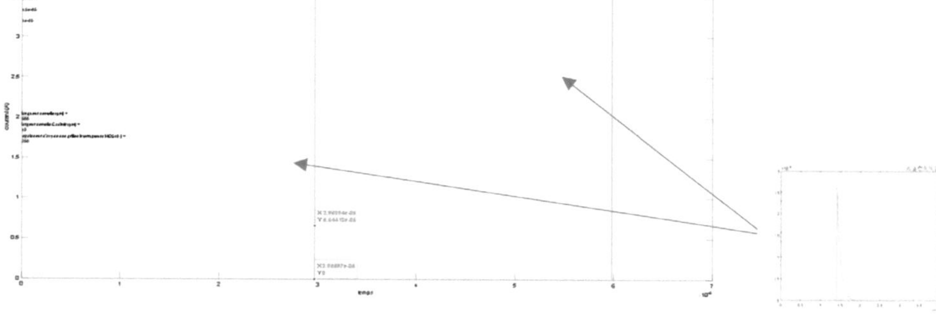

Figure 33.
Materials = PMN-PT: current peak as a function of time obtained over 2 cycles. Starting interface = 200 A °. Ratio p = F_{CO} / F_{CA} = 1000.

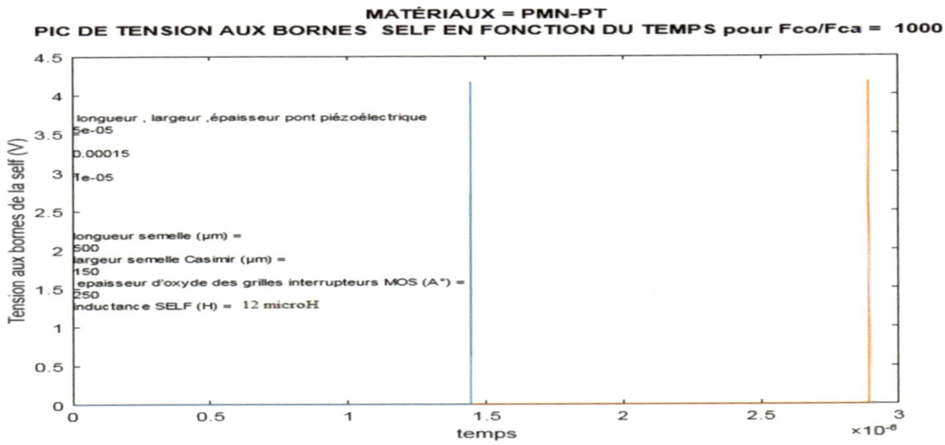

Figure 34.
*Materials = PMN-PT: Voltage peak across the 4 * 10^{-5} H solenoid as a function of the time obtained over 2 cycles. Starting interface = 200 A °, Ratio p = F_{CO} / F_{CA} = 1000.*

4.2.1.4 Peak current as a function of time and peak voltage across the inductance for 2 periods: PMN-PT

The following figures illustrate the peak current generated by the automatic vibrating structure with an inserted magnification showing the shape of this peak as a function of time (**Figure 11**) and its exponentially decrease during about 10–9 s. This current of about 3.5 10–4 A flowing through an inductor L_{IN} of 4 10^{-5} Henri naturally generates a voltage of 4 Volts (**Figure 35**).

Note in **Figure 11** the exponential form of the current peak of a duration appearing in each period. It is the same for the voltage peaks at the terminals of the inductance (**Figure 35**).

As this current peak cross an inductor, it induces by itself a voltage peak.

The current peak that appears with each cycle of vibration is uniquely due to the homogenization of the electrical charges on the two parts of the return electrode. This current peak follows the equation.

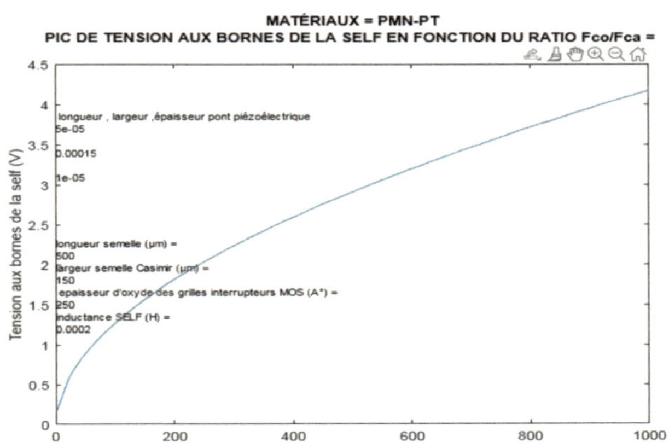

Figure 35.
*Materials = PMN-PT: Voltage peak across the 4 * 10^{-5} H solenoid as a function of the F_{CO} / F_{CA} Ratio. Start interface = 200 A °.*

$$I_{IN} = -\frac{Q_{mn2}}{R_m C_s}\left(Exp\left(-\frac{t}{R_m C_s}\right)\right) \text{ with } Q_{mn2} = d_{31} lp\ F_{CA}{}^{*}\ a_p/2.$$

This charge transferred from the electrode on the face 1 of the piezoelectric bridge to the return electrode, which is initially grounded, does not depend on the common width bp = bs = bi of the structures. This point is important and facilitates the technological realization of these structures since it limits the difficulties of a deep and straight engraving of the different structures. On the other hand, the intensity of this peak current depends linearly on the lengths lp and ls of the structures (**Figure 22**).

However, the duration of the exponential peak $t_e = R_m C_S \ \ln(2)$ is independent of the geometries of the structure. These only intervening in the frequency of vibration of the structure and in the intensity of the peak.

4.2.1.5 Peak voltage across the inductance and threshold voltage according to the desired FCO/FCA ratio: PMN-PT

We observe (**Figure 36**) that the automatically peak voltage obtained without any energy expenditure increases by a factor of 16 and goes from 0.25 V to 4 V when the ratio $p = F_{CA}/F_{CO}$ increases from 2 to 1000. Likewise, the threshold voltage MOSE and MOSD authorizing these ratios increases from 0.2 V to 3.2 V (**Figure 37**).

4.2.1.6 Vibration frequency as a function of the F_{CO}/F_{CA} *ratio and peak current as a function of the initial Casimir interval chosen: PMN-PT*

Note (**Figure 38**), that for an initial interface $z_0 = 200$ A°, the maximum vibration frequency of the structure is 3.50 MHz for a ratio Fco/Fca = 2. It falls to 750 kHz for a ratio of 1000. These frequencies remain lower than that of the first resonance of the structure which is the order of 7.94 Megahertz! For an $p = F_{co}/F_{ca}$ ratio = 500, the maximum current delivered by the structure drops as a function of the initial Casimir interval (**Figure 39**).

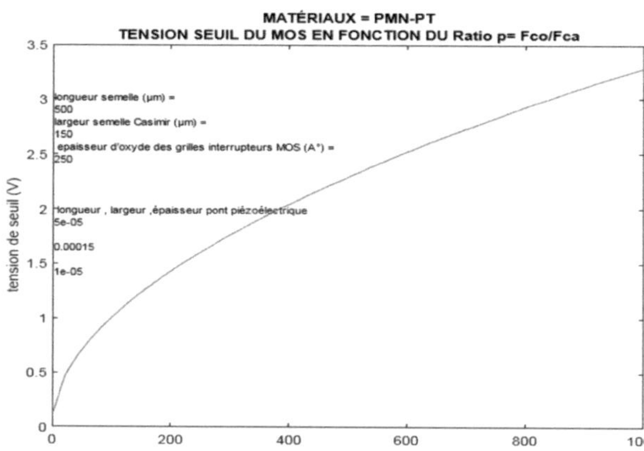

Figure 36.
Materials = PMN-PT: Threshold voltage of the Enriched or Depleted MOTS according to the F_{CO} / F_{CA} *ratio. Start interface = 200 A °.*

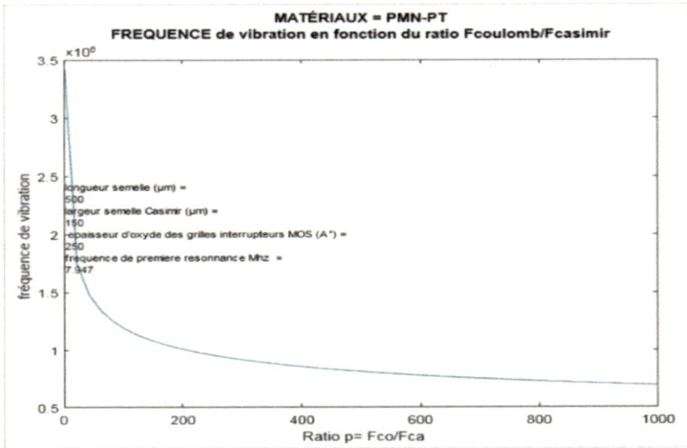

Figure 37.
Materials = PMN-PT: vibration frequency as a function of the F_{CO} / F_{CA} ratio. Start interface = 200 A °. Start interface = 200 A °.

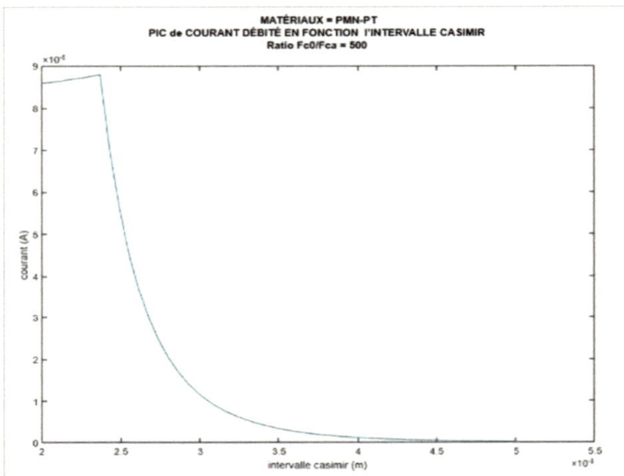

Figure 38.
*Materials = PMN-PT: Current peak across the 2 * 10-4 H inductance as a function of the starting interval between Casimir electrodes. Start interface = 200 A °.*

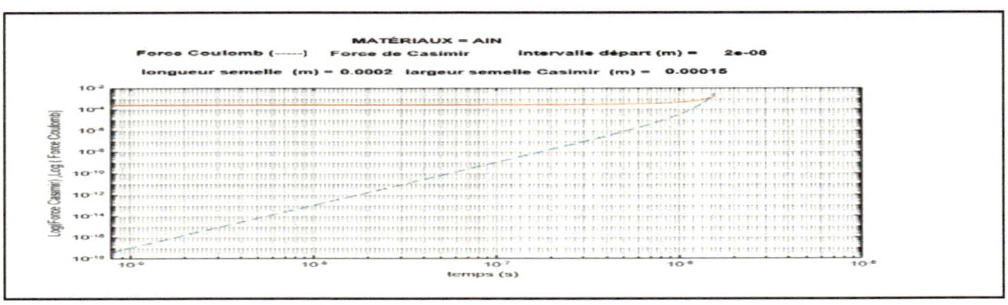

Figure 39.
Piezoelectric Material = AlN Casimir, Coulomb Force = f (Time) starts Interface = 200 A °.

This vibration frequency of the Casimir structure approaches that of the first resonance for weaker interfaces and less than 200 A°. We are then unfortunately confronted with the technological possibility of mastering such a weak interface.

It seems that the piezoelectric material PMN-PT coupled with a conductor like aluminum is an interesting couple for our vacuum energy extraction structure!

4.2.2 Piezoelectric material = AlN

With the MATLAB simulation of the behavior of the structure for piezoelectric Aluminum Nitride (AlN), we obtain the evolution with the time of the Casimir and Coulomb forces as well as the F_{CO}/F_{CA} ratio of **Figures 40** and **41** below. For a ratio F_{CO}/F_{CA} of 10, the maximum current delivered by the vibrating structure, the threshold voltage of the MOS and the vibration frequency of the structure is respectively 1.85 10^{-7} A, Vt = 3.7 V and 667,000 Hertz.

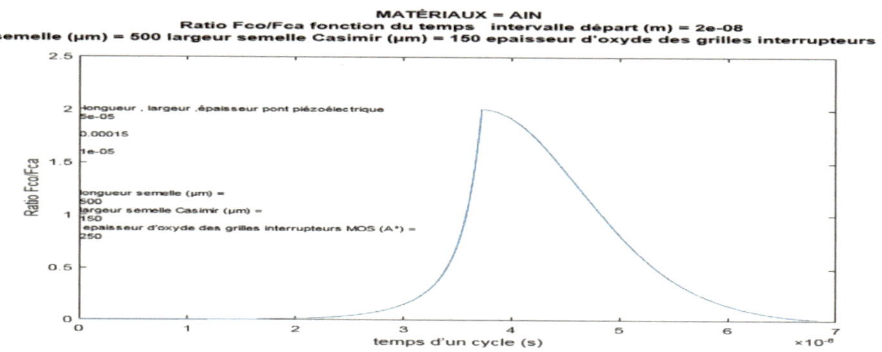

Figure 40.
Piezoelectric Material = AlN. Ratio F_{CO} / F_{CA} = f(Time). Start interface = 200 A °.

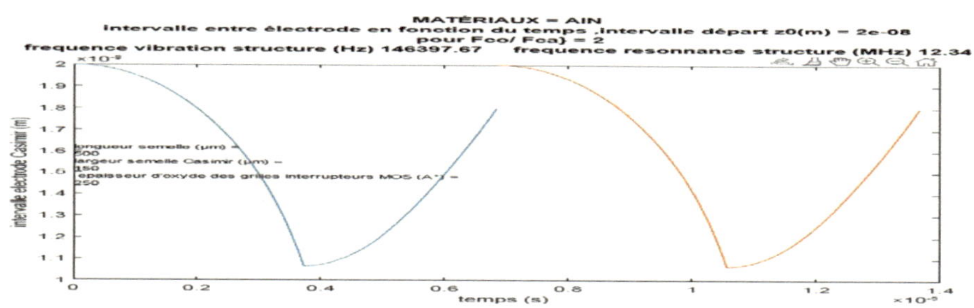

Figure 41.
Material = AlN Interval between Casimir electrodes = f (time) during two complete cycles: Interface between starting electrodes = 200 A °.

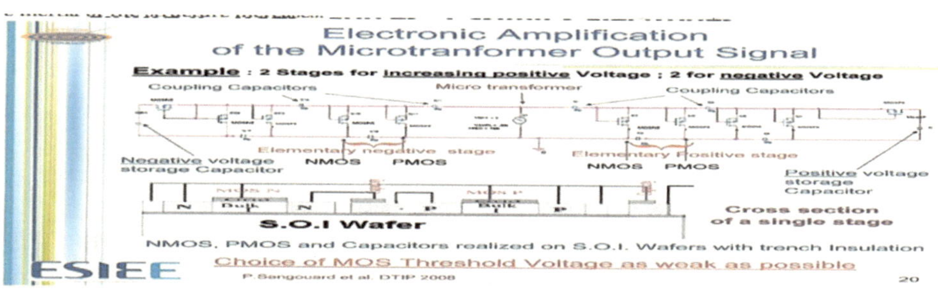

Figure 42.
Principle of electronics for amplifying and rectifying a week AC signal.

We observe (**Figure 42**) that the ratio $p = F_{CO}/F_{CA}$ barely equals 2, and that the time of "rise" of the mobile Casimir electrode is relatively slow, it is a consequence of the low value of the piezoelectric coefficient d_{31} of AlN.

In conclusion, the use of AlN does not seem suitable for this vacuum energy extraction application.

4.3 Conclusions

It seems that for the piezoelectric material we used, the most suitable piezoelectric material for this vacuum energy extraction device is PMN-PT with a peak current of 350 m A, at least for the materials we used for the previous simulations (**Figures 7, 31–39**).

In order to convert these alternating current peaks into an alternating voltage without input of energy, this current passes through a L_{IN} inductor coil which converts these current peaks without input of external energy into voltage peaks of several volts and of a duration of the order of the nanosecond.

Inductors L_{IN} for printed circuits of the order of 100 µH or less are conventional and are commercially available.

The next chapter proposes to convert these peaks of alternating voltage, to amplify them to obtain a direct voltage of several volts without any external power supply!

5. Transformation and amplification electronics without external supply of a periodic signal of a few millivolts in a continuous voltage of a few volts

Consider the diagram in **Figure 5** which shows the location and configuration of the electronic circuits for collecting the current generated during a small fraction of the period of hypothetically inductance-sustaining vibrations of the structure. The mobile charges Q_{mn} at the terminals of the electrode S_{S2} are variable over time since they vary cyclically from null to $Q_{mn}/2$. A part of these electrical charges, in the ratio of the input impedance between the inductor L_{IN} and that of the transformer electronics of this signal, generate during the short durations of their homogenization time, a peak of current which go inside this inductor and create between the two terminals of this solenoid, a voltage peak.

In **Figure 5**, the gray and red surfaces are stationary, the others are free to move. The green metallic connector connects face n°1 (Sp1) of the piezoelectric bridge to the return electrode. The red metallic connector connects face n°2 (Sp2) of the piezoelectric bridge to the MOS gates. For circuit n°1: The MOSEN and MOSEP sources in // are connected to the metallic electrode of face n°1 of the piezoelectric bridge. The MOSE drains are connected in parallel to the electronics without external power supply for transforming the AC signal, and to a terminal of the L_{IN} solenoid. The other terminal of the inductance L_{IN} is connected to the return electrode if circuit 2 is open or to ground via circuits n°2 if it is conductive.

When, the enriched MOSNE and MOSPE of circuit n°1, connect face 1 to the return electrode via the inductance L_{IN}, as already seen, during this period the circuit composed of MOSND and MOSPD of circuit n°2 is then blocked. This connection occurs when their gates have a voltage making one of them ON.

When circuit n°1 is blocked, circuit n°2 is conducting and connects the return electrode to ground, which eliminates the charges present on this electrode and then prevents any electrostatic attraction.

We will describe these electronics designed and successfully tested at ESIEE with SPICE when I was studying abandoned sensors. This electronic without external energy gave very encouraging SPICE simulation results and delivers in its output an exploitable direct voltage when it have in its input an alternating and small signals!

In these SPICE simulations, the micro transformer was assimilated to a voltage source delivering a power U * I limited to a few nW (voltages of a few mV and current much lower than the microampere).

Now retired and no longer having sufficient means of simulations, I am simply describing the results of the SPICE simulations obtained in 2008.

The principle used to amplify and transform a weak signal without power supply derives from that of the diode bridge rectifier of Graetz or the doubler of Schenkel and Marius Latour.

The crippling problem is that the diodes of these rectifiers are conductive only with a minimum voltage of around 0.6 V at their terminals. As the alternating signal from the vacuum energy extraction device can be weaker, it is necessary to have switches that are triggered with a lower control voltage.

The principal diagram of this electronics is presented in **Figures 43–45**.

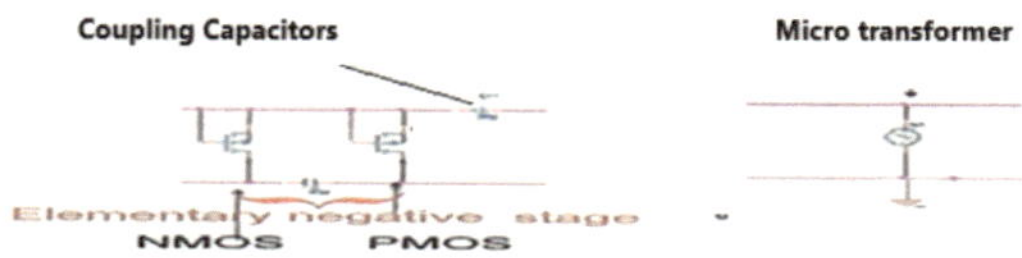

Figure 43.
Elementary stage for obtaining a negative voltage from the alternative signal of the transformer (inductance). Start interface = 200 A °.

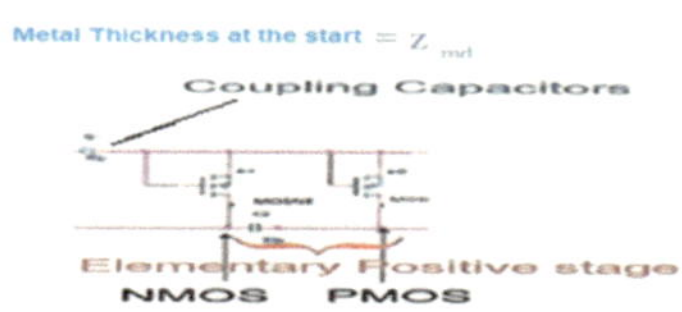

Figure 44.
Elementary stage for obtaining a positive voltage from of the alternative signal of the transformer (inductance) Start interface = 200 A °.

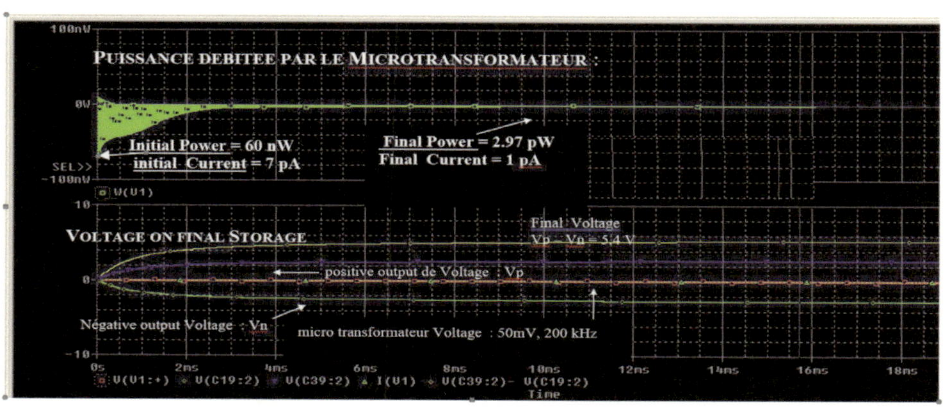

Figure 45.
SPICE simulations of voltages, current, power of the transformation electronics into a direct voltage (5.4 V) of an alternating input signal of 50 mV, frequency= 150 kHz, number of stage =14, coupling capacities = 20 pF, stocking capacity = 10 nF.

The MOSE N and P transistors of this rectifier circuit must have a technologically defined threshold voltage as close as possible to zero. The precision of nullity of these threshold voltages will depend on the values of alternating voltages at the terminals of the inductor L_{IN}, therefore on the second derivative of the temporal variations of the charges appearing on S_{p2} during the time of their homogenization which is of the order of: $t_e = R_m.C_s.Log$ (2) In the circuit of **Figure 8**, a micro transformer replaced the inductance L_{IN}. But this inductance plays the same role as this micro transformer since it delivers a limited power U.I.

The left part of the micro-transformer takes care of the negative voltages of the input signal, while the right part takes care of the positive voltages. The circuit is composed of several stages without no power supply which rectify and amplify, on the one hand the negative parts of the weak input signal and on the other hand the positive parts.

The number of elementary stages depends on the desired DC voltage, but this Dc voltage saturates with the number of stages in series (**Figure 46**). The results obtained from SPICE simulation are shown in **Figure 8**.

We observe an important point in **Figures 8** and **47**, the very low power and current consumption on the source since:

1. In **Figure 48** the power delivered by the source begin at the start with 60 nW and ends at 2.97 pW for an input current starting at 7 pA and finishing at 1pA.

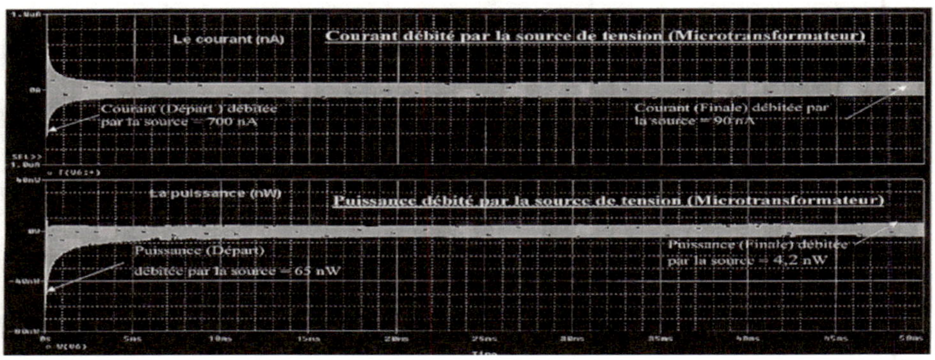

Figure 46.
SPICE simulations of the currents drawn by the transformer and the power consumed by this transformer. Input signal = 100 mV, frequency = 150kHz, number of stage = 30, coupling capacities = 20 pF, stocking capacity = 10 nF.

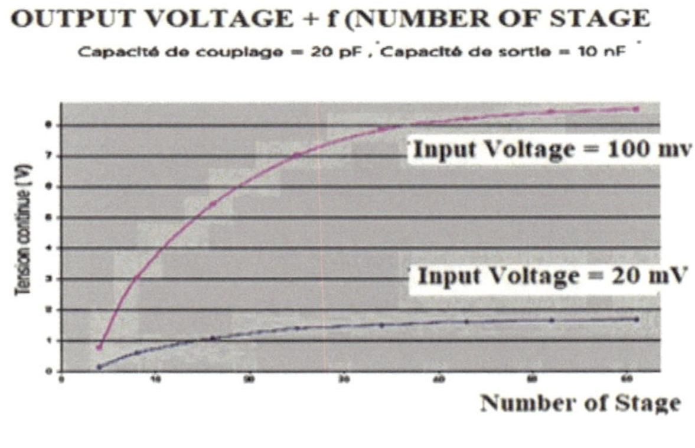

Figure 47.
DC output voltages as a function of the number of elementary stages for AC input voltages of 20 mV and the other of 100 mV. Start interface = 200 A °.

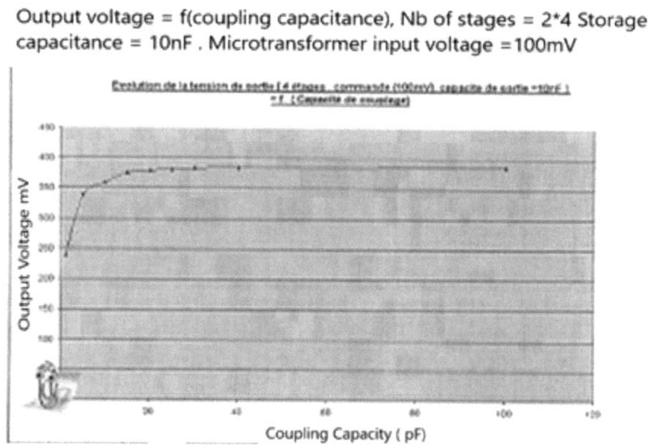

Figure 48.
Influence of the coupling capacitance on the amplification of the input signal. Start interface = 200 A °.

The negative component of the alternating signal is transformed in 10 ns into a negative direct voltage of V_n = -2.7 V. Likewise the positive component the positive alternating part is transformed into a positive direct voltage of V_p = 2.7 V We obtain therefore a direct voltage Vp-Vn = Vt = 5.4 V.

2. In **Figure 47** the power delivered by the source begin at the start with 65 nW and ends at 4.2 nW for an input current starting at 700 nA and finishing at 90 nA. The negative component of the alternating signal is transformed in 10 ns into a negative direct voltage of V_n = -3.9 V. Likewise the positive component the positive alternating part is transformed into a positive direct voltage of V_p = 3.9V. We obtain therefore a direct voltage Vp-Vn = Vt = 7.8 V.

An important point is the need to have a high circuit output impedance of several 10^7 ohms, so typically the input impedance of an operational amplifier.

The DC voltage obtained depends on the number of stages constituting these electronics without electrical power for transforming an AC signal of a few millivolts into a DC signal of a few volts. However, this transformation saturates with the number of floors, as shown in **Figure 49** Note in **Figure 49** that the DC output

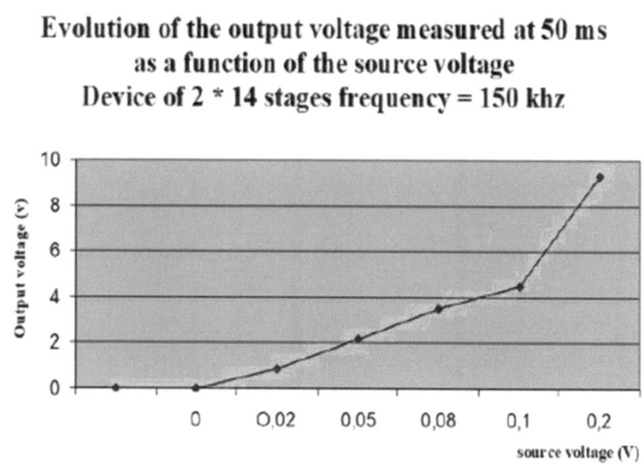

Figure 49.
Evolution of the DC output voltage as a function of the amplitude of the AC input signal for a frequency of 150 kHz. Start interface = 200 A °

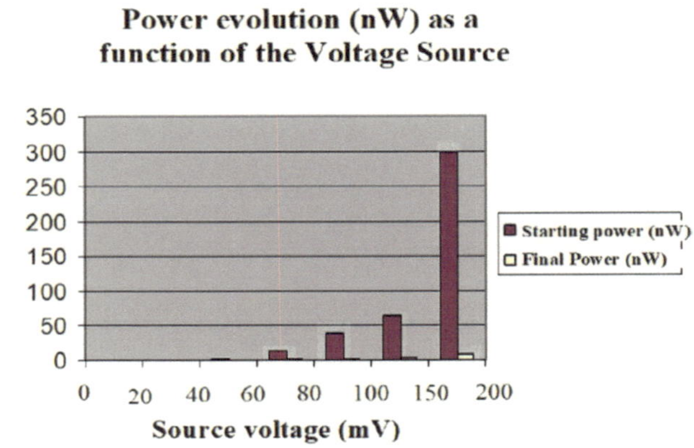

Figure 50.
*Evolution of the power supplied in nW by the source as a function of the amplitude of the voltage supplied in mV. with 2 * 14 stages and a frequency of 150 kHz Start interface = 200 A °.*

voltage saturates with the number of elementary stages and that the optimal number of stages is of the order of 40. We also looked at the influence of the coupling capacitance on the amplification of an input signal of 100 mV with a storage capacity of 10 nF. This amplification saturates and a coupling capacity of 20 pF which seems to be optimal signal (**Figure 48**).

The following **Figure 50** shows the influence of the value of the input AC voltage, with a frequency of 150 kHz, on the DC voltage obtained at the output of a 2 * 14 stage device.

Figure 46 shows the power in nW delivered by the source at the start of the amplification and at the end of this amplification.

A summary of the performance of this low "voltage doubler" device is shown in **Figure 51** below.

The interesting points for the presented electronics' device are:

1. the low alternative input voltages required to obtain a continuous voltage of several volts at the output

CHARACTERISTICS OF OUTPUT VOLTAGES (V), POWERS (nW) CURRENTS (nA)
AS A FUNCTION OF THE NUMBER OF STAGES
INPUT SIGNAL FREQUENCY = 150 kHz OUTPUT VOLTAGE MEASUREMENT FOR t = 50 ms

	Vg=50mV					Vg=100mV				
number of stages	Output Voltage	Current (nA) start	Current (nA) end	Power (nW) start	Power (nW) end	Output Voltage	Current (nA) start	Current (nA) end	Power (nW) start	Power (nW) end
2*3	550mV	300nA	26nA	15nW	1.3nW	1.1v	800nA	46nA	75nW	5nW
2*6	1	300nA	29nA	13nW	1.3nW	2V	700nA	67nA	60nW	6.5nW
2*14	2.2v	300nA	40nA	14nW	2.6nW	4,5v	700nA	50nA	65nW	4.8nW
2*21	2.8v	250nA	38nA	13nW	860pw	6v	600nA	80nA	60nW	2.7nW
2*30	3.3	250nA	43nA	12nW	1.2nW	6.5V	750nA	85nA	61nW	4nW
2*39	3.5v	250nA	45nA	12nW	900pW	7.5V	750nA	95nA	64nW	3.5nW
2*48	3.6v	250nA	46nA	12nW	1nW	7.6V	750nA	100nA	60nW	4.2nW
2*60	3.8	270nA	47nA	12nW	1.1nW	7.9V	700nA	90nA	65nW	4.2nW
2*61	3.8	270nA	48nA	12.1nW	1.3nW	8V	700nA	90nA	65 nW	42nW

Figure 51.
Summary of transformations from low alternating voltages to direct voltage frequency of 150 kHz. Start interface = 200 A °.

Figure 52.
S.O.I technology for making the elements of the "doubler".

2. the low power and current consumed by this conversion and amplification circuit on the source which in this case is only an inductor supplied by the current peaks generated by the autonomous vibrations.

3. the rapid time to reach the DC voltage (a few tens of milliseconds)

The technology used to fabricate the MOSNE and MOSPE transistors with the lowest possible threshold voltages is CMOS on intrinsic S.O.I. and the elements are isolated from each other on independent islands. This technology, represented in the following **Figure 52**, strongly limits the leakage currents.

We note that, the coupling capacities of 20 pF this electronic, like that of storage of the order of 10 nF, have relatively high values witch will require a square surface of:

- 120 μm^2 for a thickness of 250 A°, if the use silicon oxides SiO_2 with its relative permittivity of order of 4, which is a lot!

- Alumina Al_2O_3 obtained by oxidizing aluminum has only a relative permittivity of the order of 9 and would require alumina squares of 79 μm^2.

But if we use titanium dioxide as insulator, it has a relative permittivity of the order of 100 and is one of the most important for a metal oxide then the size of the capacity passes to 23 mm for a thickness of TiO2 = 500 A°, which is more reasonable!

6. Technology of realization of the current extractor device using the forces of Casimir in a vacuum

It can be seen in **Figures 25** and **26** that if PZT is used as the piezoelectric material, then the peak current output goes from 2.10^{-8} A to 6.10^{-8} A, when the width of the piezoelectric layer bp changes from 50 μm to 150 μm (for a length of the piezoelectric layer lp = 50 μm, a thickness ap = 10 μm, a length of the Casimir electrode ls = 200 μm, starting interface $z_0 = 200$ A° and a Ratio of $F_{CO}/F_{CA} = 10$).

As a result, it will be necessary, using micro-technology techniques on S.O.I silicon, to machine devices with on a high thickness while maintaining exceedingly small spaces between structures Casimir. The microelectronics laboratory of the ESIEE has acquired a great experience in the plasma etching of deep sub-micron structures by etching remarkably parallel layers of silicon of 100 mm separated by intervals of 0.8 mm and well parallel [6, 7] and **Figure 53**.

However, for the structures presented above, the space between the two surfaces of the reflectors must be of the order of 200 A°, almost 4000 times less wide … . which is not technologically feasible by engraving!

Yet it seems possible, to be able to be obtained this parallel space of the order of 200 A° between Casimir reflectors, not by etching layers but by making them thermally grow!

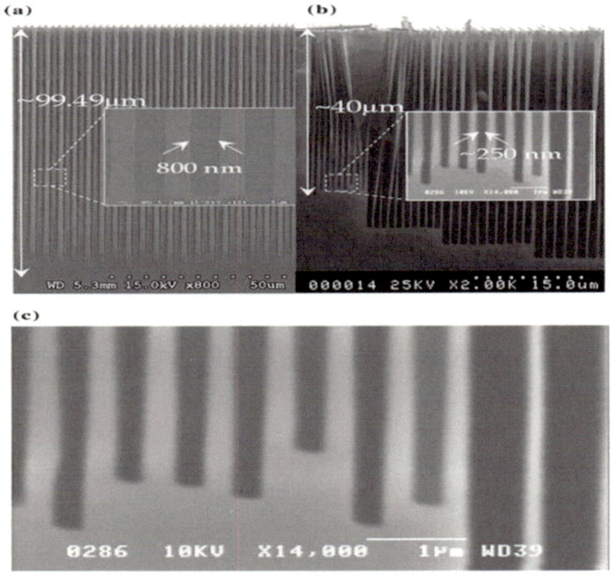

Figure 53.
Extract de [7]: High aspect ratio (HAR) structures manufactured using the Bosch process: (a) 800 nm-wide trenches with a depth of 99.5 lm (aspect ratio 124:1) and (b) 250 nm-wide trenches with a depth of 40 lm (aspect ratio 160:1). Some of the walls collapsed during the dicing procedure. © is a magnified view of the inset shown in (b) [8].

Indeed, the S_{S3} and S_{S2} surfaces of the Casimir reflector must;

- be metallic to conduct the mobile charges
- insulating as stipulated by the expression of Casimir's law who established for surfaces without charges! This should be possible if we grow an insulator on the z direction of the structure, for example Al_2O_3 or TiO_2 or other oxide metal which is previously deposited and in considering the differences in molar mass between the oxides and the original materials (see **Figure 33**).

For example, silicon has a molar mass of 28 g/mol and silicon dioxide SiO_2 has a molar mass of 60 g/mol. However, it is well known that when we grow a silicon dioxide SiO_2 of one unit we "attack" a silicon depth of the order of 28/60 = 46.6% (**Figure 54**).

The initial silicon layer reacts with the oxidizing element to form SiO_2. We will thus "consume" Silicon. The Si/SiO_2 interface will therefore end up "below" the initial surface. A simple calculation shows that the fraction of oxide thickness

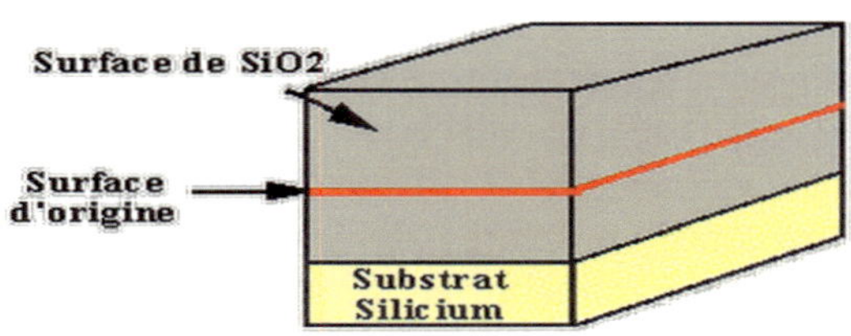

Figure 54.
Growth of SiO2 oxide on silicon.

"below" the initial surface is 46% of the total oxide thickness; the fraction "above" therefore represents 54% according to S.M. Sze. We therefore moved the original silicon surface by 46% [9].

The same must happen, for example for thermal growth of alumina. As the molecular masses of Alumina and aluminum are M_{Al2O3} = 102 g/mole and M_{Al} = 27 g/mole, we obtain an aluminum attack ratio of 27/102 = 26%, which implies that the original surface of this metal has shifted by 26% so that 74% of the alumina has grown out of the initial surface of the aluminum .... .

Likewise, if titanium is used for thermal growth of TiO_2, the molar mass ratio being M_{TiO2} = 79.9 g/mole and M_{Ti} = 47.8 g/mole we obtain a titanium attack ratio of 59.8% which implies that the original surface has moved by 59.8%.

So, this growth covers up the initial interface and it can be finely controlled! As a result, it should be possible to define finely the interface between the two Casimir reflectors using the oxidative growth conditions of a metal such as titanium or aluminum.

As regards the technological manufacture of electronics and structure, it therefore seems preferable:

1. For electronics to choose Titanium Oxide because of its high relative permittivity ε_r = 114 allowing to minimize the geometries required for the different capacities

2. For the Casimir structure, the choice of aluminum seems preferable, because its low density increases the resonant frequency of the structure and that 74% of the Alumina is outside the metal, allowing to reduce the interface between Casimir electrodes and that of the return electrode by Al_2O_3 alumina growth. A quite simple calculation shows for example that for aluminum (**Figure 55**):

We obtain: $z_{od} = 2^* (z_{md} + z_{of} - z_{ma}) + z_0 = 2^* (z_{md} + z_{of}(1 - .26)) + z_0$.

For example, if we start from an opening $z_{od} = 3\,\mu m$ and deposit a metal layer of aluminum that is etched leaving a width $z_{md} = 1\mu m$ on each side of the reflector. Then an Alumina Al_2O_3 can grow, the thickness of which is precisely adjusted, simply by considerations of time, temperature, and pressure to increase a necessary thickness to have a desired interface z_0.!

For example, if $z_0 = 200A°$, $z_{od} = 3\mu m$, $z_{md} = 1\mu m$, then $z_{of} = 0.662\mu m$. So, we obtain a Casimir interface of 200 A°. The final remaining metal thickness will be $z_{mf} = 0.338\,\mu m$ and will act as a conductor under the aluminum oxide.

Obviously, the growth of this metal oxide between the electrodes of the Casimir reflector modifies the composition of the dielectric present between these electrodes, therefore of the mean relative permittivity of the dielectric.

Let: ε_0 be the permittivity of vacuum and $\varepsilon_0 . \varepsilon_r$ the metal oxides one (ε_r = relative permittivity = 8 in the case of Al_2O_3), z_{of} the final oxide thickness on one of

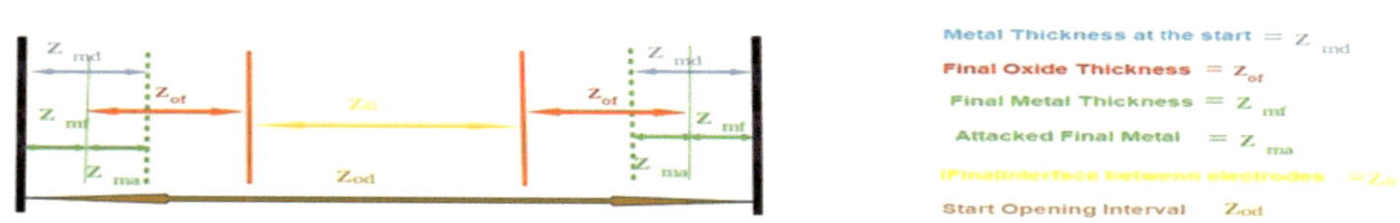

Figure 55.
Distribution of thickness.

the electrodes and z the thickness of the vacuum present between electrode (initially we want $z = z_0$).

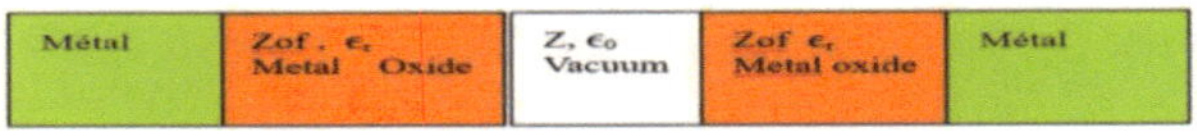

Then the average permittivity e_{0m} of the dielectric is:

$$\begin{aligned} \varepsilon_{0\,m} &= (z_{of}.\varepsilon_0 \cdot \varepsilon_r + z \cdot \varepsilon_0 + z_{of} \cdot \varepsilon_0 \cdot \varepsilon_r)/(2z_{of} + z)\varepsilon_{0m} \\ &= \varepsilon_0 \cdot (2 \cdot z_{of} \cdot \varepsilon_r + z)/(2.z_{of} + z) \cong \varepsilon_0 \cdot \varepsilon_r \end{aligned}$$

because z is $< < z_{of}$.!! For example, $z_{of} = 6620$ A° is large compared to $z < =$ 200A° therefore $\varepsilon_{0m} \cong 8^* \varepsilon_0$ in the case of Al_2O_3.

We have taken this change in permittivity into account in the preceding simulations.

7. Steps for the realization of the structure and its electronics

We start by the voltage "doubler". These electronics are produced using CMOS technology with 8 ion implantations on an S.O.I wafer with an intrinsic silicon layer above the oxide:

1. To make the drains, sources of the MOSNE, MOSND of the "doubler" circuits, of the MOSNE and MOSND of the Coulomb force trigger circuit and of the grounding switches of the S_{S3} electrode of the Casimir reflector

2. To make the source drains of the MOSPE, MOSPD of the "doubler" circuits, of the MOSPEs and MOSPDs of the Coulomb force trigger circuit and the grounding switches of the S_{S3} electrode of the Casimir reflector

3. To best adjust the zero-threshold voltage of the MOSNE of the "doubler" circuit

4. To best adjust the zero-threshold voltage of the MOSPE of the "doubler" circuit

5. To define the threshold voltage of the MOSNE of the parallel circuit triggering the Coulomb force

6. To define the threshold voltage of the MOSPE of the parallel circuit triggering the Coulomb force

7. To define the threshold voltage of the MOSND of the series circuit for grounding the S_{S3} electrode of the Casimir reflector

8. to define the MOSPD threshold voltage of the series circuit for grounding the S_{S3} electrode of the Casimir reflector.

   Once this electronics is done, we are interested in the realization of the structure of CASIMIR with the following technology proposal:

9. engrave the S.O.I. silicon to the oxide to define the location of the Casimir structures (**Figure 56**)

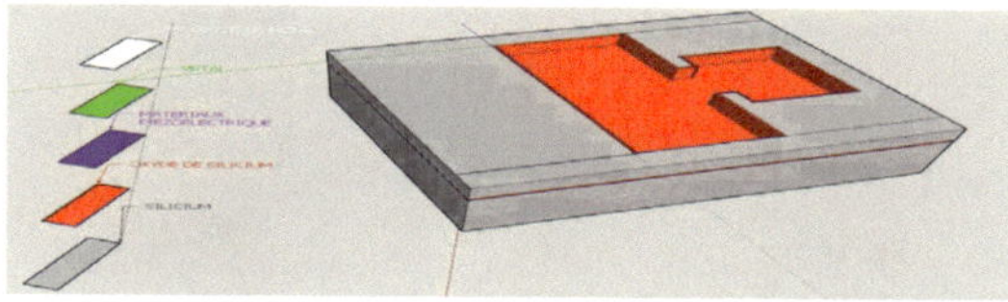

Figure 56.
Etching of S.O.I silicon.

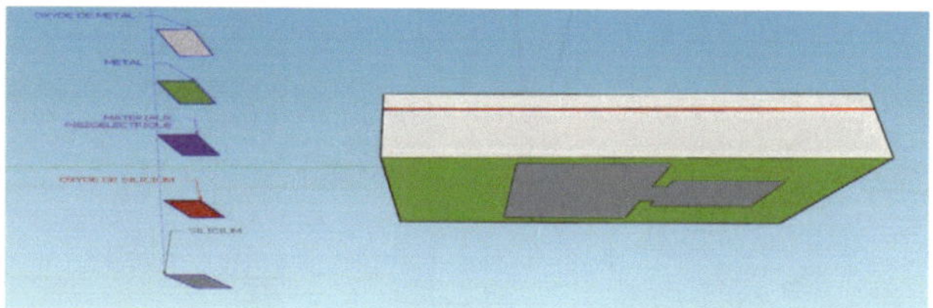

Figure 57.
Engraving of the protective metal rear face of the S.O.I. silicon.

10. Place and engrave a protective metal film on the rear faces of the S.O.I wafer (**Figure 57**)

11. Deposit and engrave the piezoelectric layer (**Figure 58**)

12. Depose and etch the metal layer of aluminum (**Figure 59**).

13. Plasma etching on the rear side the silicon of the Bulk and the oxide of the S. O.I wafer protected by the metal film to free the Casimir structure then very finely clean both sides (**Figure 60**)

14. Place the structure in a hermetic integrated circuit support box and carry out all the bonding necessary for the structure to function.

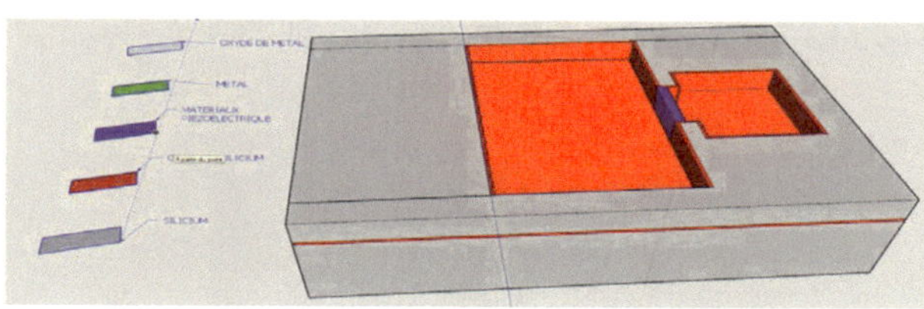

Figure 58.
Deposition and etching of the piezoelectric layer e 61 deposition and etching of the piezoelectric layer.

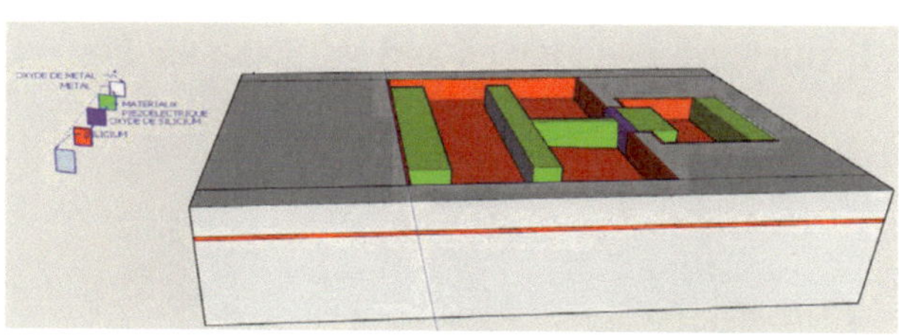

Figure 59.
Metal deposit, metal engraving etching of the piezoelectric layer.

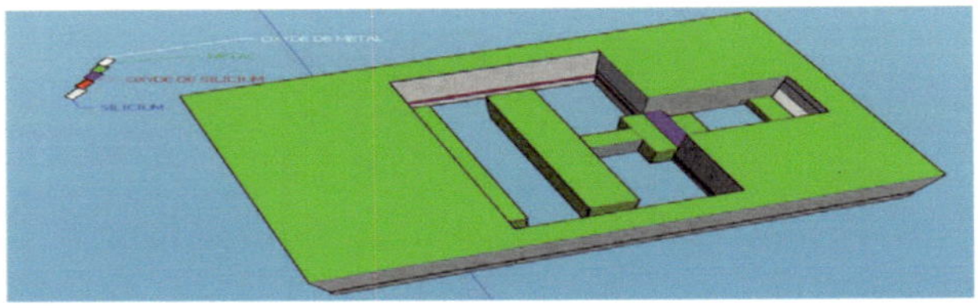

Figure 60.
View of the Casimir device on the rear face, engraving on the rear face of the structures.

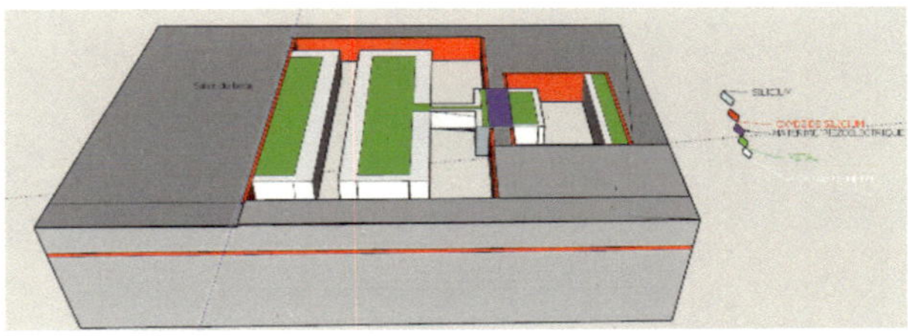

Figure 61.
Adjusted growth of metal oxide under the electronic control, front view of the Casimir device.

15. Carry out the thermal growth of aluminum oxide Al_2O_3 with *a measurement and control of the circuit under a box which should generate a signal when the interface between the Casimir electrodes becomes weak enough for the device to vibrate* ... and then stop the oxidation (**Figure 61**).

16. Create a vacuum in the hermetic box

In the case where the 2 metal electrodes of Casimir which are separated by a very weak interface, of the order of 200 A°, adhere to one another, then these two surfaces can be separated by the application of an electrical voltage on the electrodes on the other side of the piezoelectric bridge!

In order to obtain a current peak greater in intensity and duration, the Casimir cells can be positioned in a series and parallel network at the 2 terminals of a single inductance. For example, 20 Casimir cells can be placed in parallel and 10 in series.

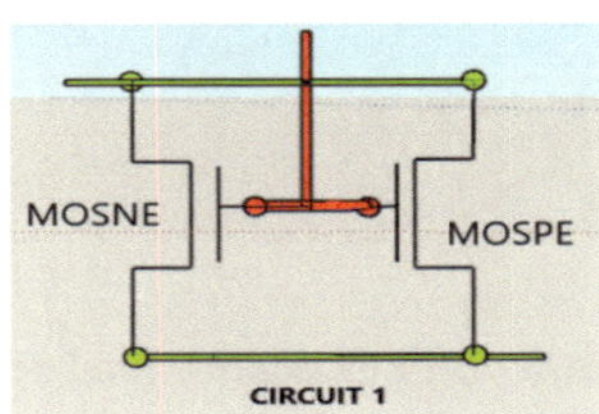

Circuit 1: parallel MOSPE and MOSNE ; see **Fig. 6**.

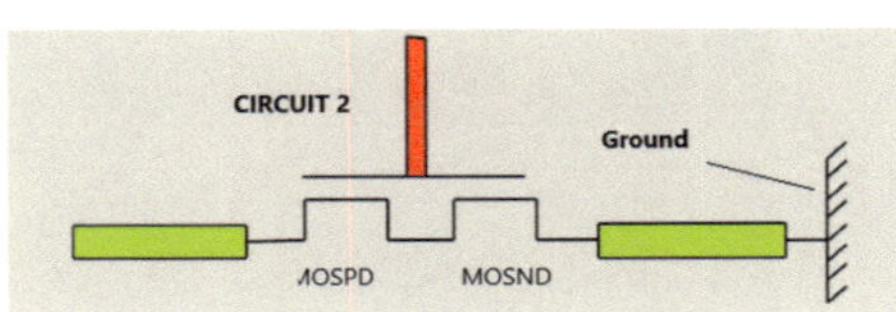

Circuit 2: serial MOSPD and MOSND ; see **Fig. 6**.

8. Energy balance

1. We remark that there is no bending of the free Casimir reflector electrodes. This only moving parallelepiped metallic electrode is in parallel to a fixed metal electrode defined on an S.O.I wafer and stay parallel to it. It transmits its motion to a piezoelectric bridge that deforms by bending.!

2. So the entropy expulsion ΔS from this Casimir vibrating structure and transmitted to the piezoelectric bridge is done by heat expulsion out of this bridge, but causes an extremely low temperature increase of the piezoelectric bridge!

A quick order of magnitude calculation gives: ΔQ_{vib} = heat transmitted by the vibrations of the piezoelectric bridge and evacuated outside follow, in first approximation, the well-known formula $\Delta Q_{vib} = \Delta S \Delta T \Delta Q_{vib} = \Delta S\, \Delta T$, with ΔS entropy variation ($J°K^{-1}$) and ΔT = temperature variation (°K).

However, we know that: (cf: M. BARTHES, M. Colas des Francs SOLID MECHANICAL VIBRATIONAL PHYSICS, ESTP: (Special School of Public Works).

$\Delta Q_{vib} = 1/2^* \ m(2\pi f_{vib})^2 x_{max^2}$; with: f_{vib} = Vibration frequencies of the piezoelectric bridge, m = mass of this bridge, x_{max} = maximum deflection of the bridge.

This heat, expended at the level of the piezoelectric bridge, causes a temperature increase.

As a first approximation we can say: $\Delta Q_{vib} = mC_{piezo}\Delta T$, with C_{piezo} = Specific heat capacity of the piezoelectric bridge ($J\ Kg^{-1}°K^{-1}$), $\Delta T =$ T emperature variation (°K).

Consequently $\Delta T = 2(\pi\ f_{vib})^2 x_{max^2}/C_{piezo}$ = Temperature variation.

We have for example:

$C_{piezo} = C_{PMN\text{-}PT} = 310(\ JKg^{-1°}K^{-1}), f_{vib} \cong 10^6$ Hz, $x_{max} \cong 100 * 10^{-10}$ m, we obtain: $\Delta T \cong 10^{-3°}K$ which is negligible!!

So, the expulsion of entropy from the vibrating Casimir Electrode is negligible!

From 0 to z_e:

In a cycle from 0 to z, the energy $E_{Casimir}$ is: $E_{Casimir} = \int_0^{ze} F_{ca} dz = \int_0^{ze} \left(\frac{\pi^2 \hbar c}{240 z_S^4}\right) dz.$

No mobiles electric charges appear on the face of return electrode which is connected to the mass and isolated of the piezoelectric bridge, and so the Coulomb force disappears!

During this displacement "going" from 0 to ze the deformation of the piezoelectric bridge, generates a potential energy W_{Bridge} accumulated in the capacity of this bridge which follows the equation:

$$W_{Bridge} = \int_0^{Q\ e} \frac{Q_F}{Cpi} dQ_F = \left[\frac{Q^2{}_F}{2^*\ Cpi}\right]_0^{Q_e} = \frac{a_p}{2\ l_p\ b_p \varepsilon_0 \varepsilon_{pi}}.$$

$$\left(\frac{d31\ lp}{2ap}\right)^2 F_{CA}^2 = \frac{a_p}{2 l_p\ b_p \varepsilon_0 \varepsilon_{pi}} * \left(\frac{d_{31}\ lp\ l_S b_S \pi^2 \hbar c}{480.ap}\right)^2 \left(\frac{1}{z_e}\right)^8$$

With Q_F the naturally creating fixed charges on this piezoelectric structure.

$Q_F = \frac{d_{31} F_{CA} l_P}{a^P}$ Eq. (3),

$Q_e = -Q_F$ the accumulated mobile charges on both the two surfaces of the "return" electrode when coulomb's force triggered at $F_{CO} = p\ F_{CA}$, C_{pi} = electrical capacity of the piezoelectric bridge, z_e the position of appearance of the Coulomb force.

From z_e to 0:

For this very quick "returning" from position z_e to position 0, the associated energy $E_{Returning}$ is:

$$E_{Returning} = \int_{ze}^{0} (F_{co} + F_{ca})dz = E_{Coulomb} - E_{Casimir}$$

Thus, we see that in the balance $E_{Casimir} + E_{Returning,} = E_{Coulomb}$. So, over a complete cycle, the energy $E_{Casimir}$ is conservative.!

The Coulomb energy $E_{coulomb}$ is the net energy appearing during a cycle. It is not due to any electrical energy applied, but to the consequence of vacuum energy!

If we choose $z_r = z_0$, the expression of $E_{coulomb}$ is,

$$E_{coulomb} = \left(l_S b_S \frac{\pi^2 \hbar c}{240} \frac{d_{31} l_P}{a_P}\right)^2 \left(\frac{1}{4\pi\varepsilon_0\varepsilon_r}\right) \int_{ze}^{0} \left(\left(\frac{1}{z_s^4} - \frac{1}{z_0^4}\right)\right)^2 \left(\frac{1}{z_r + z_0 - z_S}\right)^2 dz$$

A Coulomb force then appears between these two electrodes.

$$F_{CO} = \left(S_S \frac{\pi^2 \hbar c}{240} \frac{d_{31} l_P}{2^* a_P} \left(\frac{1}{z_s^4} - \frac{1}{z_0^4}\right)\right)^2 \left(\frac{1}{4\pi\varepsilon_0\varepsilon_r}\right) \left(\frac{1}{z_r + z_0 - z_S}\right)^2 \Bigg)$$

The position z_e of appearance of this force is such that $F_{CO} = p\, F_{CA}$, so if we choose $z_r = z_0$.

Then, the z_s position of appearance of this force is when:

$$\left(l_S b_S \frac{\pi^2 \hbar c}{240} \frac{d_{31} l_P}{2^* a_P} \left(\frac{1}{z_s^4} - \frac{1}{z_0^4}\right)\right)^2 \left(\frac{1}{4\pi\varepsilon_0\varepsilon_r}\right) \left(\frac{1}{2z_0 - z_S}\right)^2 \Bigg) = p l_S b_S \left(\frac{\pi^2 \hbar c}{240}\right) \frac{1}{z_s^4}$$

This position and evolution z_s are illustrated in the following **Figures 62** and **63.** As a result, the movement of the movable electrode shown in **Figure 62.**

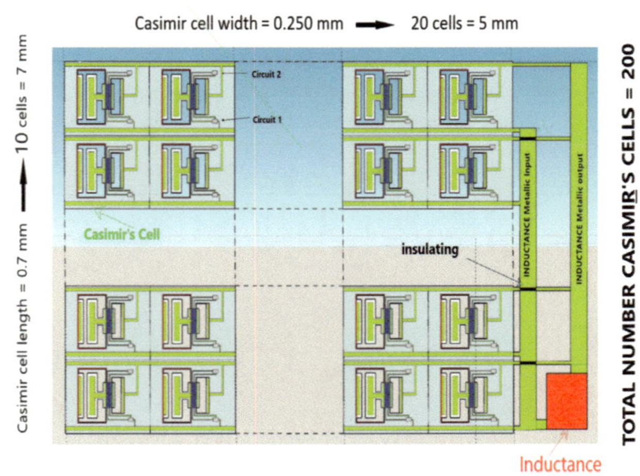

Figure 62.
Positioning of 20 Casimir cells in parallel and 10 in series. Circuit 1, Circuit 2 and Switches of circuit n ° 1 and n ° 2. Total of Casimir cells delivering a periodic current during a small part of the vibration frequency of the devices = 200!. Total des cellules = 200, width = 5 mm, length = 7 mm.

Figure 63.
Position of the mobile casimir electrode when the coulomb force occurs ls = 500 μm; bs = 150 μm; lp = 50 μm; bp = 150 μm; ap = 10 μm.

Note that the displacement of this mobile Casimir electrode is extremely small since it goes from 2 A° for an F_{CO}/F_{CA} ratio = 2 to 105 A° for a ratio of 1000.

When the electrical potential on the gate of the switches is greater than their threshold voltage, then this switch commute and the accumulated energy in the piezo electric's bridge will be used for the homogenization of the mobile charges of the bridge's electrodes connected to sources or drains of the MOS switch n°1 and the Coulomb return electrode.

During the time of this homogenization appears a current peak for a short time. So, an electrical voltage at the terminals of the self in series between switch 1 and the return electrode appears also!

So, an important point of obtaining current peaks related to the homogenization of charges appears.

His expression is: $I_{IN} = -\frac{Q_{mn2}}{R_m C_s}\left(Exp\left(-\frac{t}{R_m C_s}\right)\right.$.

With defining $t_c = R_m{}^* Cs^* \; ln\;(2)$ the duration with which the mobile charges equalize.

between the electrode of face 1 and the return electrode and R_m = the resistance (Ω) of the devices in series (Self + switch n°1); C_s the capacity (F) of the device constructed on face n°2 of piezoelectric bridge, return electrode).

These cyclic current peaks induce at the terminals of the coil voltage peaks whose expression is:

$$U_{IN} = L_{IN}\frac{d(I_{IN})}{dt} = L_{IN}\frac{Q_{mn2}}{R_m C_s}\left(Exp\left(-\frac{t}{R_m C_s}\right)\right) = L_{IN}\frac{Q_{mn2}}{R_m C_s}\frac{ln\;(2)}{t_e}\left(Exp\left(-\frac{t\,ln\;(2)}{t_e}\right)\right)$$

With L_{IN} the value of the inductance, Q_{mn2} the charges on the return electrode and the face n°1 of the bridge.

The only energy which is effectively used outside, during one cycle, is associated with these current and voltage peaks becomes:

$$W_{electic} = Abs\left(\left(\int_0^{t_e}\frac{I_{IN}\;UIN}{}d(t) = \frac{L_{IN}}{2}\left(\frac{d_{31}F_{CA}l_P}{a_P}\right)^2\left(\frac{\ln\;(2)}{te}\right)^2(1 - Exp(-2\;ln\;(2)))\right.\right. \quad (20)$$

For example, we obtain, for an interface between Casimir's electrodes $z_0 = 200$ A°, dimensions of the Casimir electrodes (length = 500 μm, width = 15 μm, thickness = 10 μm), dimensions of the piezoelectric bridge in PMN-PT (length = 50 μm, width = 15 μm, thickness = 10 μm), a proportionality factor $p = F_{CO}/F_{CA}$ = 1000, an inductance $L_{IN} = 1.10^{-6}$H:

- $Ze = 9.46\ 10^{-09}$ (m) i.e., a displacement of the mobile Casimir electrode of about 105A° (**Figure 63**)
- $W_{CA} = E_{Casimir} = 3.4\ 10^{-11}$ (Joule) = Energy of vacuum = Energy dispensed by the force of Casimir
- $W_{bridge} = 2.7\ 10^{-11}$ the potential energy accumulated in the piezoelectric bridge
- Peak current = $350\ 10^{-6}$A
- Voltage peak across the coil = 1.2 V
- Structure vibration frequency = 750 kHz
- Threshold Voltage of enriched MOSE = Threshold Voltage of in depletion MOSD = 3.25 V
- $W_{electric} = 2.7\ 10^{-11}$ (Joule) = Usable energy associated with current and voltage peaks.

We remark that this energy $W_{electric}$ is equal to the energy W_{bridge} accumulated in the bridge and is not brought by an external electrical source but is caused by the omnipresent and perpetual force of Casimir, itself controlled by a Coulomb force of opposite direction.

This Energy is less than that developed by Casimir's strength.

- The intensity of the Coulomb force is defined and technologically adjustable by adjusting the threshold voltage of the MOS transistors of circuits 1,2,3,4.
- ΔQ_{vib} = heat transmitted by the vibrations of the piezoelectric bridge = $7.8\ 10^{-14}$ J and is very small.
- We notice that $\Delta Q_{vib} + W_{electric} < E_{Casimir}$ which is consistent with Noether's theorem.

This energy $W_{electric}$ is equal to the energy W_{bridge} accumulated in the bridge and not brought by an external electrical source, but is caused by the omnipresent, timeless and perpetual force of Casimir, itself controlled by a Coulomb force of opposite direction.

Remember that Energy is defined as the "physical quantity that is conserved during any transformation of an isolated system." However, the system constituted by simply the MEMS device is not an isolated system.

On the other hand, the system constituted by the MEMS device plus the vacuum becomes an isolated system.

A system of 200 structures (**Figure 64**) gives a usable energy by second and so 750,000 peaks, $W_{electric} \cong 60\ 10^{-3}$ (Joule) for a coefficient of proportionality $p = F_{CO}/F_{CA} = 10^6$ and all switch transistors (Width = Length = 100 µm, SiO_2 grid thickness = 250 A°), thresholds voltage = 3 V.

This energy is not brought by an external electrical source but is caused by the omnipresent and perpetual force of Casimir, itself controlled by a Coulomb force of opposite direction.

This Coulomb force appears by the automatic switching of MOS transistors when its intensity is greater than a predetermined value and opposite to that of the Casimir force.

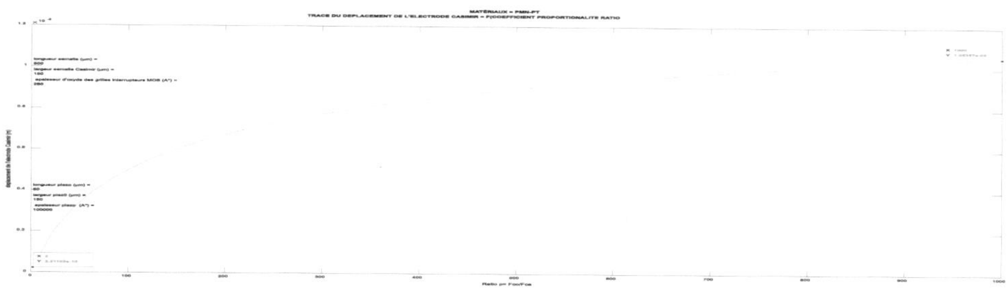

Figure 64.
Displacement of the mobile casimir electrode during the appearance of the coulomb force ls = 500 μm, bs = 150 μm, lp = 50 μm, bp = 150 μm, ap = 10 μm.

This technologically programmable switching of the MOS switches induces the spontaneous appearance of current peaks during a few nanoseconds, themselves inducing voltage peaks at the terminals of an inductor. When the system returns to its starting position, the Coulomb force disappears, leaving the Casimir force to deform the structure again.

The system then spontaneously enters into vibrations.

The energy balance of a cycle therefore seems to satisfy Emmy NOTHER's theorem!

9. Conclusions

1. the proposal to use isotropic and perpetual energy of Casimir called Energy of the vacuum, to obtain a variation of electric charges of a piezoelectric bridge which generates current peaks at the frequency of inductance-sustaining vibrations of the structure, usable without no energy input

2. the system which should allow the conversion of this vacuum energy into alternating current peaks at the vibration frequency of the system. This current passes through an inductor which converts these alternating current peaks into alternating voltage peaks.

3. The current and the voltage at the terminals of this choke feed electronics which rectifies and amplifies, without any external power supply, these voltage peaks in a direct voltage of a few usable volts.

4. a proposal for micro and nano electronic technology giving hope for a possible realization of this set.

According to this study, it would seem we can extract energy from the vacuum by the use of the Casimir force thwarted at the appropriate time by a temporary Coulomb force which makes the system return to its initial position and makes enter into vibrations the structure!!

I am fully aware that this concept may sound like incredible, but it does not seem to contradict Emmy Noether's theorem and, mathematical calculations and simulations give encouraging results and merit further study of this concept by a thesis. I am looking for a microelectronics laboratory with sufficient technological and design resources to confirm or deny this idea of a retired dreamer. The dreamer would be happy to participate in this dream of a new source of energy.

A. Appendices

A.1 A few reminders from RDM

A.1.1 Calculation of the deflection of a bridge embedded at its 2 ends

Note: We take the case of pure bending, the shear force T is such that With M the bending moment applied to the piezoelectric bridge. The Casimir force in the z axis is applied in lp/2 at the center of the bridge (**Figure 65**).

Recall that the neutral axis G (x), is the place where the normal bending stress is zero. G is a point of this neutral axis which in the case of a symmetrical bridge is also on the median line of the center of gravity of the bridge. Let us name:

- Mz (x) = Bending moment depending on the position x on the bridge,
- Gz a point of the neutral axis, z = f (x) the deformation of the bridge
- $I_P(x)$ the bending moment of inertia of the bridge section
- Ep the Young's modulus of the piezoelectric material
- ρ = the radius of curvature of the deformation z = f (x) the deformation of this bridge.

We will assume that the strains of the bridge subjected to the weak Casimir force are themselves small, so the induced strain angles d (ρ) are also small. Thus as it is stipulated in all the works of R.D.M. we can assimilate the radius of curvature ρ with Eq. (21) (**Figure 66**)

$$tan\ (d\omega) = \frac{dx}{\rho} \approx \omega \Rightarrow \frac{1}{\rho} = \frac{d\omega}{dx} = -\frac{M_Z(x)}{E_P I_P(x)}$$

Since the bridge is parallelepiped in shape, the bending moment of inertia along the z axis of the section of this bridge is Eq. (22) (**Figure 67**)

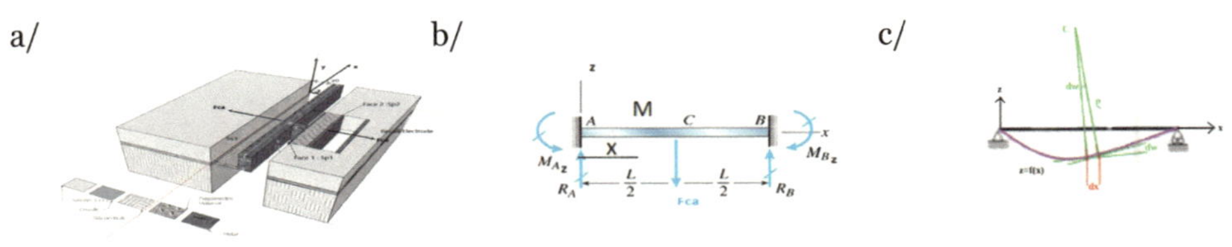

Figure 65.
General appearance a/ b/ of the device studied, forces and applied moments, c/ of the deformed bridge.

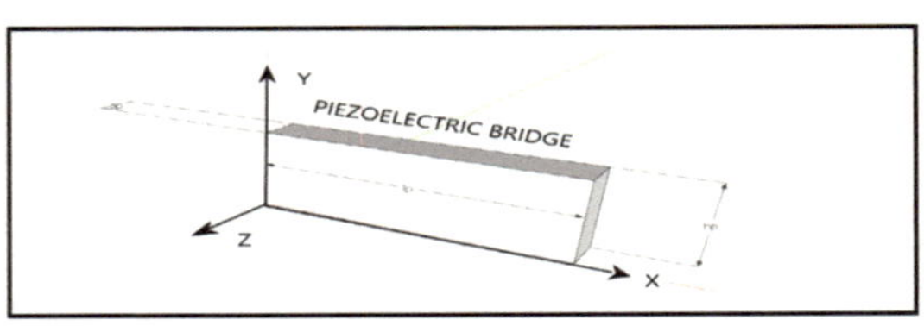

Figure 66.
Piezoelectric Bridge Nomenclature.

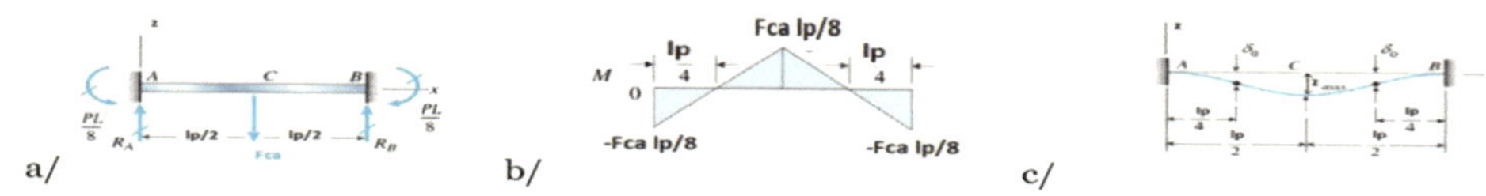

Figure 67.
(a) Forces, shear forces and moments applied on the bridge. (b) Variation of bending moment. (c) Shape and arrow of the bridge embedded at both ends. With: δo = inflection points, z_{max} = arrow of the bridge.

$$I_{GZ} = \frac{b_P a_P^3}{12} = Cte$$

$$In\,x = \frac{l_P}{2} \Rightarrow \frac{dz}{dx_{x=\frac{lP}{Z}}} = 0 \Rightarrow MAZ(x = lp/2) = -\frac{F_{CA}}{8} lP \Rightarrow MAZx = 0) = -\frac{FCA}{8} lP$$

The maximum deflection is in x = lp/2 which gives an arrow: $z_{\max} = F_{CA}.l_p{}^3/(192\ E_P.I_P)$ in $x = l_P/2$ (Eq. (23)). See RDM nomenclature

A.1.2 Calculation of the resonant frequency of the piezoelectric bridge

It is demonstrated (*see for example: Vibrations of continuous media Jean-Louis Guyader (Hermes)*) that the amplitude z (x, t) of the transverse displacement of a cross section of the beam is given by the partial differential equation: $\frac{\delta^4 z}{dx^4} + \frac{\rho S}{E_P IP} \frac{d^2 z}{dt^2} = 0$ if one neglects the internal damping!

The linear mass m(x) being equal to ρ_S (Kg/m), with ρ the density (Kg/m^3) and S the section (m^2), hence the differential equation of free vibrations deduced is: (Eq. (28)) $\frac{\delta^4 z}{dx^4} = -\frac{\rho S}{E_P IP} \frac{d^2 z}{dt^2}$ With $k = \left(\rho S \omega^2 / E_p I_p\right)^{1/4}$, the solution of this differential equation is.

written in the general form: $Z(x) = A1 \exp(k\,x) + A2 \exp(-k\,x) + A_3 \exp(i\,k\,x) + A_4 \exp(i\,k\,x)$ and in the more convenient form:

$$Z(x) = a \cdot \sin(k \cdot x) + b \cdot \cos(k \cdot x) + c \cdot sh(k \cdot x) + d \cdot ch(k \cdot x) \qquad (21)$$

A.1.2.1 Eigen modes and frequencies

In this part, we will assume that the moving part of the structure (**Figure 28**) vibrates at its resonant frequency. The only series of discrete pulsations w_i (proper pulsations of vibration) will be authorized, these pulsations being obtained in the general form: with: w_{pi} = resonance pulsation and therefore f_{pi} = frequency resonant

$$\omega_{PR} = \left(\frac{\alpha_I}{lP}\right)^2 \sqrt{\left(\frac{E_P IP}{\rho S}\right)} \Rightarrow f_{PI} = \frac{1}{2\pi} \left(\frac{\alpha_I}{lP}\right)^2 \sqrt{\left(\frac{E_P IP}{\rho S}\right)}$$

A.1.2.2 Boundary conditions

In our situation we have a recessed-recessed bridge. $Z(x) = a \cdot \sin(k \cdot x) + b \cdot \cos(k \cdot x) + c \cdot sh(k \cdot x) + d \cdot ch(k \cdot x)$ (30).

In this case for x = 0 and for x = lp, we have z(x) = 0 and dz./dx = 0 (zero elongations and slopes). Let:

$$\text{For}: x = 0: 1: a + c = 0 a = -c 2: b + d = 0 b = \text{-}d$$
$$\text{For } x = lp: 3: a \sin(k\,lp) + b\cos(k\,lp) + c\,sh(k\,lp) + d\,ch(k\,lp) = 0;$$
$$4: a\ \cos(k\,lp)\text{-}b\ \sin(k\,lp) + c\,ch(k\,lp) + d\,sh(k\,lp) = 0$$

We deduce from the preceding Eqs. (1–4) that: c. (sh (k lp) – sin (k lp)) + d (ch (k lp) – cos (k lp)) = 0; and c. (ch (k lp) – cos (k lp)) + d (sh (k lp) + sin (k lp)) = 0.

These last 2 equations lead to a transcendent equation in k l_p: so, the determinant of this system is zero! $\left[sh\ (k.lp)^2\text{-} \sin\ (k.lp)^2\right] - [ch\ (k.lp) - \cos\ (k.lp)]^2 = 0\ \ \cos\ (k.lp) = 1/ch\ (k.lp)$.

The solutions of this equation can be solved graphically or numerically.

The numerical resolution (for example by the dichotomy method in **Figure 68** of this equation gives for the first 5 solutions: a1 = 4.7300; a2 = 7.8532; a3 = 10.9956; a4 = 14.1317; a5 = 17.2787.

So, the first resonant frequency of the piezoelectric bridge is.

$\omega_{P1} = (4.73)^2\sqrt{\frac{E_P I_P}{M_S l_P^3}} \Rightarrow f_{P1} = \frac{1}{2\pi}(4.73)^2\sqrt{\frac{E_P I_P}{M_S l_P^3}}$ (31).

For example, for a bridge embedded at both ends with the following characteristics,

For geometries:

$$lp = 50\,\mu m, bp = 150\,\mu m, ap = 10\,\mu m; Sp = 1{,}510^{-9}\ m^2,$$
$$Ip = bp \cdot ap3/12 = 1.2510^{-20}\ m^4$$
$$li = 10\,\mu m; bi = 150\,\mu m; ai = 10\,\mu m; ls = 1000\,\mu m; bs = 150\,\mu m; as = 10\,\mu m$$

For the material: PZT: density

$$r = 7600\left(kg/m^3\right), \text{Young's modulus } Ep = 6^*\ 10^{10}(Pa)\ \left(Kg\ m\ s^{-2}\right)$$

For the section inertia: $Ip = bp^*\ ap^3/12 = 1.25\,10^{-20}(m^4)$.

Then the calculated first resonance frequency is for the PZT material: fp1 = 1.1553 * 10^7 hertz. For this embedded bridge, an ANSYS simulation (**Figure 69**) gives a resonant frequency of f1 = 1.02. 10^7 Hz which is close to that calculated in the draft calculation presented in this report and validates the orders of magnitude obtained with the equations for these preliminary calculations. If one carries out the calculation of the resonant frequency of the structure of **Figure 5** which comprises a free sole of Casimir S_{S2} parallel to a fixed surface S_{S3} and transmitting by a mechanical link finger the force of Casimir, one finds that the resonant frequency have the same form but with Ms. a fixed mass applied in the middle of the bridge. M_s = the total mass of the structure! $M_s = r\left(S_p{}^* a_p + S_i{}^* a_i + S_s{}^* a_s\right) = r\left(l_p{}^* b_p{}^* a_p + l_i{}^* bi^* a_i + l_s{}^* b_s{}^* a_s\right)$.

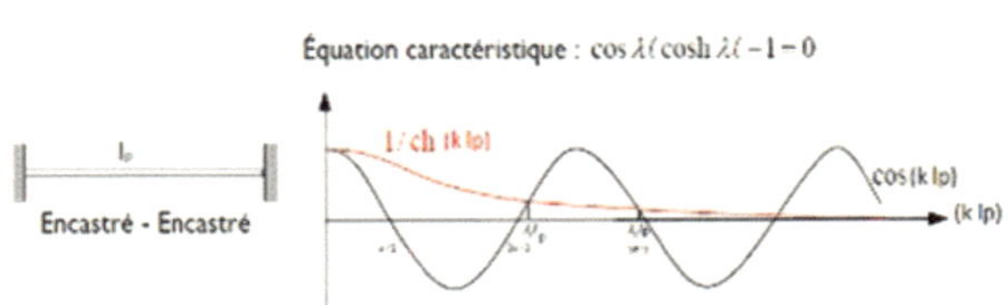

Figure 68.
Numerical solution of Eq. (18).

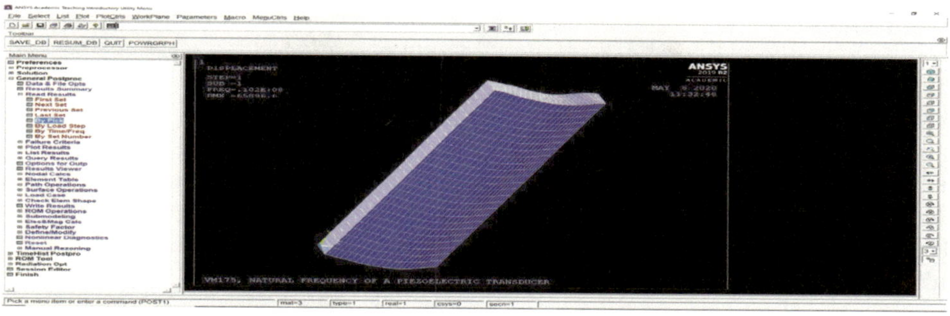

Figure 69.
ANSYS simulation of the resonant frequency of the piezoelectric.

With r = the medium density of the piezoelectric material, of the connecting finger, of the Casimir electrode sole and S_p,S_i,S_s the longitudinal surfaces of this bridge. Indeed, the presence of the Casimir sole connected by the Casimir force transmission finger in the middle of the piezoelectric bridge, modifies the resonant frequencies of this bridge.

The calculated resonance frequency then becomes for the same geometries and materials. fs1 = 2.509 10^6 hertz. With these characteristics, an ANSYS simulation of this structure gives a close resonance frequency: fs1 = 2.62 10^6 Hertz. This approach greatly simplifies these preliminary calculations because otherwise the curvature of the piezoelectric bridge makes the Casimir force strongly depend on the longitudinal and transverse positions x and z of the facing surfaces!

Author details

Patrick Sangouard
Mathematization, Simulation, Electronics, and Technology, Dijon, France

*Address all correspondence to: patrick.ps.patrick@gmail.com

References

[1] Reynaud S, Lambrecht A, Jaekel MT. Fluctuations du vide quantique. In: Laboratoire Kastler Brossel UPMC case Jussieu F Paris Cedex 05. F 75231 Paris Cedex 05: Laboratoire de Physique Théorique de l'ENS 24 rue Lhomond; Juin 2001

[2] Casimir HBG. On the attraction between two perfectly conducting plates. Proceedings of the Koninklijke Nederlandse Akademie van Wetenschappen. 1948;**51**:793

[3] Deriagin BV, Abrikosova II. Direct measurement of the molecular attraction of solid bodies. 2. Method for measuring the gap. Results of Experiments. Soviet Physics JETP. 1957; **3**:819

[4] Lifshitz EM, Pitaevskii LP. Landau and Lifshitz course of theoretical physics: Statistical physics part 2 ch VIII (Butterworth-Heinemann, 1980). Soviet Physics JETP. 1956;**2**:73

[5] Techniques de l'Ingénieur 14/12/ 2012: l'expertise technique et scientifique de référence « Applications des éléments piézoélectriques en électronique de puissance » Dejan VASIC: Maître de conférences à l'université de Cergy-Pontoise, Chercheur au laboratoire SATIE ENS Cachan, François COSTA: Professeur à l'université de Paris Est Créteil, Chercheur au laboratoire SATIE ENS Cachan

[6] Wachel JC, Bates CL. Techniques for controlling piping vibration failures. ASME Paper. 76-Pet-18. 1976

[7] Parasuraman J, Summanwar A, Marty F, Basset P, Angelescu DE, Bourouina T. Deep reactive ion etching of sub-micrometer trenches with ultra-high aspect ratio. Microelectronic Engineering. 2014;**113**:35-39

[8] Marty F, Rousseau L, Saadanya B, Mercier B, Français O, Mitab Y, et al. Advanced etching of silicon based on deep reactive ion etching for silicon high aspect ratio microstructures and three-dimensional micro- and nanostructure. Microelectronics Journal. 2005;**36**: 673-677

[9] Semiconductor Devices, Physics' and Technology S. M. SZE Distinguished Chair Professor College of Electrical and Computer Engineering National Chiao University Hsinchu Taiwan, M.K. LEE Professor Department of Electrical Engineering, Kaohsiung Taiwan

Section 2

Energy Efficiency Evaluation

Chapter 6

Performance Evaluation of Desalination Technologies at Common Energy Platform

Muhammad Wakil Shahzad, Kim Choon Ng, Muhammad Burhan, Chen Qian, Doskhan Ybyraiykul, M. Kumja, Muhammad Ahmad Jamil, Yinzhu Jiang, Nida Imtiaz and Ben Bin Xu

Abstract

A major fraction of secondary energy consumed for our daily activities, such as electricity and low-grade heat sources, emanates from the conversion of fossil fuels in power plants. In the seawater desalination processes, the energy efficiency is usually expressed in kWh electricity or kWh of low-grade heat per unit volume of water produced. Although kWh energy unit provides a quantitative measure of input energy, it has subtly omitted the embedded quality of supplied energy to desalination plants. In assuming the equivalency across dissimilar energy forms, it results in a thermodynamic misconception that has eluded the desalination industry hitherto, i.e., not all units of derived energy are created equal. An incomplete energy efficacy approach may result in the inferior selection of desalination processes to be deployed; —a phenomenon observed in the trend of installed desalination capacity globally. Operating a less efficient desalination plant over its lifespan would create much economic burdens including a higher unit cost of water, higher CO_2 emissions and greater brine discharge to the environment. This book chapter clarifies the key concept and a thermodynamic framework to rectify the misconception in energy consumption, permitting energy planners and designers to optimize deployment of future desalination plants for energy sustainability. We have derived conversion factors to convert assorted derived energies into standard primary energy for fair comparison.

Keywords: sustainable desalination, thermodynamic limit, universal performance ratio, primary energy

1. Introduction

The global demand for potable water to meet all activities of mankind, in the industrial, domestic and agricultural sectors has been increasing rapidly due primarily to three growth factors, namely (i) an increasing world' population in developing countries, (ii) the quest for higher economic growth in all economies and (iii) the over-abstraction of ground water and the degradation of existing natural water sources on land. Much of fresh water found on land, namely lakes, wetlands and rivers, is gradually being polluted by indiscriminate discharge of man-made pollutants. By the year 2030, Global Water Intelligence [1] has projected an increase in annual potable-water demand from the current level of 5300–6900 billion cubic meters (bcm), equivalent to a compound annual growth rate of over 2%, as shown in **Table 1**. Yet, the existing sustainable potable water supply, mainly from natural precipitation sources, remains constant at 4200 bcm annually. Such a shortfall in the supply-demand of greater than 2700 bcm annually can only be met by reliable desalination methods [2]. Many ad-hoc measures to conserve water consumption and better manage the supply infrastructure can improve the water use inventory in water stressed countries [3]. However, ground water extraction rates are far greater than the rates at which they are replenished and there is over extraction from rivers [4]. Even with a degree of water re-use there will be a deficit between consumption and sustainable supply. Thus, the only practical means of meeting the future global potable water needs is by seawater desalination [5].

For seawater desalination at ambient temperature, the minimum work needed to separate dissolved salt ions of 3.5% by weight from the brine (within the solution) is termed as the thermodynamic limit (TL) of the normal seawater. Invoking the Gibbs equations for the separation process where the mass fractions of dissolved salts, the activity coefficients of water and solute are known, the theoretical work can be readily found to be 0.78 $kWh_{\_primary\ energy(pe)}$ per cubic meter of potable water or alternatively, the amount of potable water could be theoretically attained at TL is 1.282 m^3 per $kWh_{\_pe}$ consumption [6]. The primary energy (PE) is the naturally available work and it is equivalent to the respective calorific value of fuel burned. It implies that the $kWh_{\_pe}/m^3$ of energy consumption at TL is totally devoid of dissipative losses as the processes are deemed ideal, i.e., the available work as described by classical thermodynamics. Unfortunately, such a concept has been grossly misinterpreted in the literature. The recent reports indicated that energy efficacy of exiting methods in seawater desalination have achieved merely 13% of the TL [7–11]. This shows that currently desalination processes are not consuming fossil fuel energy sources efficiently. Thus, there is a great motivation to improve the energy efficacy of desalination processes to meet the sustainable goals of future water supplies.

In this chapter, the authors attempt to address two challenges facing the desalination industry: Firstly, there is a need to have a common thermodynamic framework to define the absolute value of energy supplied to separation processes. The energy consumed by assorted desalination processes must incorporate both quantity and quality aspects at the respective input conditions of processes. Unfortunately, the quality of dissimilar energy supplied to assorted desalination methods hitherto has been inadvertently omitted. We accentuate that a meaningful efficacy comparison of dissimilar desalination methods can be achieved with a common thermodynamic platform of high to low temperature reservoirs. All derived energy consumption of desalination processes is equivalently transformed to the

Water consumption by sectors	**Potable water from natural precipitation 10^9 m^3 or bcm**	**2010 bcm/year**	**2018 bcm/year**	**2030 bcm/year**	**Predicted deficit in water bcm/year**
Industry	4200	3100	3600	4500	2700
Agriculture		800	1028	1500	
Domestic		600	705	900	
Total		4500	5333	6900	

Table 1.
The projected demand and supply of potable water for the industry, agriculture and domestic sectors, as reported by global water intelligence (GWI) [1–3].

consumption of primary energy. Such procedures are predicated on either the same equivalent Carnot work output or input depending on the nature of desalination methods used. More importantly, the proposed methodology provides a direct apportionment, in the form conversion factors (CF), in the existing cogeneration power plants setting producing electricity and heat for desalination processes. Secondly, the authors opined that an optimally-designed desalination system can readily attain up to 35% of the TL, as reflected by the many plausible heat engines operating currently in other industries. A quantum improvement in the efficiency of separation processes of seawater desalination is most likely to realize either by (i) developing better performing work-driven systems such as thin-film composite materials or (ii) a higher thermodynamic synergy between the heat-driven processes. In the later section, the authors will highlight a hybrid heat-driven cycle, that were successfully tested at KAUST, attained the best energy efficacy for seawater desalination of 20% [12].

2. Limitations of current evaluation methodologies

At present, the conventional secondary or derived energy units, expressed in kWh of electricity or thermal heat source, are used inadequately for energy efficiency comparison between all types of desalination processes [13, 14]. This practice is insufficient because it has omitted a key aspect of energy quality embedded in the supply fuels. As demonstrated later, the assumption that all derived energy, no matter how dissimilar in forms, are deemed directly equivalent to each other which is thermodynamically inadequate. The units of energy measurement, namely kWh or 3.6 MJ, expresses merely its quantitative aspect but it has ignored the qualitative aspect of the energy used.

For example, same heat input, Q(1 kWh), is supplied to two processes, as depicted on a temperature versus entropy diagram of **Figure 1**. States 1–2 shows higher temperature process and states 3–4 at a lower temperature, i.e., $T_1 >> T_3$. Using the concept of an ideal Carnot cycle at the same heat input, the available work that could be extracted from the former process is higher than the latter. Being an isothermal cycle in a T-S diagram, i.e., $\oint_{12dc} Q = \oint_{12dc} W \ and \ \oint_{12dc} U = 0$, the energy input from a higher temperature source yields a larger amount of useful work due primarily to the better quality of heat input. This is reflected by the dissimilar

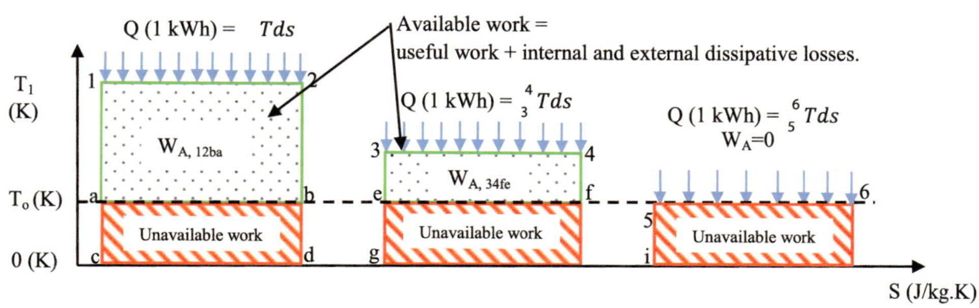

Figure 1.
A graphical demonstration of the energetic quantity and quality to thermodynamic cycles. Despite having same quantity of energy input, say, Q = 1 kWh, a higher available work could be produced from the process of a higher input temperature that is, $W_{A,12ba} >> W_{A,34fe}$. Note that the available work constitutes the useful work, the internal and external dissipative losses incurred by the processes of cycle. At near to ambient temperature, T_o. The cycle-56Ji has zero available work, even it is supplied with the same heat quantity. Despite the same heat input quantity, the cycles have dissimilar amount of unavailable work, and it is attributed to the quality of energy (defined by temperature and pressure) being supplied over the limit of "dead state".

unavailable work that were demarcated by the ambient temperature and the entropy change. Such unavailable work is also commonly known as the dissipation trapped by the "dead state". This aspect of diminishing available work with lower heat source temperature can be observed to reduce to zero at the limit, $T_{source} \rightarrow T_o$. A second aspect of **Figure 1** is cascading of processes in which exhaust of processes at higher temperature can be used as a heat source for a second process operating at relatively lower temperature to optimize the cycle efficiency.

Over many decades, decision makers within the desalination industry have failed to notice the above-mentioned misconception. Should it remain uncorrected, sub-optimal decisions will be made and, in a world, seeking to become carbon neutral the implications are serious. The consequences will be inferior selection of desalination methods for the supply of large quantity of potable water in many water-stressed countries. Operating a non-optimal desalination plant over its lifespan not only burdens consumers economically with a higher unit water cost but the associated carbon dioxide emissions will be higher and probably there will be a higher discharge of chemically laden brine into the sea.

Those interested in pedagogy might care to ask the following rudimentary question, why despite many decades of advancement in science and engineering, how is it possible for the desalination industry to treat dissimilar energy quantities as if they were the same or, as the English would say, compare apples with oranges? We will seek to give an answer at the end of the paper.

3. A level playing field across the processes

Currently, the best available power plants for co-generation of electricity and heat sources are the combined cycle gas turbines plant or CCGTs in short to utilize fossil fuel primary energy optimally. It consumes conventional primary energy, by burning the natural gas or liquid fuels in the combustor, to generate secondary or derived energy. Such derived energy is used conveniently for powering the work and heat-driven processes for treating impaired water to produce potable water, as shown in **Figure 2**.

One notable point is the relative rates of primary energy consumption in producing the derived energy types. A detailed analysis indicates a disproportionate distribution of primary energy use by the assorted processes of a conversion plant. For example, the electricity generation from the gas and steam turbines incurred almost 96 ± 1% of the total input exergy, and followed by a minor portion in the form of bled-steam at low pressures at 2 ± 0.1% for powering desalination processes, whilst the remaining 2 ± 0.1% of input exergy are traced to the heat rejection to ambient by exhaust gases and to the condenser.

The stark differences in exergy destruction nullifies any implicit long-held assumption that some might have of direct parity or equivalence between dissimilar derived energy consumptions. Many would recognize that 1 $kWh_{\_elec}$ is not equal to 1 $kWh_{\_thermal}$ but would not be able to establish a weighting between them. Thermodynamically, not all derived energy is created equal. Unfortunately, the current practice of quantifying energy efficacy across assorted seawater desalination methods is based on "equal parity between all types of derived energy". Obviously, this is a flawed assumption. A simple analogy is found in monetary currencies conversion in between countries. For example, a unit US dollar is not equivalent to another currency such as the Australian dollar. Economists employed a method of purchasing power parity (PPP), based on a basket of essential consumer products that normalized the necessary conversion factors in between all currencies globally. Here we seek to achieve the same for desalination processes.

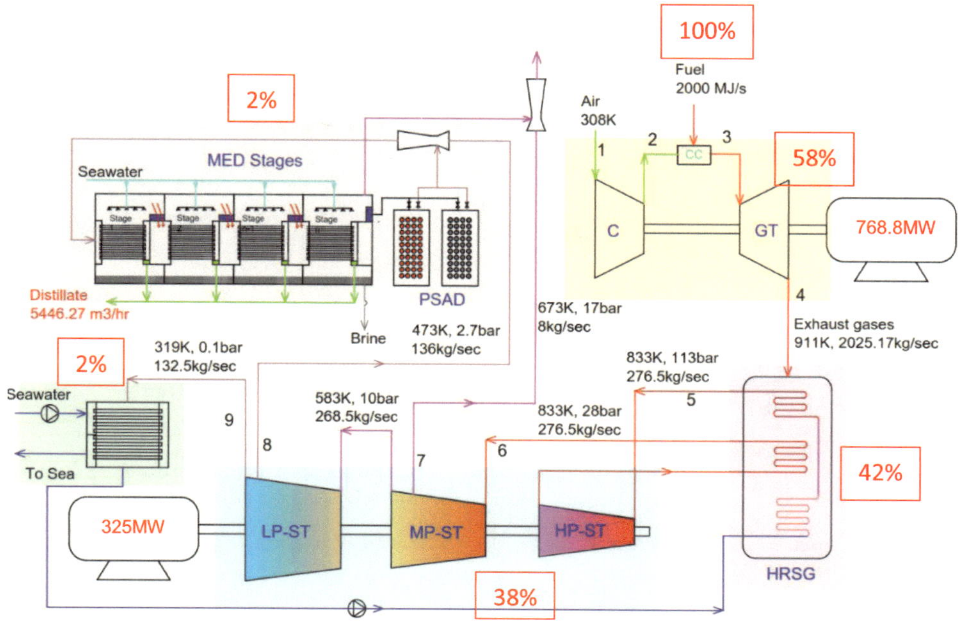

Figure 2.
Typical primary energy consumption in a combined power and water cycle.

4. Thermodynamic framework

The term exergy was devised by Zoran Rant [15] in 1956 by using two Greek words, i.e., "ex" and "ergon" meaning "from work". However, the main concept was first studied by Willard Gibbs in 1873 [16]. The term exergy is defined as "the available work" and it constitutes the maximum useful (shaft) work that could be extracted from of a cycle. Recently, many researchers published on exergoeconomic and thermoeconomic analysis of desalination process [17–20].

To provide any misconceptions across the various type of seawater desalination processes, the thermodynamics of heat engines, representing the desalination methods, are invoked. The amount of ideal or Carnot work (W_C) that can be extracted from a flow of heat input Q_H, emanating from a higher temperature (T_H) heat source to an engine, producing an ideal work W_C, whilst rejecting heat Q_L into a low temperature (T_L) reservoir, is depicted schematically in **Figure 3**.

Due to incipient dissipative losses, the actual useful work ($W_{act,i}$) produced by an engine is lower than the ideal or Carnot work and thus, the Second Law efficiency ($\eta_i^{''}$) of engine defines the work ratio, i.e., $\eta_i^{''} = \frac{W_{act,i}}{W_C}$. We invoke the derived corollary of Second |Law of Thermodynamics relationship, i.e.,

$$\frac{W}{T_H - T_L} = \frac{Q_H}{T_H} = \frac{Q_L}{T_L}, \quad (1)$$

where T_H and T_L are the process average temperatures corresponding to any desalination methods. For a given Carnot work (W_C,) output from a cycle, the corresponding amount of heat supply (Q_H at T_H) to the engine is deemed as the primary energy input. This can be expressed as the product of Carnot work and the ratio of T_H to the temperature difference (T_H–T_L) between the reservoirs:

$$Q_H = W_C\left(\frac{T_H}{T_H - T_L}\right). \quad (2)$$

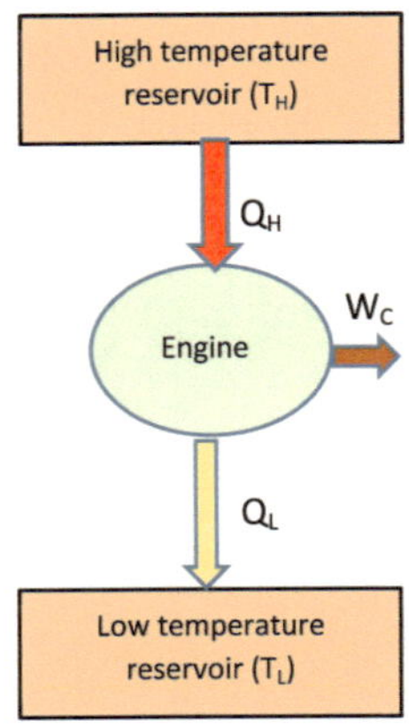

Governing Thermodynamic Laws and Corollary:-

First Law of Thermodynamics: $Q_H = W_C + Q_L$.

Corollary of Second Law: $\frac{Q}{T} = constant$.

Invoking the above Second Law, i.e.,

$$\frac{Q_H}{T_H} = constant = \frac{Q_L}{T_L} \quad or \quad Q_L = \frac{Q_H T_L}{T_H}$$

And substitute into the First Law;

$$Q_H - \frac{Q_H T_L}{T_H} = W_C,$$

hence, the relationship between ideal work, heat transfers and reservoir temperatures is given by:

$$\frac{W_C}{(T_H - T_H)} = \frac{Q_H}{T_H}$$

Figure 3.
A heat engine driven by heat transfers at high and low temperature reservoirs and Carnot (ideal) work that can be emanated by it.

Assuming the same work output were to be derived from an adiabatic flame (T_{adia}) of a fuel burned with ambient air and operating between maximum temperature difference across the reservoirs (T_{adia}-T_o), the heat supply to the engine is equivalent to the work potential (exergy) of heat engine. A common thermodynamic platform across the temperatures, T_{adia} and T_o, is proposed where an equivalent heat transfer (Q_{SPE}) at the referenced platform would deliver the same Carnot work, i.e.,

$$Q_{SPE} = W_C\left(\frac{T_{adia}}{T_{adia} - T_o}\right). \tag{3}$$

Given the temperature platform, i.e., $(T_{adia} - T_o)$, Eq. (3) implies the input exergy, Q_{SPE}, is equivalent to a fraction of supplied fuel to generate the Carnot work. Should there be "n" number of engines operating synergistically across the same referenced temperature reservoirs, then the total standard primary energy consumption by all engines is given by the summation of the right hand terms of Eq. (3), i.e.,

$$\sum_{i=1}^{n} Q_{spe,i} = \left(\frac{T_{adia}}{T_{adia} - T_o}\right) \sum_{i=1}^{n} W_{C,i}. \tag{4}$$

where "i" refers to a process in a combined machine. Eq. (4) depicts an important observation of decomposition of total input exergy (work) into fractions as accrued by a host of sequential machines. This is similar to the equivalent primary energy input, i.e., $Q_{\_spe}$, contributed by all processes in a CCGT plant. At known heat transfer rates corresponding to each set of inlet and outlet temperatures of a cycle, the total Carnot work can be cumulatively summed to yield the primary energy of the fuel burned as presented in case example in following sections. Equivalently, the apportionment of standard primary energy consumption incurred by the processes of CCGT, namely the generation of electricity and low-grade steam energy, can now be accurately determined using a conversion factor for ease of application.

Extending the Eq. (4) by taking the ratio of standard primary energy and the Carnot work of a process to their respective total in the cycle gives their equivalency. Also the temperature ratios ($1-T_o/T_{adia}$) are eliminated.

$$\frac{Q_{SPE}}{\sum_{i=1}^{n} Q_{SPE,i}} = \frac{W_C}{\sum_{i=1}^{n} W_{C,i}} \tag{5}$$

Herein a conversion factor CF_i is defined as the standard primary energy to the actual derived energy. It can be expressed as

$$CF_i = \frac{Q_{SPE}}{W_a} = \left(\frac{\sum_{i=1}^{n} W_{C,i} / \left(1 - \frac{T_o}{T_{adia}}\right)}{\sum_{i=1}^{n} \left(W_{C,i} \eta_i''\right)} \right) = \frac{1}{\left(1 - \frac{T_o}{T_{adia}}\right) \eta''} \tag{6}$$

where the Second Law efficiency of a process is defined as $\eta_i'' = \frac{W_{a,i}}{W_{C,i}}$, and T_{adia} is the adiabatic flame temperature of fuel burning in air which characterizes the highest temperature difference (T_{adia}–T_{o}) across the reservoirs of the heat engine.

5. Results and discussion

For clarity, a typical CCGT plant of nominal primary energy input of 2000 MW is considered as presented in **Figure 2**. By analyzing the heat transfer rates at the respective temperature reservoirs for each of the cascaded processes, the ideal or Carnot work can be determined with a selected common temperature platform, defined by the adiabatic flame and ambient temperatures. By summing all the standard primary energy ($Q_{\_SPE}$), as described by Eq. (4), it yields the equivalent primary energy of fuel or the fuel exergy supplied to the CCGT plant. In terms of the useful output, the total electricity generation from both turbines of CCGT amounts to 1094.37 MW_{elec} and a steam-powered multi-effect distillation (integrated MED_TVC) produces 5445 m^3/h potable water. To sustain the dissimilar derived energy, a steady heat rate of 2000 MW is needed by burning a fossil fuel such as the natural gas at the combustor of gas turbines (GT) cycle. The detailed thermodynamic states and the mass flow rates of working fluids operating in key components of CCGT, either the products of combustion or steam at all state points of key components, are presented in Appendix 1.

This procedure offers a means of apportionment of the Q_{spe} into fractions that generate all types of useful derived energy to power the assorted desalination plants, as summarized in **Figure 4**. Based on these fractions of primary energy dissipation, the appropriate conversion factors are derived which forms a basis for level platform to normalize the primary fuel to derived energy or vice versa. For example, the conversion factor for electricity is simply expressed as the ratio of $Q_{\_SPE}$ to the electricity generated or alternatively, it can also be determined from the Second Law and temperature ratios as shown below:

$$\text{CF}_{\text{elec}} = \frac{\left(Q_{SPE\_GT} + Q_{SPE\_ST}\right)}{\left(W_{SPE\_GT} + W_{SPE\_GT}\right)} = \frac{1}{\left(1 - \frac{T_o}{T_{adia}}\right)\eta_i''} = 1.7328 \tag{7}$$

Similarly, the conversion factor for low-grade steam input to MED_TVC is expressed by the ratio Q_{spe} to the thermal energy input or it can also be determined from the appropriate temperature ratios:

$$\text{CF}_{\text{thermal}} = \frac{\left(Q_{SPE\ of\ bled\ steam}\right)}{\left(Q_{actual\ bled\ steamat\ low\ pressure}\right)} = \frac{1 - \frac{T_o}{T_{MED}}}{\left(1 - \frac{T_o}{T_{adia}}\right)} = 0.1250 \tag{8}$$

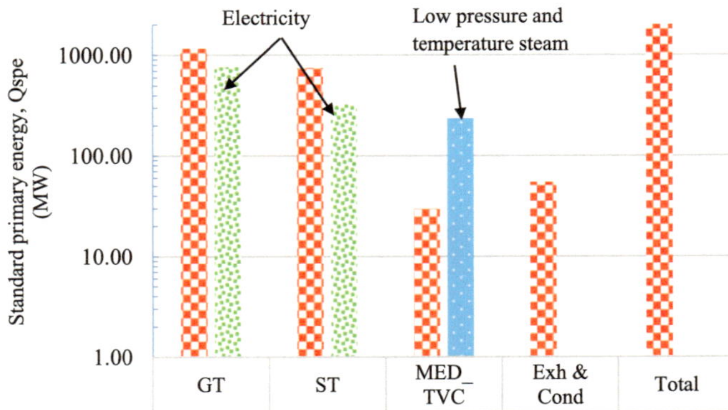

	GT	ST	MED_ TVC	Exh & Cond	Total
Q_spe (MW)	1165.00	750.00	29.75	54.93	1998.66
Derived energy (MW)	768.87	325.00	237.54	0.00	

Figure 4.
The consumption of standard primary energy and the production of useful derived energy by the major components of a combined cycle gas turbines (CCGT) power plant. The units of accompanying table are in MW.

The thermodynamic limit of 0.78 kWh_{spe}/m^3 is engaged to determine the Carnot work and the temperature reservoirs of the ideal states. Thus, the above calculations demonstrated that a common standard primary energy platform could resolve the long-held implicit misconception of equivalency that were assumed between different types of derived energy, namely that between electrical and thermal energy. Such a thermodynamic fallacy has unfortunately persisted in the desalination industry for over 5 decades.

Figure 4 present the standard primary energy consumptions and the production of useful derived energy by the major components of a CCGT power plant based on derived conversion factors.

It is noticed that at ideal conditions, the maximum potable water production per unit primary energy consumed is 1.282 m^3/kWh_{spe} or minimum specific energy consumption is 0.943 kWh_{spe}/m^3. Being an ideal process, no conversion of primary energy to derived energy is needed. However, a common misconception, often seen in literature where the graph of specific energy consumption for desalination processes is presented against the various recovery ratios. Conventionally, it showed a curve of gradual increase of the derived energy consumption with increasing recovery ratio (RR) from zero to more than 60%. This depiction of specific derived energy consumption has omitted the dissipative losses incurred by the conversion plants in producing the derived energy when the RR is other than zero. A similar concept is found in the Carnot efficiency of a heat engine when the actual work output is deemed zero at the ideal limit, although the available Carnot work from the cycle is at its highest. Using the proposed common platform of standard primary energy consumption for all desalination processes, the cross comparison of energy efficiency amongst all desalination methods can now be accurately resolved. **Figure 5** shows the energy efficacy from about 60 seawater desalination plants powered by assorted desalination methods, stretching from 1983 to the present [21].

For a fair comparison, all conventional specific derived energy consumption in these plants is transformed to their equivalent primary energy with the relevant conversion factors, where the embedded quantitative and qualitative aspects of the derived energy are now incorporated. It can be seen that SWRO is has a slightly better energy efficacy than MED and MSF, achieving around 13% of TL.

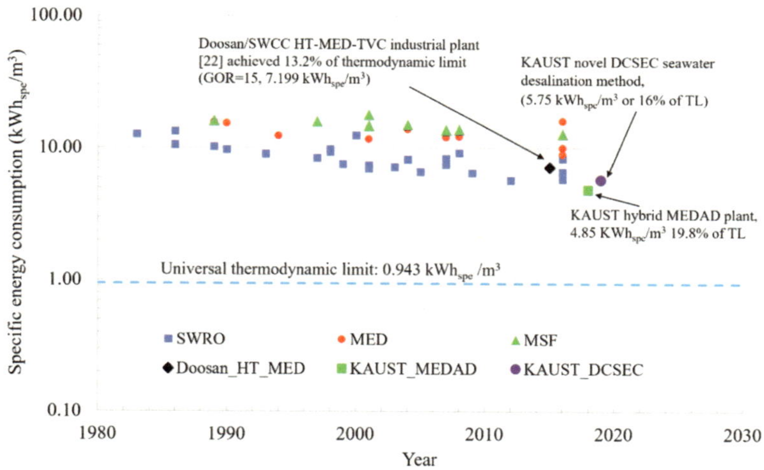

Figure 5.
Energy efficiency of seawater desalination processes based on standard primary energy. The MED-TVC shows a higher energy efficacy as compared to MSF and the SWRO.

Nevertheless, all practical methods available hitherto are still far below the TL, hovering less than 10–13% of the ideal.

The authors have conducted an experimental study at KAUST of a hybrid approach involving the well-proven heat driven MED-TVC processes with an adsorption (AD) cycle, arranged in a back-to-back manner [12, 22–31]. A quantum jump in the energy efficiency is achieved through the thermodynamic integration of two thermally-driven cycles with two salient consequences, namely (i) an increase in the available temperature differences between the top to bottom brine temperatures and hence more MED stages could be inserted, and (ii) an opportunity to scavenge more enthalpy from the seawater feed by liquid flashing in the lower stages of MED where the corresponding stage temperatures were below ambient. The recent pilot-scale experiments, conducted with hybrid design of MED-AD plant at KAUST, have attained a lowest brine temperature of 5°C. The vapor generation in these MED stages maximized both the effects from the thermally-driven film evaporation and the liquid flashing from the excess enthalpy embedded available in the feed spray [32–35]. Consequently, the thermodynamic synergy between MED-TVC and AD cycles have boosted distillate production by more than two folds with the same energy input to the top brine stage, attaining a specific energy consumption level of 4.85 kWh_{spe}/m^3 that shows a quantum jump in energy efficiency from current 13 to 20% of TL, as indicated in **Figure 5**.

6. Summary

In summary, the common platform of standard primary energy consumption is thermodynamically the most rigorous method for the cross comparison of energy efficiency of assorted desalination processes. The outward acceptance of equivalency between electricity and low-grade thermal energy has led to a long-held indifference to the quality of derived energy supply to utilize more optimally. This attitude has afflicted the desalination industry for more than 5 decades. The consequence from such a fallacy has led to some inferior decisions by leaders of desalination industry particularly regarding the adoption of less energy efficient desalination processes and hence non-optimal energy consumption. Such poor selection has burdened the future economy of many water-stressed countries with higher unit water costs over the decade-long life-span of plants. In concluding it is noted, firstly that the energy efficiency of all practical desalination methods available hitherto have been shown, on a standard primary energy platform, to be far below the ideal limit, typically hovering between 10 and 13% of the TL. Secondly, the design experiences accrued by scientists and engineers have demonstrated, in some other disciplines, that a plausible energy efficiency target of an engine operating between 35 and 40% of ideal limit is tenable for the cascaded designs of assorted desalination plants. Only at these higher efficacy levels will the desalination processes be poised to meet the future goals of sustainable seawater desalination. Hence, there is motivation to strive for higher efficiency with better thermally-driven distillation techniques or thin-film composite membranes [36–40]. The caveat is that a common platform for energy efficacy comparison is desirable, and it is anchored to the best available conversion technology known. For example, in the past three decades, the CCGT has the highest conversion efficiency in the production of convenient derived energy that powers the desalination processes. In future when making appropriate comparison, the same thermodynamic-rigorous methodology of using a standard primary energy platform is equally valid.

Acknowledgements

Authors would like to thank Northumbria University, Newcastle Upon Tyne NE1 8ST, United Kingdom POC grant for Solar Water project awarded to Dr. Muhammad Wakil Shahzad, and King Abdullah University of Science and Technology, Saudi Arabia for the research support of this research. This work was also supported by the Engineering and Physical Sciences Research Council (EPSRC, UK) grant-EP/N007921.

Appendix 1: CCGT plant schematic and summary of state points

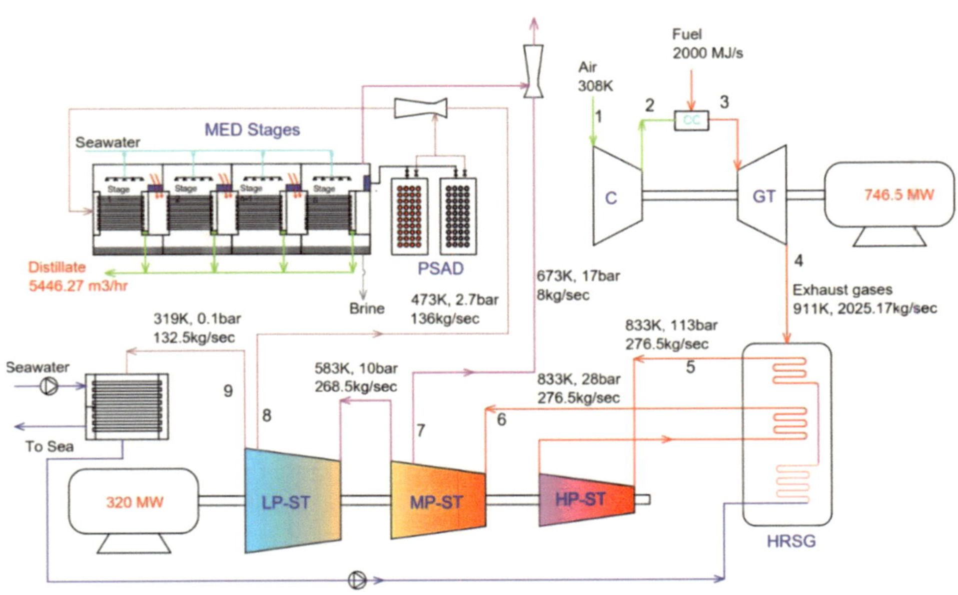

Thermodynamic states of air and steam of CCGT cycle under investigation.

State points	ṁ (kg/s)	T (K)	P (bar)	h (kJ/kg)	s (kJ/kg-K)
1	2021.64	305	1.0	305.6	6.88
2	2021.64	592	8.0	599.3	6.98
3	2056.00	1470	8.0	1559.1	8.30
4	2056.00	911	1.2	945.0	8.0
4a	2066.00	370	1.05	371.8	7.08
5	295.16	833	113	3514.3	6.7
5a	295.16	683	28	3360.0	7.0
6	295.16	833	28	3600.0	7.5
6a	287.16	703	10	3380.0	7.8
P1	8.0	768	17	3490.0	7.55
7	287.16	703	10	3380.0	7.8
7a	151.16	319	0.1	2590.0	8.16
P2	136.0	573	2.7	3090.0	7.75
8	151.16	308	0.1	189.0	0.639

Author details

Muhammad Wakil Shahzad[1*], Kim Choon Ng[2], Muhammad Burhan[2], Chen Qian[2], Doskhan Ybyraiykul[2], M. Kumja[2], Muhammad Ahmad Jamil[1], Yinzhu Jiang[3], Nida Imtiaz[1,4] and Ben Bin Xu[1]

1 Department of Mechanical and Construction Engineering, Northumbria University, Newcastle Upon Tyne, UK

2 Water Desalination and Reuse Centre, King Abdullah University of Science and Technology, Thuwal, Kingdom of Saudi Arabia

3 School of Materials Science and Engineering, State Key Laboratory of Clean Energy Utilization, ZJU-Hangzhou Global Scientific and Technological Innovation Centre, Zhejiang University, Hangzhou, Zhejiang, P.R. China

4 School of Mechanical Engineering, Universiti Teknologi Malaysia, Johor Bahru, Johor, Malaysia

*Address all correspondence to: muhammad.w.shahzad@northumbria.ac.uk

References

[1] Market Insight: Global Water Market, Global Water Intelligence Report. Oxford, United Kingdom: Global Water Market; 2018

[2] Odhiambo GO. Water scarcity in the Arabian Peninsula and socio-economic implications. Applied Water Science. 2016;**6**:21-35

[3] Wehn U, Collins K, Anema K, Carrera LB, Lerebours A. Stakeholder engagement in water governance as social learning: Lessons from practice. Water International. **43**(1, 2017):34-59

[4] Vorosmarty CJ, Green P, Salisbury J, Lammers RB. Global water resources: Vulnerability from climate change and population growth. Science. 2000; **289**(5477):284-288

[5] Shahzad MW, Burhan M, Ang L, Ng KC. Energy-water-environment nexus underpinning future desalination sustainability. Desalination. 2017;**413**: 52-64

[6] Van Der Ham F. Eutectic Freeze Crystallization. Technische Universiteit Delft, Delft, Netherlands; 1999 (Eutectic Freeze—TU Delft Repositories, repository.tudelft.nl)

[7] Shahzad MW, Burhan M, Ng KC. A standard primary energy approach for comparing desalination processes. npj Clean Water. 2019;**2**:1

[8] Shahzad MW, Ng KC. On the road to water sustainability in the Gulf. Nature Middle East. 2016;**4**:1-11. DOI: 10.1038/nmiddleeast.2016.50

[9] Shahzad MW, Burhan M, Son HS, Oh SJ, Ng KC. Desalination processes evaluation at common platform: A universal performance ratio (UPR) method. Applied Thermal Engineering. 2018;**134**:62-67

[10] Ng KC, Shahzad MW, Son HS, Hamed OA. An exergy approach to efficiency evaluation of desalination. Applied Physics Letters. 2017;**110**(18): 184101. DOI: 10.1063/1.4982628

[11] Shahzad MW, Burhan M, Ybyraiymkul D, Ng KC. Desalination processes' efficiency and future roadmap. Entropy. 2019;**21**(1):84

[12] Son HS, Shahzad MW, Ghaffour N, Ng KC. Pilot studies on synergetic impacts of energy utilization in hybrid desalination system: Multi-effect distillation and adsorption cycle (MED-AD). Desalination. 2020;**477**:114266

[13] Ghaffour N, Missimer TM, Amy GL. Technical review and evaluation of the economics of water desalination: Current and future challenges for better water supply sustainability. Desalination. 2013;**309**:197-207

[14] Yip NY, Elimelech M. Thermodynamic and energy efficiency analysis of power generation from natural salinity gradients by pressure retarded osmosis. Environmental Science & Technology. 2012;**46**(9): 5230-5239

[15] Zoran R. Exergie, Ein neues Wort für "technische Arbeitsfähigkeit". Forschung Auf dem Gebiete des Ingenieurwesens. 1956;**22**:36-37

[16] Gibbs JW. A method of geometrical representation of thermodynamic properties of substances by means of surfaces: Reprinted in Gibbs, collected works. In: Longley WR, Van Name RG, editors. Transactions of the Connecticut Academy of Arts and Sciences. New York: Longmans, Green; 1873. pp. 382-404

[17] Jamil MA, Shahzad MW, Zubair SM. A comprehensive framework for thermoeconomic analysis of

desalination systems. Energy Conversion and Management. 2020;**222**: 113188

[18] Abid A, Jamil MA, us Sabah N, Farooq MU, Yaqoob H, Khan LA, et al. Exergoeconomic optimization of a forward feed multi-effect desalination system with and without energy recovery. Desalination. 2021;**499**:114808

[19] Jamil MA, Goraya TS, Shahzad MW, Zubair SM. Exergoeconomic optimization of a shell-and-tube heat exchanger. Energy Conversion and Management. 2020;**226**:113462

[20] Jamil MA, Goraya TS, Ng KC, Zubair SM, Xu BB, Shahzad MW. Optimizing the energy recovery section in thermal desalination systems for improved thermodynamic, economic, and environmental performance. International Communications in Heat and Mass Transfer. 2021;**124**:105244

[21] Ng KC, Burhan M, Chen Q, Ybyraiymkul D, Akhtar FH, Kumja M, et al. A thermodynamic platform for evaluating the energy efficiency of combined power generation and desalination plants. npj Clean Water. 2021;**4**(1):1-10

[22] Shahzad MW, Ng KC. On the Road to Water Sustainability in the Gulf, Nature Middle East. Berlin/Heidelberg, Germany: Springer; 2016. DOI: 10.1038/nmiddleeast.2016.50

[23] Shahzad MW, Ng KC, Thu K. Future sustainable desalination using waste heat: Kudos to thermodynamic synergy. Environmental Science: Water Research & Technology. 2016;**2**(1): 206-212

[24] Thu K, Kim YD, Shahzad MW, Saththasivam J, Ng KC. Performance investigation of an advanced multi-effect adsorption desalination (MEAD) cycle. Applied Energy. 2015; **159**:469-477

[25] Shahzad MW, Thu K, Kim YD, Ng KC. An experimental investigation on MEDAD hybrid desalination cycle. Applied Energy. 2015;**148**:273-281

[26] Ng KC, Thu K, Oh SJ, Ang L, Shahzad MW, Ismail AB. Recent developments in thermally-driven seawater desalination: Energy efficiency improvement by hybridization of the MED and AD cycles. Desalination. 2015; **356**:255-270

[27] Shahzad MW, Ng KC, Thu K, Saha BB, Chun WG. Multi effect desalination and adsorption desalination (MEDAD): A hybrid desalination method. Applied Thermal Engineering. 2014;**72**(2):289-297

[28] Shahzad MW, Ybyraiymkul D, Burhan M, Ng KC. Renewable energy-driven desalination hybrids for sustainability. In: Desalination and Water Treatment. London: InTechOpen; 2018

[29] Shahzad MW, Burhan M, Ybyraiymkul D, Ng KC. Desalination with renewable energy: A 24 hours operation solution. In: Water and Wastewater Treatment. London: InTechOpen; 2019

[30] Shahzad MW, Burhan M, Ghaffour N, Ng KC. A multi evaporator desalination system operated with thermocline energy for future sustainability. Desalination. 2018;**435**:268-277

[31] Shahzad MW, Ybyraiymkul D, Burhan M, Oh SJ, Ng KC. An innovative pressure swing adsorption cycle. In: AIP Conference Proceedings. Vol. 2062. Melville, NY: AIP Publishing LLC; 2019. p. 020057

[32] Alrowais R, Qian C, Burhan M, Ybyraiymkul D, Shahzad MW, Ng KC. A greener seawater desalination method by direct-contact spray evaporation and condensation (DCSEC): Experiments. Applied Thermal Engineering. 2020; **179**:115629

[33] Chen Q, Alrowais R, Burhan M, Ybyraiymkul D, Shahzad MW, Li Y, et al. A self-sustainable solar desalination system using direct spray technology. Energy. 2020;**205**:118037

[34] Chen Q, Muhammad B, Akhtar FH, Ybyraiymkul D, Muhammad WS, Li Y, et al. Thermo-economic analysis and optimization of a vacuum multi-effect membrane distillation system. Desalination. 2020;**483**:114413

[35] Chen Q, Burhan M, Shahzad MW, Ybyraiymkul D, Akhtar FH, Li Y, et al. A zero liquid discharge system integrating multi-effect distillation and evaporative crystallization for desalination brine treatment. Desalination. 2021;**502**:114928

[36] Najar BA, Peters CD, Albuflasa H, Hankins NP. Pressure and osmotically driven membrane processes: A review of the benefits and production of nano-enhanced membranes for desalination. Desalination. 2020;**479**:114323

[37] Peters CD, Hankins NP. Osmotically assisted reverse osmosis (OARO): Five approaches to dewatering saline brines using pressure-driven membrane processes. Desalination. 2019;**458**:1-13

[38] Ostergaard P, Duic A, Noorollahi N, Mikulcic Y, Kalogirou SA. Sustainable development using renewable energy technology. Renewable Energy. 2020; **146**:2430-2437

[39] Kalogirou SA. Introduction to Renewable Energy Powered Desalination, Ch 1, Renewable Energy Powered Desalination Handbook, Application and Thermodynamics. Oxford, United Kingdom: Butterworth-Heinemann; 2018

[40] Nguyen T, Kook S, Lee C, Field RW, Kim IS. Critical flux-based membrane fouling control of forward osmosis: Behavior, sustainability, and reversibility. Journal of Membrane Science. 2019;**570–571**:380-393. DOI: 10.1016/j.memsci.2018.10.062

Chapter 7

Contributing by Monitoring Energy Efficiency to the Development of Optimization Measures to Improve Energy Performance in an Industrial Platform

Laurentiu Constantin Lipan

Abstract

Greenhouse gas emissions and climate change are currently major international problems. This makes it important to increase energy efficiency. The implementation of energy efficiency improvement measures has reduced energy demand for industrial platforms, so that energy plans designed before these measures are no longer appropriate for current tasks (extensions are not considered at this time). I intended to give a clear image and a better understanding of the factories' power consumption (the industrial area in question is located near a city). A power plant-specific power system is quite disruptive, as can be seen from the monitored data at the power plant users and from the general power supply voltage bars (110 kV, 20 kV) as well as from the voltage bars of the adjacent users. Based on real-time measurements and monitoring devices, the following characteristic curves have been extracted for local energy systems. That is because data analysis is easy to be used for monitoring energy efficiency and elaborating optimization measures to improve power performance.

Keywords: greenhouse gas emissions, intelligent platforms, advanced metering infrastructure (AMI), intelligent networks, U-THD, I-THD

1. Introduction

The experimental study of the energy action/behavior of an industrial user allows to highlight its energetic performance, practical ways to improve its energy use, as well as validating the theoretical hypotheses formulated in order to assess this user. Assessing the power quality supplied in the public system and observing the quality indicators for compliance with accepted standard limits plays an important role in order to establish the optimal functioning parameters for this industrial power user. In this paper, the results of the experimental study achieved at a large-scale/industrial power user are exposed, along with the problems that emerged from this research. Moreover, solutions for solving these kind of problems are indicated. The analysis of the power supply quality of the three-phase general power supply circuit of the analyzed user was performed on the 110 kV side in the measuring cells in the secondary measuring circuits of the T1 110/20 kV transformer, 16 MVA.

2. Analysis of the quality of the electrical power on the bars of 110 kV

"Analysis of the quality of the electrical power supply of the three-phase general power supply circuit of the user on 110kV" is based on "Study on the quality of power supply including load analysis, problems regarding voltage/current fluctuations, interruptions in the company's power supply, flicker monitoring, asymmetries, harmonic levels on voltage and current, for the general supply on the 110kV side of the company in the Intelligent Electrical Connections Station, without personal workers, with Advanced Metering Infrastructure (AMI)".

The electrical installations (**Figure 1**) of the analyzed industrial user consist of:

- two 110/20 kV 16 MVA substations (one being in reserve) – **Figure 2** and **Table 1**;

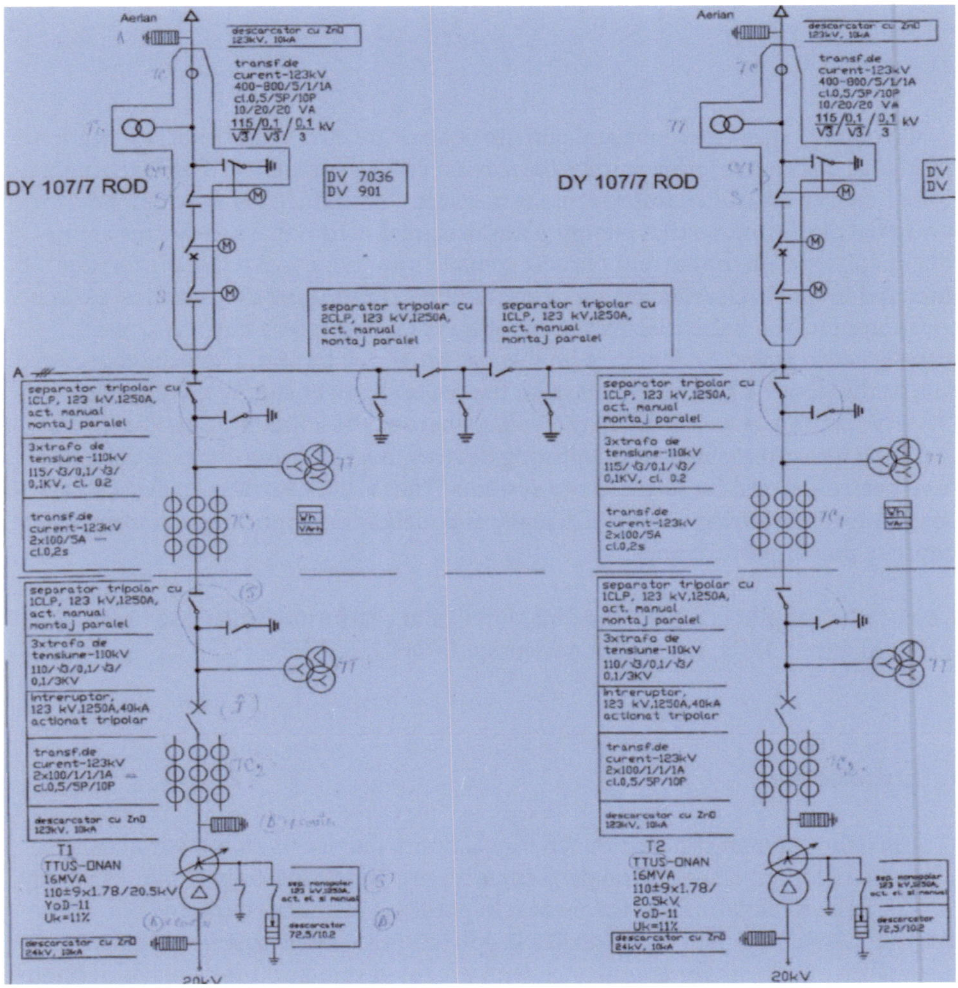

Figure 1.
110 kV single-wire power supply diagrams of the analyzed company with transformers 1 and 2, 110/20 kV of 16 MVA [copyright © 2020 PLT – Adapted working scheme].

Figure 2.
Real photo - transformer 1 (T1), 110/20 kV of 16 MVA.

Transformer	S_n	ΔP_0	ΔP_{sc}	Tension		Fabrication year	
				U_{MT}	U_{MT}		
	MVA	kW	kW	kV	V		
TTUS-ONAN	16	38	300	110		20.5	2008
TTUS-ONAN	16	38	300	110		20.5	2008

Table 1.
Characteristics of the transformer in the transformer station.

- four 20/0.4 kV substations - PT1, PT2, PT3 and PT4;
- electrical distribution boards (intelligent electrical networks platforms);
- 0.4 kV power receivers (users).

The measuring point for the analyzed substation, in which the monitoring equipment was mounted, is represented by the entry point on the 110 kV (**Table 1**) side of the 110/20.5 kV, 16MVA transformer, or the exit point from the 110 kV electrical connection station of the conveyor (the common part with that of the distributor – the IT cells are actually located in the same premises). **Table 2** presents some samples of measurements performed for the analyzed transformer in the electrical connections station.

Thus, an ION7600 Class A type analyzer was installed (**Figure 3**), approved and verified up to date (for monitoring the various electrical quantities that will be presented below), in order to make an energy analysis of the monitored electrical measures for the technological process in the general connections power station, at the supply point inside the company.

The analysis of the data registered in the period 26/06/2021 at 16:20–20/08/2021 at 19:00, for the power supply of the receivers within the company allows for the highlighting of the electrical characteristics of the power supply system.

Measurements performed on the HV bars (High Voltage) in the above mentioned period have been achieved in order to carry out the electric power study

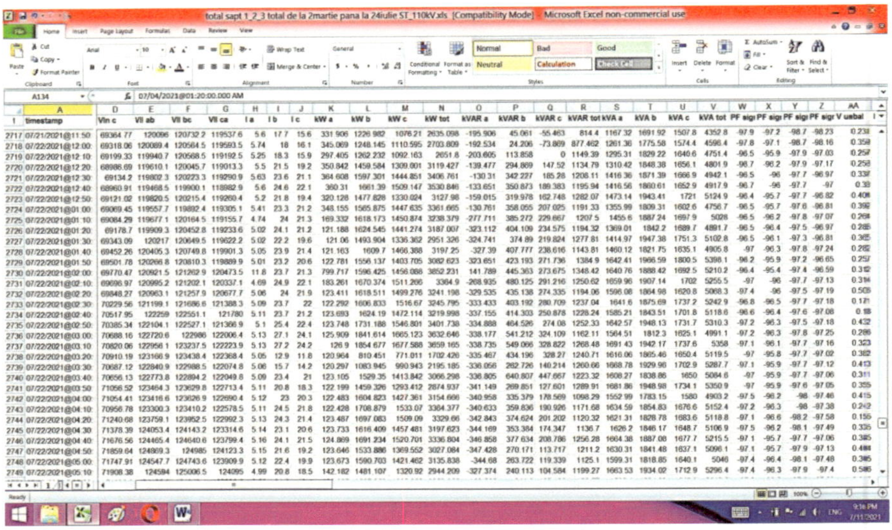

Table 2.
Measurement sheet for the transformation station on the first day of the monitoring period 13/03/2021 at 18:00–22/03/2021 at 16:10.

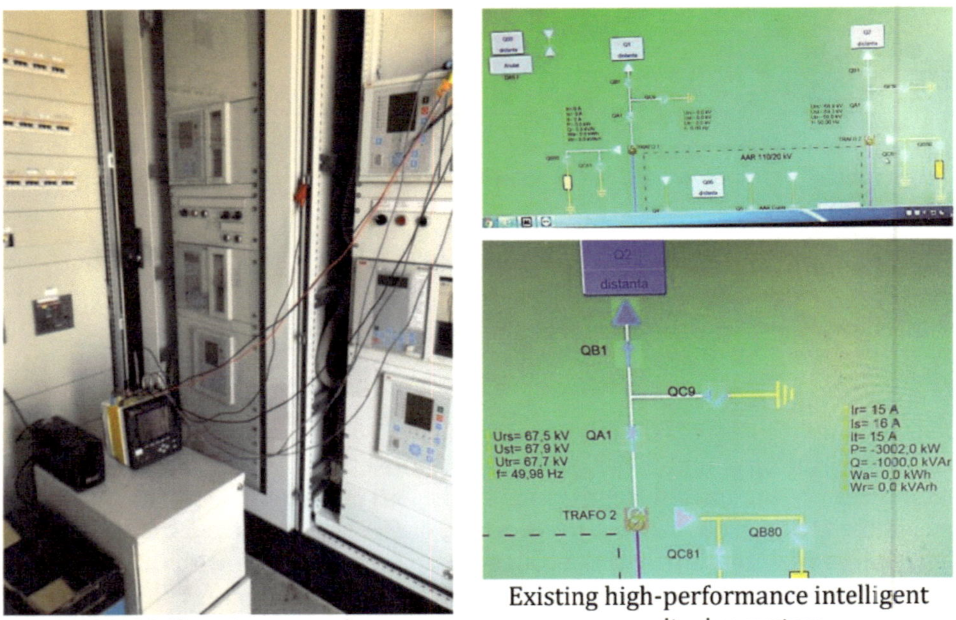

ION7600 Class A type analyzer

Existing high-performance intelligent monitoring system

Figure 3.
Real photo – Measure cell for transformer 1 (T1), 110/20 kV of 16 MVA in Consumer's Power Station [copyright © 2020 PLT – Working scheme & photo].

within the company. The measurements that were recorded are calculated as the average values of the electrical quantities, at an interval of 10 minutes, according to current quality standards IEC 61000–3-4 [1].

The voltage at the supply bars of the receivers shows variations (60.136 ... 72.53) kV voltage per phase/(104.254 ÷ 126.17) kV voltage between phases, for TGD (**Figure 4a** and **b**), which must fit within the limits of 60 ± 10% / 110 ± 10% kV

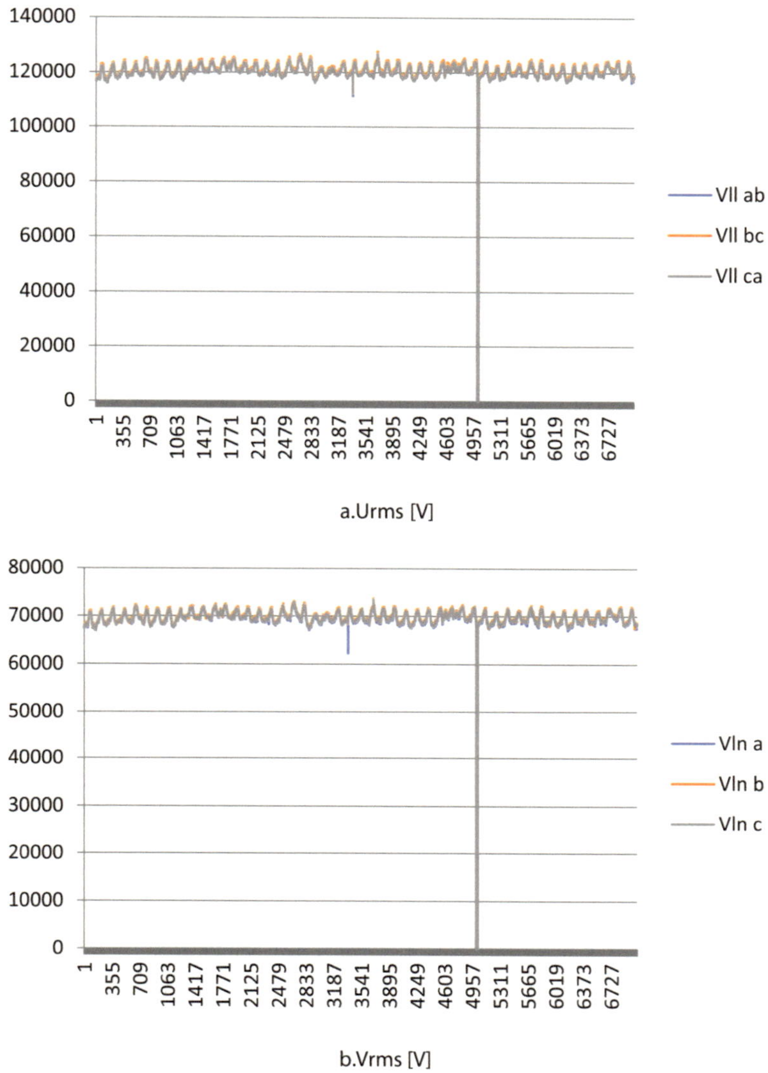

Figure 4.
Voltage curve at the supply bar for TGD: a - voltages between phases, b - voltages per phase.

(54 ... 66/100 ... 120 kV). Rapid variations in voltage values caused by specific processes within the company (factory) lead to the recording of voltage fluctuations, accompanied by the flicker effect. The 3-phases are relatively symmetrically charged. It has been observed that the values of the supply voltage during the measurements did not exceed the normal (normed) limit values.

The voltages at the supply bars have a shape close to a sinusoid, being characterized by a relatively low total THD distortion factor (**Figure 5**), which fits within the values allowed at the low voltage supply bars (THD admitted = 8%). The total voltage distortion factor - THD U (**Figure 5**) is relatively low (it fits between the values (1 ÷ 2.1%).

Figure 6 shows the variation of the unbalance factor during recording. It is observed that the values of the negative unbalance factor ($k_s = V_{unb}$):

$$k_s = \frac{U^-}{U^+} \tag{1}$$

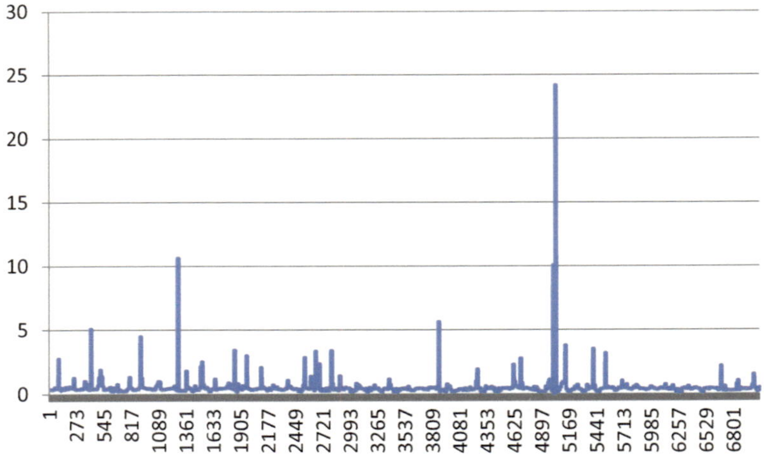

Figure 5.
Voltage distortion factor on supply phases a, b, c (THD U) - monitored values.

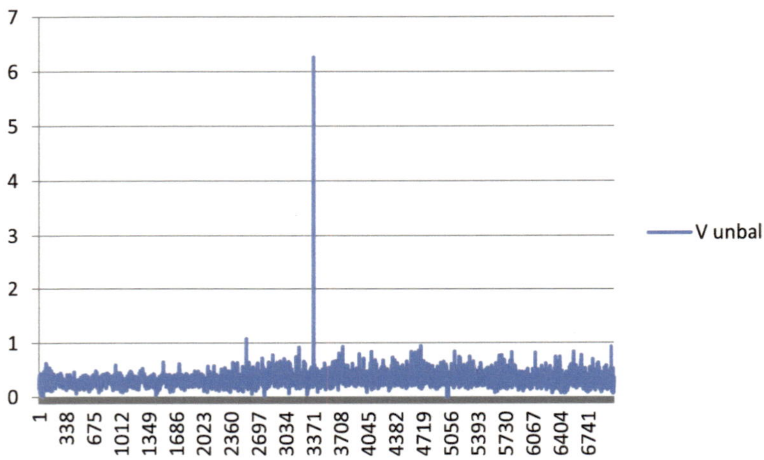

Figure 6.
Negative voltage unbalance factor at TGD supply bars (ks).

where *U* is the negative sequence voltage (inverse) and U+ is the positive sequence voltage (direct), does not exceed the permissible values of the negative unbalance factor (<2%) at TGD (power supply point) in the monitored enclosure.

The analysis of the data in **Figure 6** highlights an unbalance factor of about 5%, which indicates the existence of unbalanced and non-linear three-phase users connected in the low voltage network, unevenly, fact which causes different voltage drops on the three phases. It is important that the unbalance factor of the voltages between the phases has values within the accepted limits so that the three-phase motors connected in a triangle in the low voltage network are not affected by the unbalance of the supply voltages. Connecting star motors could lead to a reduction in their operating performance [2].

The variation of the measured values of the negative unbalance factor ks, determined as the ratio between the negative sequence component of the voltage curve U- and the positive sequence component of the voltage curve U+, is indicated in **Figure 6** [3].

The analysis of the data in **Figure 7** highlights the fact that, at the low voltage bars of the receiver supply system, the admitted levels of voltage fluctuations are exceeded in many cases, in the public network (Pst_admitted = 0.9, and

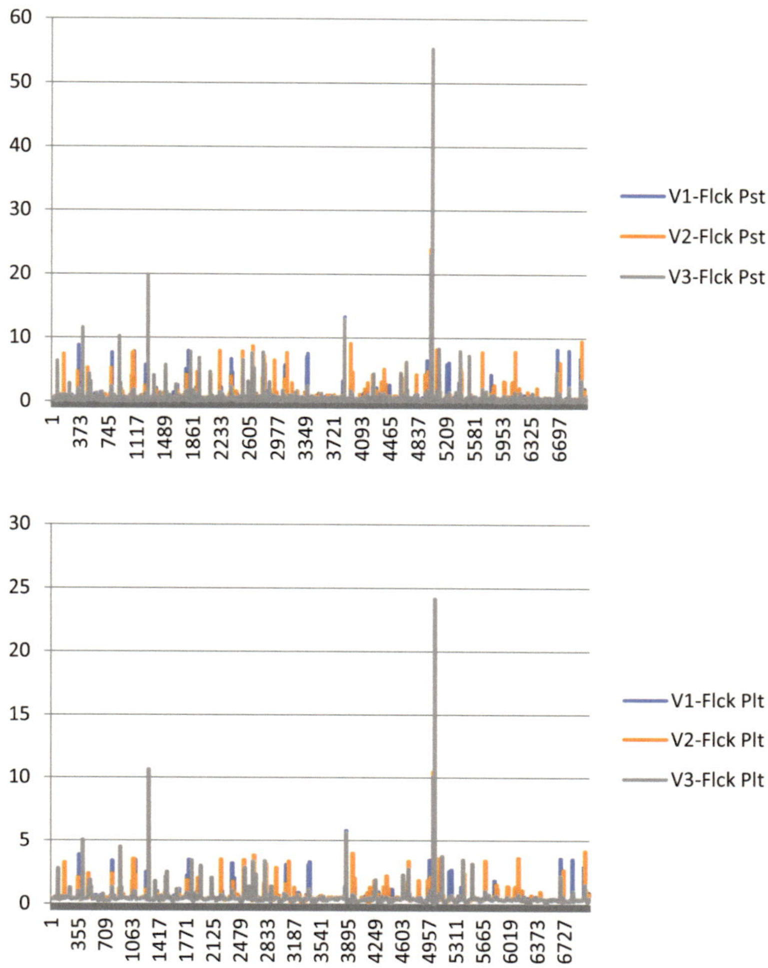

Figure 7.
Variation in the level of voltage fluctuations at the supply bars of the TGD.

Plt_admitted = 1.0) [3–6]. The number of events that lead to values higher than those admitted in the public network is small and is due to specific processes. In addition, in the energy system of other users, the admitted values are different from those in the public network, established according to the effects of the process carried out. The data in **Figure 7** indicate that higher voltage variations occur in phase c, accompanied by a higher level of flicker indicators. It is noted that the curves Pst and Plt in the general connections table (TGD) inside the studied company (factory), have almost the same shape (allure).

Moreover, based on the measurements, it is observed that at the feed bars of the TGD, the flicker level (Pst, Plt) is within the normalized parameters (limit). The disturbances determined at the supply bar are due, first of all, to the variation, in wide limits, of the reactive power necessary for the operation of the furnace. In the case of AC voltage supply, the efficient operation of the oven requires the existence of a low power factor, which requires the adoption of measures to improve it.

The variability specific to the technological process may require adequate compensation of the reactive power. The wide variation of the reactive power absorbed from the mains supply causes rapid voltage variations (voltage fluctuations) at the supply bars (usually within the accepted limits of slow voltage variations, +10%, in mains networks high voltage) accompanied by the phenomenon of flicker in users connected in the same area of use. The complexity of the phenomenon, as well as its propagation in the power supply network, requires experimental determinations to be based on the analysis of the level of disturbances in users in the area and, if necessary, to evaluate measures to reduce the level of disturbances.

The technical study carried out comprises a wide area of use in which it has been identified and analyzed, in particular, the electromagnetic disturbance in the form of voltage fluctuation which causes flicker phenomenon. The determinations in the area followed a large number of parameters and quality indicators, but the detailed analysis focused on the short-term and long-term flicker indicators Pst, measured according to the recommendations of IEC 61000.

Also, the electrical quantities determined in the representative points of the scheme in the area were analyzed, namely at the high voltage supply bar of the analyzed user, at the high voltage supply bar of other users in the area and at the low voltage bar, at which are powered by the affected receptors.

The determined values enable the comparison with the admitted values recommended or imposed by the international CEI standards and the RET&RED performance standard (RET = Electrical transmission networks; RED = Electrical distribution networks).

In order to ensure the quality of electricity, limit values are indicated in the standards and norms in force, technical energy norms for limiting voltage fluctuations, including the flicker effect, in electricity transmission and distribution networks - NTE 012/14/00 and standards international IEC 61000–4-30 ver. III, IEC 61000–3-7: 2008 and IEC 61000–2-2. The national standards NTE 012/14/00, Ord_12_2016_RET and international performance standard IEC 61000–4-30 ver.3, IEC 61000–3-7: 2008 and IEC 61000–2-2 indicate, for the low voltage level, the limits of compatibility in **Table 3**. For the range of medium, high and very high voltages, the planning values of the flicker indicators on short-time (Pst) and long-time (Plt) are indicated in **Table 4**.

For the low voltage level, in any time interval of one week, the long-term flicker indicator must meet the Plt < 1 condition for 95% of the time. The same condition is imposed for the average voltage level [3–6].

Due to the fact that low voltage users are affected by voltage fluctuations (flicker phenomenon) during the operation of the electric arc furnace, experimental determinations have been proposed, in particular, to verify the compliance of the level of disturbances within the accepted limits by international standards. Also, the curves

Flicker indicator	Compatibility level
Pst	1.0
Plt	0.8

Table 3.
Compatibility limits for flicker indicators in low voltage networks.

Flicker indicator	Level of planning*	
	(Medium Tension) MT	(High Tension) IT
Pst	0.9	0.8
Plt	0.8	0.6

**The values of the flicker indicators were established considering that the transfer factor to the low voltage network is unitary. In real cases, this factor may vary and must be determined for different operating conditions. In this way, the values at the MT and IT levels may be appropriately higher. The values mentioned are for guidance only.*

Table 4.
Indicative values of the planning level for flicker indicators, in medium, high and very high voltage networks.

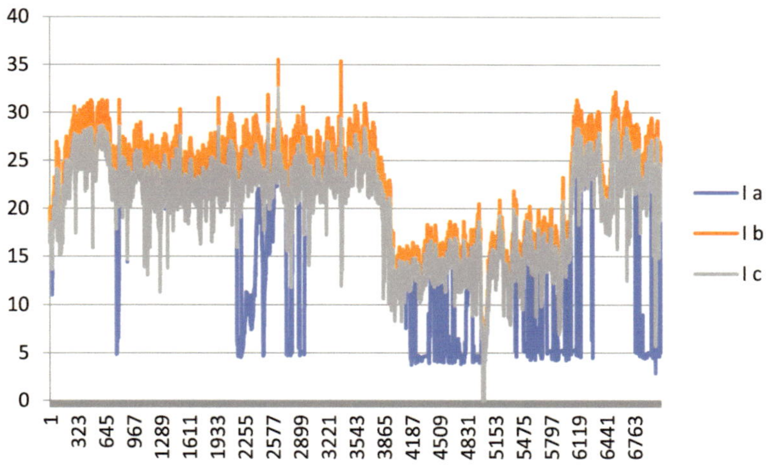

a.

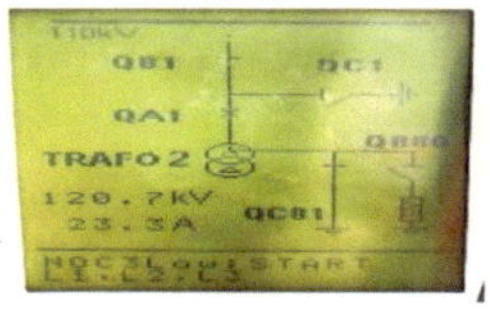

b.

Figure 8.
Electric current curve on the feed bar for TGD. a. Arms [a] – Monitored. b. Arms [a] - instantaneous values.

regarding the voltage variation in significant points of the scheme and the curves for the variation of the reactive power on the user's power supply line will be analyzed.

As electromagnetic compatibility analyzes are taken into account, in particular values with a probability of 95%, these values will be highlighted in the analysis, which must be compared with the admitted values. Also, as comparison, the values with a probability of 50% (average values) will be highlighted.

The electric currents on the three phases monitoring have quite close values indicating, however, an asymmetric phase charge. The variations of these currents are presented in **Figure 8**.

The total distortion factor of electric current - THD I (**Figure 9**), according to the shape of the graph curves, indicates the sinusoidal shape of the current in the supply circuit. At a first comparison with the data on the level of distortion of the voltage on the bars, there is no direct influence, especially in the case of phase a, in which the harmonic distortion of the electric current is superior to the other two phases. The analysis of **Figure 8a** and **b** also shows a weak imbalance between the loads of the three phases.

It should be noted that the jumps in the variation curves of the harmonic spectrum are due to sudden variations in electric current at the start of various electrical equipment - motors (in-rush current). These values are not specific for the analysis of harmonic distortion of electric current.

The variation of the unbalance factor of electric current, during the recording time, is indicated in **Figure 10**. The negative unbalance factor was determined based on the relation:

$$k_s = \frac{I^-}{I^+} \tag{2}$$

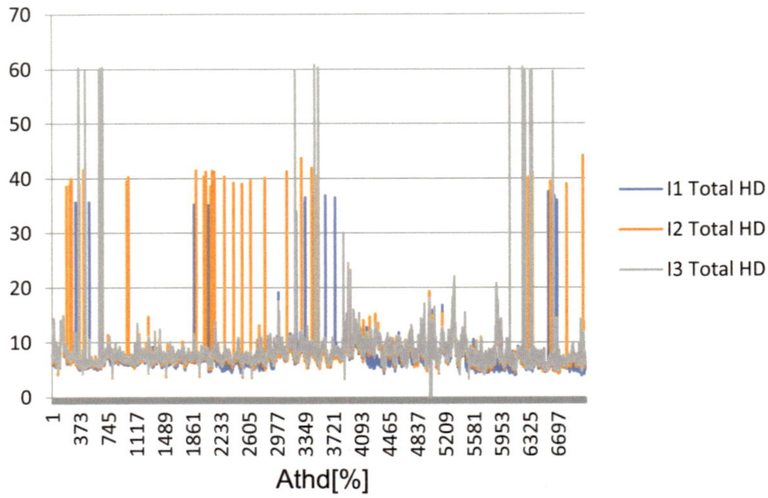

Figure 9.
Electric current distortion factor on supply phases a, b, c (THD-I or Athd) - monitored values.

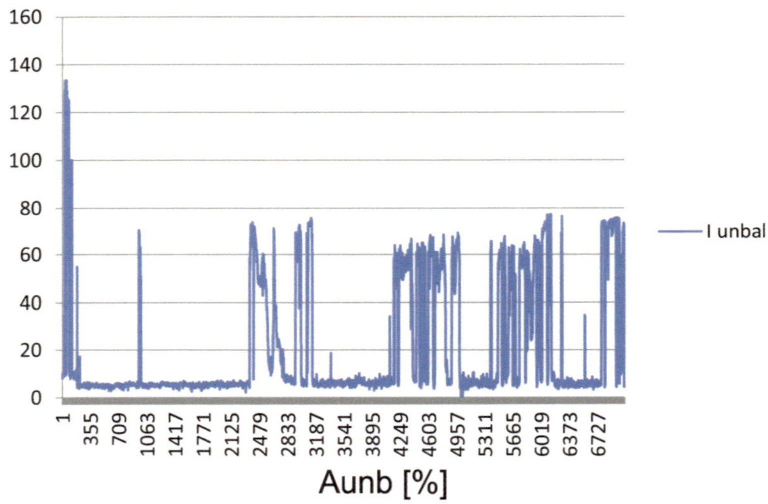

Figure 10.
Variation of the negative current unbalance factor of TGD (ks = Aunb).

where *I* is the negative (reverse) sequence component of the electric currents, and *I*+ is the positive (direct) sequence component.

The data in **Figure 10** shows that, during the recording, the unbalance factor of the electric current did not exceed the normal values in operation, these being determined especially by the voltage unbalance at the supply bars.

Special attention was paid to monitoring the values of electric currents in the supply circuits to determine both their loads and the level of losses in the user's electrical circuits.

Figures 11 and **12** show the variation P (active powers) and, respectively, Q (reactive powers) on the IT/MT (at 110 kV) bars, and **Figure 13** shows the variation of the apparent powers transited on the circuit (S), respectively, in **Figure 14** the variation of the three-phase powers comparative. The phase shift between the voltages and the electric currents monitored at the supply bar of the analyzed electrical equipment, in instantaneous values, is presented in **Figure 15**.

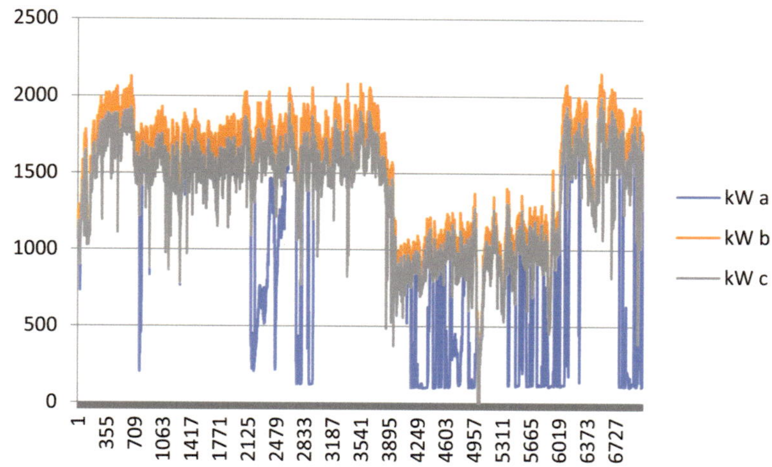

Figure 11.
P [kW] variation on the IT bars at 110 kV.

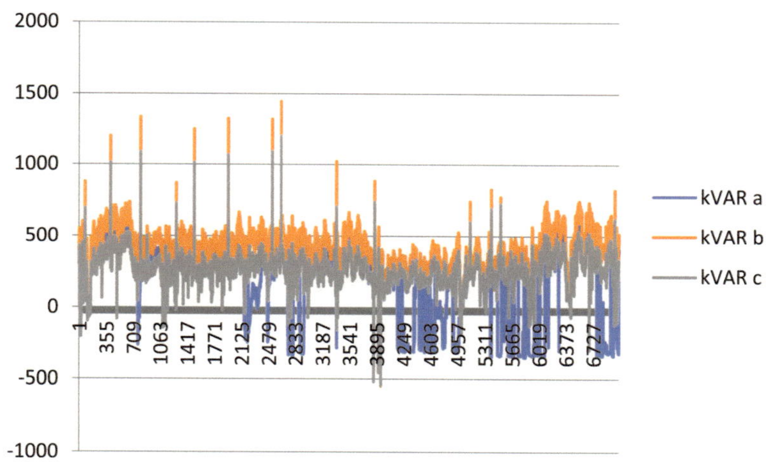

Figure 12.
Q [kVAr] variation on the IT bars at 110 kV.

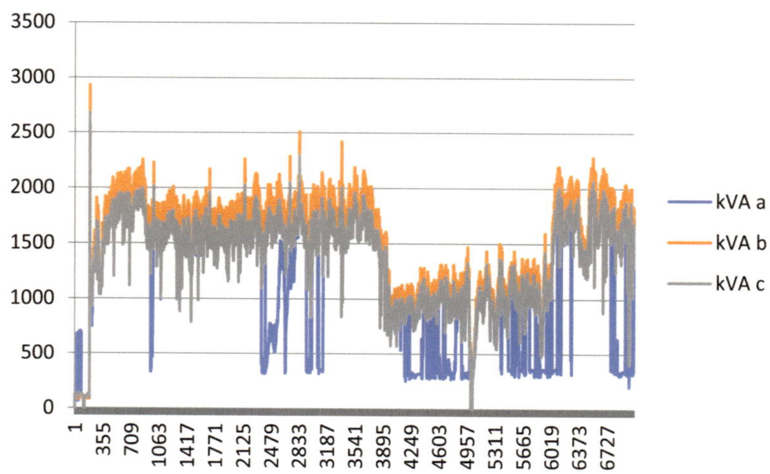

Figure 13.
Apparent single-phase powers transited through at 110 kV - S [kVA].

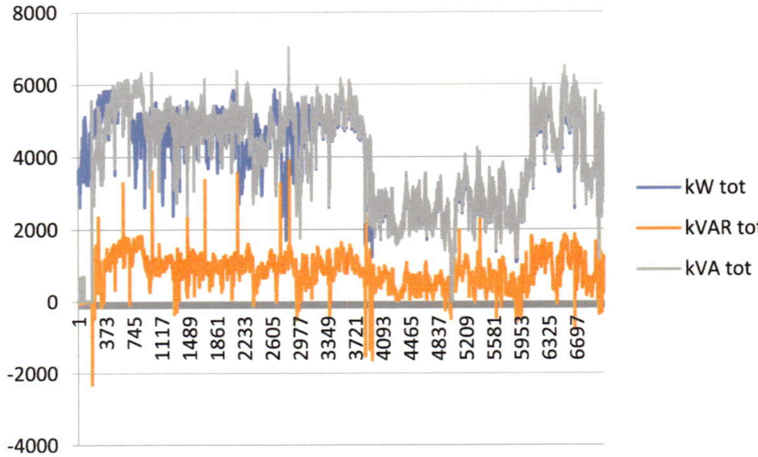

Figure 14.
Three-phase powers transited at 110 kV. SumP [kW], SumQ [kVAr], SumS [kVA].

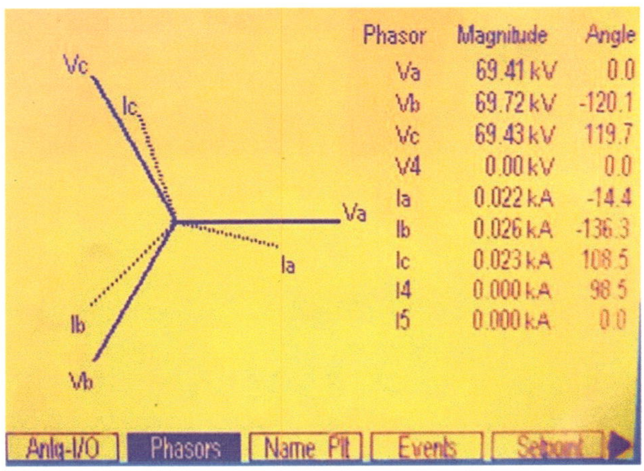

Figure 15.
The phase shift between voltages and electric currents at the supply bar at 110 kV.

The curves in **Figure 12** indicate that the user absorbs capacitive reactive power over a significant period of time. In comparison with the data in **Figure 14**, it is highlighted that the values of the capacitive power factor are within the limits accepted by the regulations in force. In order to ensure the control of the reactive power in the capacitive zone, and the limitation of some undesired increases of voltage in this interval, concrete solutions for the limitation of the capacitive regime were analyzed.

The PF power factor was determined based on the recorded energy values over a specified time interval:

$$PF = \frac{W^P_{trifazat}}{W^S_{trifazat}} \tag{3}$$

The PF value, although widely used in practical applications, provides correct information on the energy behavior of the consumer only in the case of a constant consumption during the time interval in which the power factor is evaluated.

The variation of the power factor during the recording, and in the time intervals in which the equipment was in operation, is indicated in **Figure 16** for each phase.

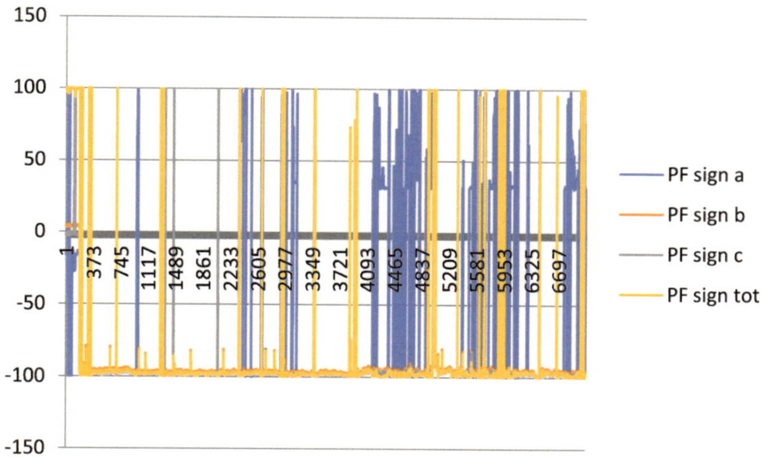

Figure 16.
PF variation (power factor) at 110 kV.

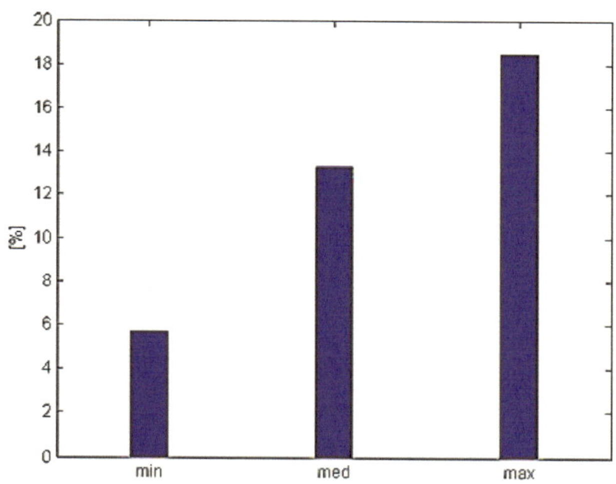

Figure 17.
Transformation station loading (load T1).

As it can be seen from the graph of the variation of the power factor (**Figure 16**), it is highlighted the fact that an important circulation of the reactive power appears in the supply circuit. The recorded values of the harmonic spectrum are affected by the specific operating mode, with frequent starts and stops of motors, as well as other electrical equipment (lighting, etc.). The real values of the harmonic spectrum can only be taken into account for stationary time intervals.

Figure 17 shows the minimum, average and maximum loads of the transformer during the monitoring period. The average load of the transformer station is about 13%, and the maximum load is 18%. Due to the low level of consumption in most operating cases, the transformer station operates poorly charged, which is an important source of energy loss. There are also times when the load is 6%.

Obs. In the power balance made for the transformer station of consumer, it is found that the transformer load does not exceed 18% of the nominal value (during 13/03/2021 at 18:00–22/03/2021 at 16:10).

Figure 18 shows the values of the voltage harmonics on the three phases (a, b and c) at 110 kV. **Figure 19** shows the values of the current harmonics on the three phases (a, b and c) at 110 kV (in Cell Measure with Ion7600 Class A).

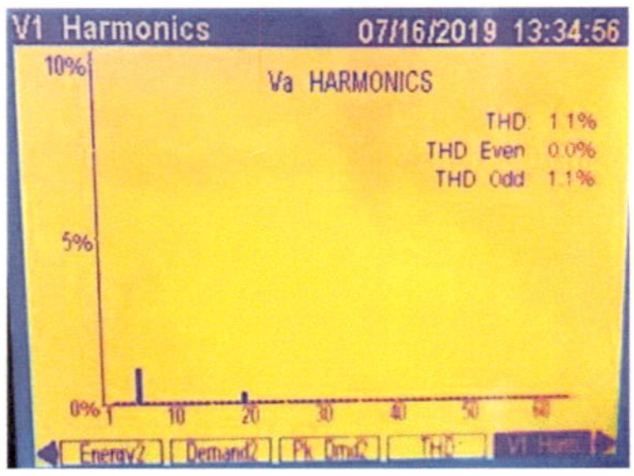

a

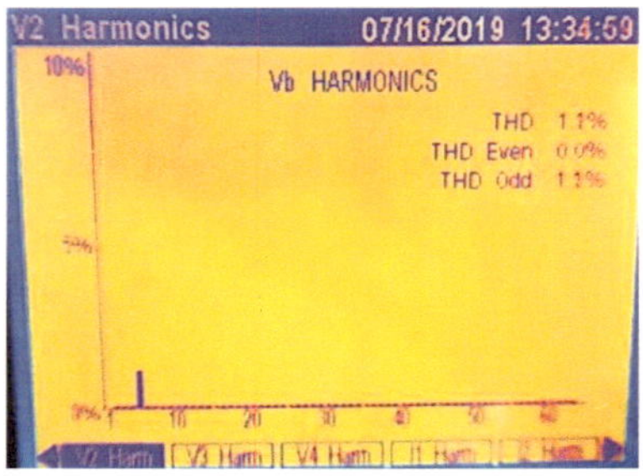

b.

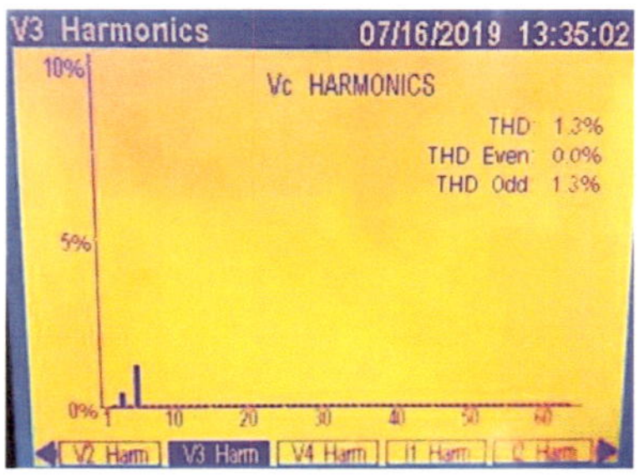

c.

Figure 18.
Voltage harmonics level on phases a, b and c at the supply bars - monitored values at 110 kV. a. Harmonic level [%] phase 1 (voltages). b. Harmonic level [%] phase 2 (voltages). c. Harmonic level [%] phase 3 (voltages).

The main observations that emerge from the analysis of the electrical measurements performed are:

- there is an imbalance of the charge of the three phases of the analyzed electrical circuit;
- the power level factor varies slightly, from negative calories to positive values;
- the flicker level varies strongly, but is maintained between optimal values;
- the level of electric harmonics is relatively high (note that electric harmonics 5 and 7 have very high values).

Under these conditions, the optimization of the power balance aims at the balanced charging of the three phases, the improvement of the power factor compensation, the improvement of the flicker level and the operation with an improved harmonic regime (harmonics of electric current).

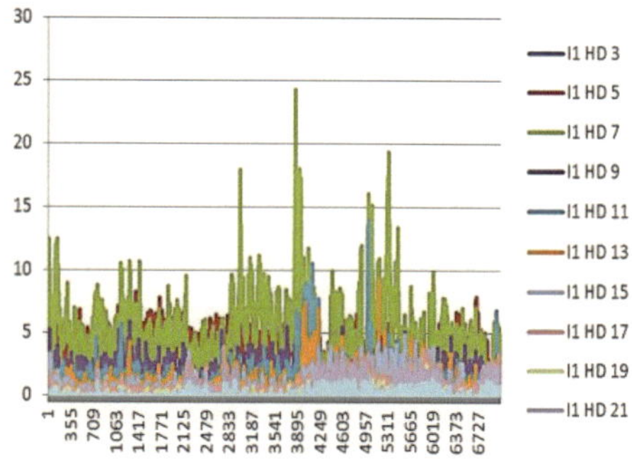

a.

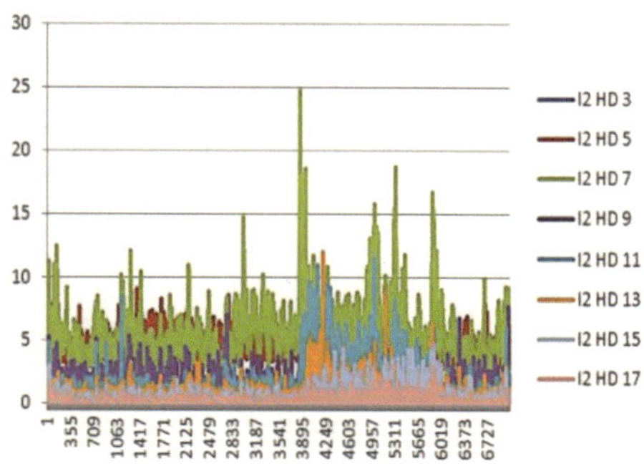

b.

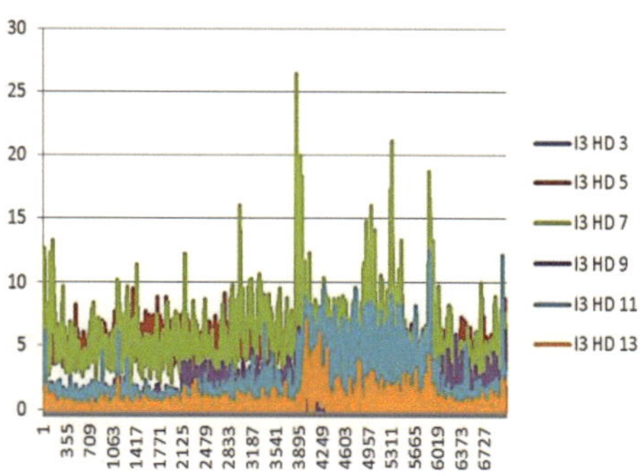

c.

Figure 19.
The level of electric harmonics on phases a, b and c at the supply bars - monitored values at 110 kV.
a. Harmonic electric currents phase1 [%]. b. Harmonic electric currents phase 2 [%]. c. Harmonic electric currents phase3 [%].

There is a good classification of the voltage level on phases and between phases in the normed parameters (framing in the normed limits).

3. Events recorded during the monitored period

This chapter presents a series of events recorded during the monitoring period (Dips&Swells) [5, 6]. These events led to production interruptions that caused significant damage to the user through recalibration, cleaning, disposal of inferior products of poor quality, the realization of additional waste and implicitly additional specific costs.

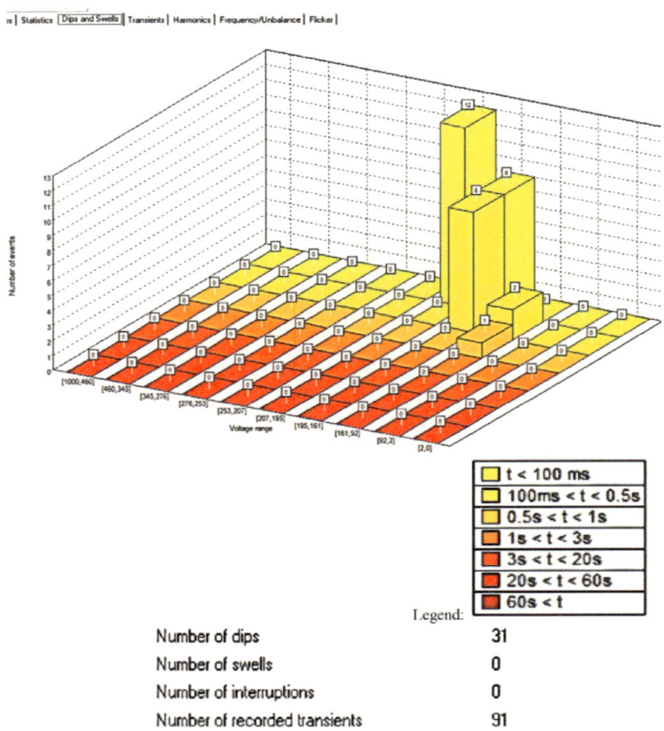

Figure 20.
Events noted in the first half of the 110 kV monitoring period in Electrical Station.

- Events noted in the first half of the 110 kV monitoring period in the Electrical Station at the consumer – **Figure 20**.
- Events noted in the last half of the 110 kV monitoring period in the Electrical Station at the consumer – **Figure 21**.

4. Technical analysis

According to the technical activity report, by downloading the monitored data by other industrial devices (such as SEPAMs), leads to the following technical statements assumed by the user [7]: “Event parity: approx. 5% of the total monitored events belong to the USER, and in approx. 95% of cases of origin of the Zonal Electricity Distributor and assumed by the Electricity Transporter.

That is, for month 1 there were 31 memorized events due to Zonal Electricity Distributor, at 2 ... 3 events of own influence - percentage, and for the 2nd month, there were 28 memorized events due to Zonal Electricity Distributor, at 2...3 events of own influence [7].

Obs. A series of additional information was analyzed that is not presented in detail in this chapter, but which competes or influences the proper functioning of the analyzed user, such as:

1. Cables, switches (with related power bars)

2. Specific operation of electrical equipment

3. Field findings

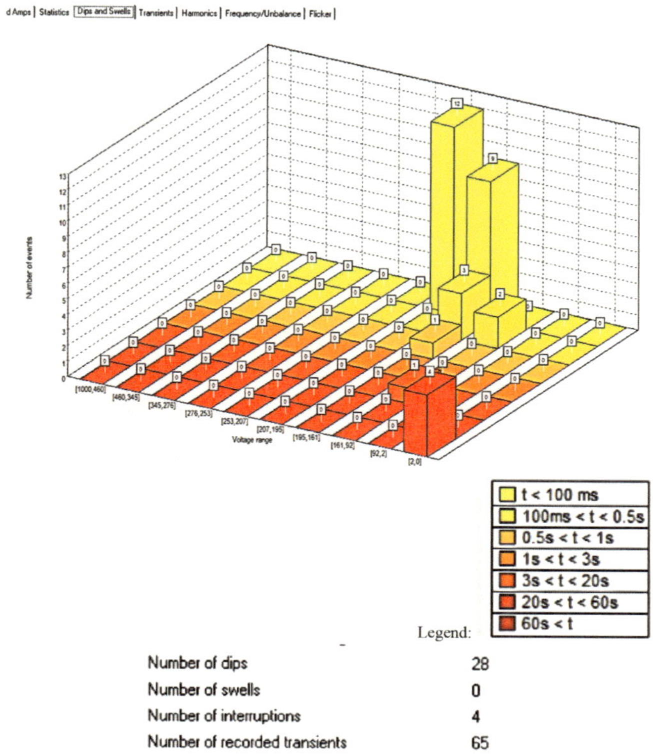

Figure 21.
Events noted in the last half of the 110 kV monitoring period in Electrical Station.

4. Protection information - provided by company representatives

5. Information specific to electrical networks

6. overall water demand, energy consumption and emission of conventional desalination processes - information such as size, emission [8] etc.

7. etc.

Thus, for the various events and power quality, the following recommendations were taken into account, presented in the **Table 5**:

The principle of operation/assembly/effects/etc. is similar to the one previously presented on specialized technical literature in the field.

The difference is given by:

1. The materials from which the insulation of various equipment will be made (cables, windings, etc.), which are much more expensive and specialized (unusual);

2. The electrical equipment/devices which is more expensive, being connected at 110 kV, compared to those connected at 20 kV, (also because of other reasons, related not only to insulation, see electromagnetic fields, etc.).

3. In the case of connection of these SVC type devices to the 20 kV bar, it is known that the reduction of the flicker level by a maximum of 2 times, is corroborated with an additional reduction by a few percent of the flicker level,

No.	Proposed operations	Equipment	Implementation Mod	Obs.
1	Internal - mandatory	PTs and transformers+users	After performing a STUDY REGARDING THE DISTRIBUTION OF USERS BY STANDARD, PRIORITY CATEGORIES, CRITICISM IN SOCIETY (distribution by PTs and transformers)	
2	Checking	Cables	With mobile technical cable testing laboratory	
		Breakers	With mobile technical testing laboratory - thorough verification inside	
		Contact resistors between IT&MT bars and related cables	Internal - with the company's electricians	
		Thorough grounding check	Check the sockets by removing the connection plates, cleaning, thoroughly checking for improvement and if necessary.	
		Transformers	With mobile technical testing laboratory - thorough verification of sensors, with cleaning of dust terminals, insulating cleaning, moisture, oxidation or oil.	
3	Purchase of equipment	Compensation of capacitive/ inductive operating modes	With specialized equipment made by established manufacturers	
			Proposal of a Rotary Compensator	
			Proposal of a PSS (Power System Stabilizer)	
			Proposal of an SVC (Static Var Compensator)	
			Proposal for a STATCOM (advanced FACTS device such as Back-to-back)	
4	Extensive inspections of the distributor's installation - Zonal Electricity Distributor	Checks for switching times and tripping and tripping times - various switching and protection equipment	With specialists or with the knowledge of the Zonal Electricity Distributor	external
		Protection settings	With specialists or with the knowledge of the Zonal Electricity Distributor	external
		Checking the operation of various switching equipment (switches, their automation etc.)	With specialists or with the knowledge of the Zonal Electricity Distributor	external
		Study and implementation of the increase of the short circuit power of the ZONAL ELECTRICITY DISTRIBUTOR power station from which the enterprise is supplied	With specialists or with the knowledge of the Zonal Electricity Distributor	external

No.	Proposed operations	Equipment	Implementation Mod	Obs.
		Checks of intervention modes in the 110 kV mains network of ZONAL ELECTRICITY DISTRIBUTOR (on the part of Medium Voltage and High Voltage) to the large users in the vicinity of Consumer who are connected to the same 110 kV network; verification of intervention times, interruptions, switching, respectively trigger times, triggering - various switching and protection equipment, plus other complaints that existed in the area (on IT and MT) during this monitoring period (but also before).	With specialists or with the knowledge of the Zonal Electricity Distributor	external

Legend: Red = with maximum priority; Yellow = priority.

Table 5.

by passing through the transformer of 110/20 kV that supplies the electric arc furnace; in case of mounting these SVC devices on the 110 kV side, it is to be taken into account only the design data of the SVC system (this additional attenuation is almost null at 110 kV).

4. However, in this case, compared to the compensation of the flicker level on the 20 kV side, we can also consider a slightly lower level of short-term and long-term flicker indicators on the 110 kV bar, compared to their level on the 20 kV, which may or may not influence the final price (depending on the manufacturer/equipment used).

5. Although the solution using SVC ensures the limitation of the flicker level to the users in the area, in the current configuration of the Electric Station at Zonal Electricity Distributor and Electric Station at Electricity Transporter with Electric Station at Consumer, the open coupling at 110 kV bus bar does not provide the necessary conditions for using a backup transformer for the user (a transformer will be dedicated to the analyzed user).

Thus, using FACTS devices (SVC or STATCOM) implies [9–11]:

- Flicker

It is a random variation caused by rapid fluctuations of the reactive power in the common power supply point of the steel enterprise. The human eye perceives voltage fluctuation as the change in brightness of light sources.

- Voltage stabilization

The operation of the electric arc can lead to a strong unbalance, especially in the initial stage of starting the melting process. The three-phase asynchronous motors are affected by voltage unbalance. The unbalance of the voltage causes the reduction of the installation's yield, overheating, loud noises, vibrations and variations of motors' speed. That is why the use of SVC in control mode on each phase of the steel company's power supply system ensures the voltage symmetry and stabilization.

- Reactive power compensation (Q)

  The transport of reactive power leads to significant voltage drops and the increase of electric currents in the supply network, decreasing the transport capacity of the active power (P). Public network operators maximize the transport capacity, encouraging users to use local reactive power compensation. SVC maintains the required reactive power (Q) within the limits imposed by the operator, thus avoiding penalties, as well as increasing the efficiency of user activities.

- Harmonic currents reduction

  Users with non-linear characteristics such as electric arc furnaces generate harmonic electric currents. Harmonic currents charge the network and lead to voltage distortion. Distorted voltages can cause IT equipment (computers, etc.), control equipment, and other sensitive equipment to malfunction. SVC filter circuits are designed to absorb harmonic load-generated currents such as thyristor-controlled coils (TCRs). THD (Distortion Factor) and the individual level of harmonic currents are limited below the specified (normed) values.

- Energy saving

  Reactive power compensation (Q) and increased electricity quality lead to increased active power transmission capacity (P) and reduced active energy losses. Thus, overloading of the power supply network can be avoided. This fact brings benefits for both the company and the environment by a more efficient use of electricity, and a reduction of the need for electricity. It is known that, for every 1 kWh saved, produced in a thermal power plant, the amount of pollutants released into the atmosphere is smaller with about 1 kg of CO2.

- Productivity increase

  The SVC system can ensure a practically constant voltage level at the company's power supply bars. This way, it decreases the duration of the melting process and increases productivity. A SVC system limits production interruptions and restarts that require long durations. The electric arc furnaces stabilized by SVC have an important effect on reducing electrode consumption, heat loss and the life of the furnace liner. By increasing the quality of electricity, the demand for equipment is reduced, the lifespan is increased and the costs of maintenance and replacement of some components are reduced.

5. Conclusions

Conclusions resulting from the analysis of the monitored data on the 110 kV – T1 side (power station) as well as the main observations that emerge from the analysis of the electrical measurements performed are:

- there is an unbalance of charge in the three phases of the analyzed electrical circuit;

- the power factor level shows variations outside the range 0.90 ... 1.00; these values are not considered optimal (correctly compensated); there are penalties to the Electricity Invoice for not falling between the mentioned values, according to ANRE (National Energy Regulatory Authority) order only if those from PCC (Common Connection Point – CCP) to the electric meter;

- the flicker level varies strongly, but is maintained between optimal values in 90% from 90% of the monitoring time (Obs. The values in the CCP matter the most);

- the level of electric harmonics is relatively high; it is noticed that the harmonics of electric current 5, 7 have high values, with significant variations.

Under these conditions, the aim is: balanced charging of the three phases, improved power factor compensation, improved flicker level and operation with an improved harmonic regime (electric harmonics).

There is a good classification of the voltage level on phases and between phases, in the normed parameters (framing in the normed limits).

The two proposed performance solutions (in this case):

A. The connection of a STATCOM to the 110 kV bus bars of the user (**Figure 22**) leads to a factor of reduction of the flicker level between 3 and 6 (they multiply). The control block BC receives the information regarding the voltage level at the 110 kV bus bars and determines its maintenance by injection of real-time capacitive reactive power from capacitor C via the appropriate controlled inverter.

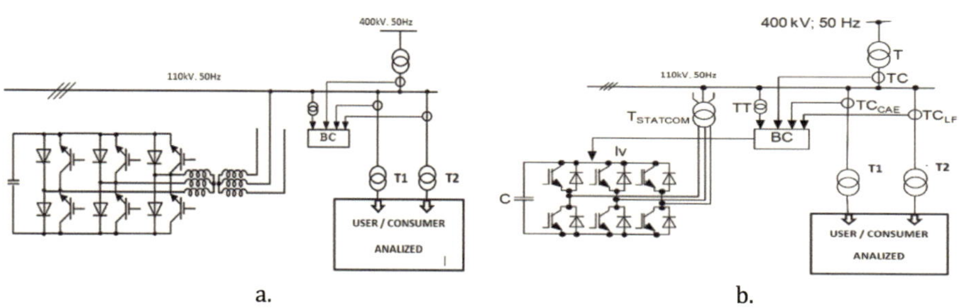

Figure 22.
Scheme of a STATCOM (for analyze) [3–7, 12]. a. Scheme of a STATCOM for voltage control at 110 kV bus bars. b. Functioning scheme/diagram of a STATCOM equipment.

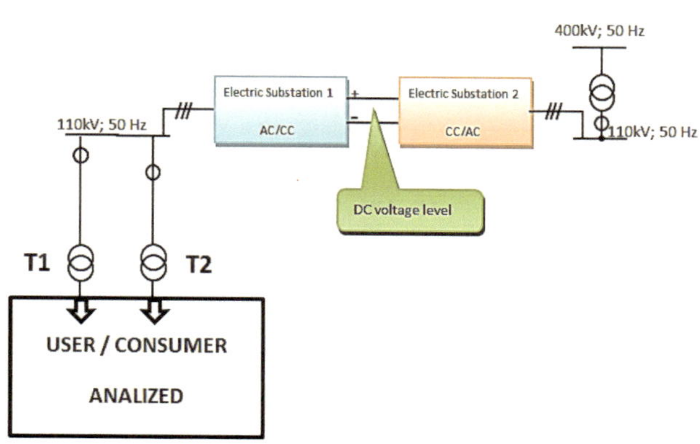

Figure 23.
The functioning diagram of a back-to-back FACTS system [15, 16].

Connecting SVC or STATCOM to the user's 110 kV bus bars, along with using a dedicated 400/110 kV transformer, enables a reduction of practically 1.5 ... 2 times (in case of using SVC), and 3 ... 6 times (in the case of using STATCOM), of the disturbance level at the 110 kV bars of the Electrical Station [3–7, 12]. In this way, the problems related to the disturbance of the users in the area are completely solved as well [13].

In CIGRE study [12, 14] regarding a STATCOM type FACTS device to improve the flicker level in the case of an ELECTRIC ARC OVEN, it is said that the reduction of the flicker level in a single installation was reduced from an initial factor of Pst95% = 3,75 without STATCOM, to Pst95% = 0.76 with STATCOM in operation, for a short-term probability of 95% [3–7, 12]. This fact indicates that the flicker level has improved about five times with the introduction of the compensator.

B. "Back-to-Back System" - By using this type of system/equipment, the level of flicker indicators on the bar on which it is mounted is eliminated, in the sense that it can be fully controlled [3–7, 12]. This is due to the well-known fact that any type of disturbance is eliminated from the arc furnace to the Power Station by passing through a DC voltage level (**Figure 23**) [15, 16]. Obs. Back-to-back equipment has high-power thyristor electronic devices, which can themselves produce/generate system disturbances, if they are not correctly chosen and calculated with great accuracy.

Author details

Laurentiu Constantin Lipan
Faculty of Power Engineering, Department of Electric Power Systems Engineering, University Politehnica of Bucharest, Romania

*Address all correspondence to: laurentiu.lipan@upb.ro

References

[1] IEC. Electromagnetic Compatibility (EMC) - Part 3–7: Limits-Assessment of Emission Limits for the Connection of Fluctuating Installations to MV, HV and EHV Power Systems. IEC 61000–3-7, second edition of CAN/CSA-C61000-3-7/2008, 2008

[2] IEC. Electromagnetic compatibility (EMC) - Part 2–2: Environment-Compatibility levels for low-frequency conducted disturbances and signalling in public low-voltage power supply systems. IEC 61000–2-2, IEC 61000-2-2: 2002+AMD1:2017+AMD2:2018 CSV, Consolidated version, 2018

[3] IEC. Electromagnetic compatibility (EMC), Part 4–30: Testing and measurement techniques- Power quality measurement methods. IEC 61000–4-30 ver. III, IEC 61000-4-30:2015/AMD1: 2021, Amendment 1, TC 77/SC 77A Additional information, ISBN 978-2-8322-2324-6, 2021

[4] SR EN 50160, Voltage characteristics of electricity supplied by public distribution systems. Standard EN 50160:1999, rev.2009; Romania, 2009

[5] IEC. Electromagnetic compatibility (EMC) - Part 3–13: Limits-Assessment of emission for the connection of unbalanced installationa to MV, HV and EHV power systems. 61000-3-13, (Adopted IEC/TR 61000-3-13:2008, first edition, 2008-02), CAN/CSA-C61000-3-13-09 (R2017), 2017

[6] IEC. Electromagnetic compatibility (EMC) - Part 3–6: Limits-Assessment of emission limits for the connection of distorting installations to MV, HV and EHV power systems. Standard IEC/TR 13, 61000–3-6, IEC TR 61000-3-6:2008. EU: IEC - International Electrotechnical Commission; 2008

[7] Golovanov N, Postolache P, Iordănescu I, Lipan L, Porumb R. Instalații Electroenergetice si elemente de audit industrial. Bucuresti: Editura N'Ergo; 2008

[8] Shahzad MW, Burhan M, Ang L, Ng KC. Energy-water-environment nexus underpinning future desalination sustainability. Desalination. 2017;**413**: 52-64. DOI: 10.1016/j.desal.2017.03.009

[9] Singh P, Senroy N. Steady-State models of STATCOM and UPFC using flexible holomorphic embedding. Electric Power Systems Research. 2021; **199**:107390, 2021. DOI: 10.1016/j.epsr.2021.107390

[10] Ufa R, Gusev A, Zaki AA, Diab AS, Ruban N, Andreev M, et al. Analysis of application of back-to-back HVDC system in Tomsk electric power system. Research paper. Energy Reports. 2020;**6**: 438-444. DOI: 10.1016/j.egyr.2020.01.017

[11] Zellagui M, Lasmari A, Settoul S, El-Sehiemy RA, El-Bayeh CZ, Chenni R. Simultaneous allocation of photovoltaic DG and DSTATCOM for techno-economic and environmental benefits in electrical distribution systems at different loading conditions using novel hybrid optimization algorithms. International Transactions on Electrical Energy Systems. 22 August 2021;**31**(8): e12992. DOI: 10.1002/2050-7038.12992

[12] CIGRE study. "Static synchronous compensator (statcom) for arc furnace and flicker compensation. Prepared by Working Group B4.19/Edited by I. Paris, EU: Arslan Erinmez; 2003

[13] Čerňan M, Müller Z, Tlustý J, Valouch V. An improved SVC control for electric arc furnace voltage flicker mitigation. International Journal of Electrical Power & Energy Systems. 2021;**129**:106831. DOI: 10.1016/j.ijepes.2021.106831

[14] Geetha Devi CE, Devanathan R. Computer simulation of linearized Statcom model. Materials Today; Proceedings. 2020;**33**(7):4243-4249. DOI: 10.1016/j.matpr.2020.07.349

[15] Fathy A, Yousri D, Abdelaziz AY, Ramadan HS. Robust approach based chimp optimization algorithm for minimizing power loss of electrical distribution networks via allocating distributed generators. Sustainable Energy Technologies and Assessments. 2021;**47**:101359. DOI: 10.1016/j.seta.2021.101359

[16] Sangeeta Sarali D, Selvi Agnes Idhaya V, Pandiyan Karuppasamy. Modeling of the three-phase converter using a predictive controller based on the dynamic voltage restorer model for improving power quality. Materials Today: Proceedings. Elsevier BV; 2021. DOI: 10.1016/j.matpr.2021.01.744

Chapter 8

Energy Efficiency, Emissions and Adoption of Biomass Cookstoves

Kailasnath B. Sutar

Abstract

Indoor air pollution due to inefficient use of solid biomass fuels in traditional cookstoves causing serious threat to human health and millions of deaths, mainly in developing countries. This chapter reports parameters for measurement of thermal as well as emission performance of biomass cookstoves. The thermal performance parameters include fire power, efficiency, specific fuel consumption and turn-down ratio whereas the emission performance parameters include emission factor or indoor concentration of a pollutant. This chapter also reports about technological improvements in the biomass cookstoves. Since early 1980s, efforts were made by the researchers for development improved cookstoves. These efforts include use of metals as cookstove materials, provision of grate for better air circulation, air preheating, provision of swirl and secondary air, provision of insulation, use of chimney, baffles etc. The improved cookstoves were found to be causing saving in biomass fuel but there was not much improvement in emission performance of these stoves as compared with their traditional versions. The research on advanced biomass cookstoves started in early twenty-first century. While designing these cookstoves, advancements in technologies such as insulating the combustion chamber, supplying correct amount of primary and secondary air at right place into the combustion chamber, use of fan to create draft, use of gasification techniques, use of high density pellets as fuel etc. are being used. Advanced biomass cookstoves are found to be highly fuel efficient and they cause negligible pollutant emissions. Various factors affecting adoption of improved biomass cookstoves such as social, functional, and cultural are discussed in detail. Recommendations for use of energy efficient and clean cooking options are also given.

Keywords: energy efficiency, emission, indoor air pollution, adoption, cookstove

1. Introduction

Access to clean cooking energy for all, is a major challenge in twenty-first century. In modern cooking practices, people across the globe are using various cookstoves with fuels such as biomass, Liquefied Petroleum Gas (LPG)/ Piped Natural Gas (PNG), kerosene, Charcoal, biogas etc. Other cooking devices such as electric, solar and induction are also being used. About one third of global population does not have access to clean energy mainly due to issue of affordability. The most commonly used cookstoves in the developing nations are the biomass cookstoves. Traditional versions of these cookstoves are highly polluting and very inefficient which results in severe health issues and millions of premature deaths globally.

According to International Energy Agency, in 2018, the global consumption of energy in residential sector was about 88 EJ (1 EJ = 1018 J) which was about 23.3% of the total energy consumption [1]. The components of residential energy are: 32% combustion of bio-fuels & waste, 24% combustion of gas, 21% combustion of coal, 4% combustion of oil and the remaining 26% energy for generation of residential electricity [1]. Since year 2000 till 2019, there is about 48% rise in global energy consumption [2]. The cost of cooking energy is also rising day by day. For example, in India the price of LPG cylinder has been doubled in last 7 years from Rs. 410.0 in 2014 to Rs. 819.0 in 2021 [3]. These statistics indicate that there is an urgent need for conservation of residential cooking energy by using energy efficient cookstoves.

Primitive humans started cooking with fire nearly 2 million years ago [4, 5]. The first method of cooking was probably roasting of a fish or a bird by holding it over an open fire [6]. The different stages of evolution in cooking process as reported in literature are: Prehistoric cooking, ancient cooking, medieval cooking, renaissance cooking, modern cooking and twentieth century cooking [6]. Since prehistoric era till present days, human beings have continued using open fires for cooking purpose. In present days, commonly used domestic cookstoves in different parts of the world can be broadly classified into two groups *viz.* combustion cookstoves and non-combustion cookstoves. The cookstoves in which direct combustion of solid, liquid or gaseous fuels occur and chemical energy of fuels is converted into thermal energy, are known as combustion cookstoves. The examples of combustion cookstoves are: Biomass cookstove, Gas cookstove, Kerosene cookstove, Charcoal cookstove and their variants. In non-combustion cookstoves, no combustion of fuels occur but solar or electric energy is converted into thermal energy. The examples of these cookstoves are: Solar cooker, Electric cookstove, Induction cook-top and their variants. With reference to the developments occurred in biomass cookstoves over last few decades, they are mainly classified into three categories *viz.* traditional cookstoves, improved cookstoves and advanced cookstoves. Biomass cookstoves are also classified as: stationary (non-metal cookstoves) and portable (metal cookstoves), natural draft (buoyancy induced) and forced draft (fan or blower driven). The advanced biomass cookstoves are of two type *viz.* combustion cookstove and a gasifier cookstove. The gasifier cookstoves are further available in four types: updraft, downdraft, cross draft and top lit up draft (TLUD). The detailed classification of biomass cookstoves can be found in literature [7–11].

According to International Energy Agency (IEA) [12], about 2.6 billion people globally (i.e. about 34% of the global population) do not have access to clean cooking energy. They still rely on solid biomass as the only cooking fuel. According to World Health Organization (WHO) [13], every year about 4 million premature deaths occur from the illnesses resulting from household air pollution due to inefficient cooking practices using solid biomass and kerosene cookstoves.

Over a long period of time, the evolutions in design and operation of cookstoves have occurred. The developments in combustion cookstoves are attributed to increase in their overall efficiencies due to improved thermal and emission performance. Also attention is being provided on user friendly designs of the cookstoves.

The present chapter reports parameters affecting thermal and emission performance of biomass cookstoves. It reports emission norms set by national and international agencies for cookstoves using biomass and fossil fuels. It reports the advancements in technologies of biomass cookstoves. It also reports factors affecting adoption of biomass cookstoves. Recommendations are also given on promotion of clean cooking energy options.

2. Performance parameters of biomass cookstoves

In biomass cookstoves, conversion of chemical energy into thermal energy takes place due to combustion of solid biomass. The parameters which affect performance of biomass cookstoves are of two types *viz*. thermal parameters and emission parameter includes emission factor of different pollutants. The thermal performance parameters include fire power, efficiency, specific fuel consumption and turn-down ratio. The emission performance parameters include emission factor of a pollutant (g/kg or g/kJ or g/MJ) or indoor concentration of a pollutant (ng/m^3 or μg/m^3 or mg/m^3 or g/m^3) [9].

The amount of thermal energy produced (kJ) per unit time (s) is known as fire power (kW). Mathematically fire power is defined as follows:

$$\text{Fire power (P)} = \frac{\text{mass of fuel burnt} \times \text{Calorific value of fuel}}{\text{Time taken for complete combustion of fuel}}\ \text{kJ/s or kW} \quad (1)$$

Fire power is the total amount of energy available for cooking the food per unit time. The energy actually used for cooking the food will be very small as compared to that of the fire power due to various losses. **Figure 1** shows energy balance for a biomass cookstove. Out of the total energy available in the form of fire power (a), some of the energy is absorbed by the cookstove body in the form of an internal energy and some energy is lost form the cookstove body to the surroundings through convection, radiation and to the ground through conduction (b). Heavier cookstoves absorb more energy in the form of the internal energy. Hence, traditional cookstoves as well as modified biomass cookstoves made of mud and brick are found to have

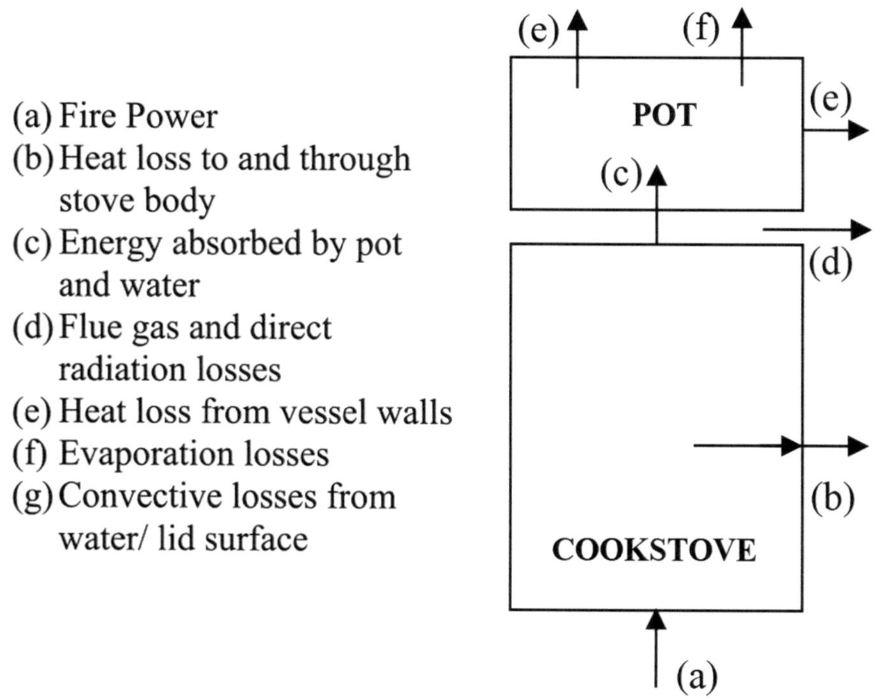

Figure 1.
Schematic of energy balance of a cookstove [9]. [Reprinted from Renewable & Sustainable Reviews, Vol. 41, Sutar K. B., Kohli S., Ravi M. R. and Ray A., Biomass cookstoves: A review of technical aspects, 1128–1166, 2015, with permission from Elsevier.].

poor efficiencies as compared to the metal biomass cookstoves. Some of the energy in the fire is absorbed by the pot and the pot contents (c). During this transfer of energy, some of the energy is lost to the atmosphere with flue gases and some energy is lost in the form of direct radiation and convection (d). Vessel walls also lose some heat to the atmosphere in the form of convection (e). From the top portion of the pot, there will always be evaporative (f) and convective energy losses (g).

From **Figure 1**, it is clear that actual energy used per unit time for cooking the food (P_u) = {(c)−[(e) + (f) + (g)]}/t. Now, thermal efficiency (η) of biomass cookstove is defined as the ratio of actual energy used by the pot and the pot contents for cooking the food per unit time to the fire power available due to combustion of fuel. Mathematically, thermal efficiency is defined as follows:

$$\text{Thermal efficiency } (\eta) = \frac{\text{Actual energy used per unit time for cooking the food}}{\text{Fire power}} = \frac{P_u}{P} \tag{2}$$

Pollutant	Averaging time	WHO [15]		USEPA [16]		CPCB India [17]	
		Value	Unit	Value	Unit	Value	Unit
CO	24 hours	07	mg/m^3	—	—	—	—
	8 hours	10	mg/m^3	09	ppm	02	mg/m^3
	1 hour	35	mg/m^3	35	ppm	04	mg/m^3
				—	—		
PM_{10}	Annual	20	μg/m^3	—	—	60	μg/m^3
	24 hours	50	μg/m^3	150	μg/m^3	100	μg/m^3
$PM_{2.5}$	Annual	10	μg/m^3	12	μg/m^3	40	μg/m^3
	24 hours	25	μg/m3	35	μg/m^3	60	μg/m^3
NO_2	Annual	—	—	53	ppb	40	μg/m^3
	24 hours	—	—	—	—	80	μg/m^3
	1 hour	200	μg/m^3	100	ppb	—	—
SO_2	Annual	—	—	—	—	50	μg/m^3
	24 hours	—	—	—	—	80	μg/m^3
	1 hour	—	—	75	ppb	—	—
Formaldehyde	30 minutes	0.1	mg/m^3	—	—	—	—
Naphthalene	Annual	0.01	mg/m^3	—	—	—	—
Benzene	Annual	Unit risk of leukemia: 6×10^{-6} per μg/m^3 of air.		—	—	05	μg/m^3
Polycyclic aromatic hydrocarbons (PAH)	Annual	Unit risk for lung cancer: 8.7×10^{-5} per ng/m^3 of B[a]P.		—	—	01	ng/m^3

USEPA: United States Environmental Protection Agency, CPCB: Central Pollution Control Board, ppm: parts per million, ppb: parts per billion, ng: Nano gram, and B[a]P: Benzo(a)Pyrene.

Table 1.
WHO guidelines for indoor air quality [15] and ambient air quality standards for USA [16] and India [17].

Specific fuel consumption (SFC) is the mass of dry fuel required (g) to produce a unit output. Here, the unit output is a mass of water remaining in the pot at the end of the test (kg). SFC is expressed in terms of g/kg [14]. Turn down ratio is the ratio of maximum and minimum power between which the cookstove can be operated satisfactorily [9].

Emission factor (g/kg or g/kJ or g/MJ) of a particular pollutant is mass of that pollutant emitted (g) per kilogram of the fuel burnt or per kJ or per MJ of energy released during the cooking task [9]. Indoor concentration of a particular pollutant (ng/m^3 or µg/m^3 or mg/m^3 or g/m^3) is defined as the amount of exposure of that pollutant (ng or µg or mg or g) to the user per m^3 of the air in the room or cooking space [9].

According to WHO guidelines [15], carbon monoxide (CO), particulate matter of size less than 10 µm (PM_{10}) and of less than 2.5 µm ($PM_{2.5}$), nitrogen dioxide (NO_2), formaldehyde, naphthalene, benzene and polycyclic aromatic hydrocarbons (PAH) are found to be major indoor air pollutants. Considering global warming potential of these pollutants, it is very important for the researchers to know the safer limits of these pollutants in ambient air as recommended by the national and international agencies. **Table 1** report WHO guidelines on indoor air pollutants resulting from combustion of fuels and also ambient air quality standards set by United States Environmental Protection Agency (USEPA) for USA [16] and by Central Pollution Control Board (CPCB) for India [17]. For a given pollutant, with increase in averaging time, values of its safe limit decrease. For example, as per WHO guidelines, permissible limit of exposure to CO emissions for 1 hour is 35 mg/m^3, for 8 hours it is 10 mg/m^3, and for 24 hours this limit is 7 mg/m^3.

3. Traditional and improved biomass cookstoves

Figure 2(**a** and **b**) shows images of traditional biomass cookstove and improved mud cookstove. Average efficiency and CH_4 emissions of traditional biomass cookstoves used in Asian countries were reported to be about 11% and 0.52 g/MJ of energy delivered by wood fuel [18]. Since early 1980s, some researchers reported ways of improving thermal performance of the traditional biomass cookstoves by modifying their designs [19–24]. These ways include: use of metals as cookstove materials, provision of grate for better air circulation, air preheating, provision of

(a) Traditional Biomass Cookstove (b) Improved Mud Cookstove

Figure 2.
Images of traditional biomass cookstove and improved mud cookstove.

swirl and secondary air, provision of insulation, use of chimney and baffles [18]. Average efficiency and CH_4 emissions of improved biomass cookstoves used in Asian countries were reported to be about 24% and 0.408 g/MJ of energy delivered respectively with the wood fuel [18]. Thermal performance of improved biomass cookstoves was found to be better than the traditional cookstoves; but there was not much improvement in their emission performance as compared with traditional ones. Researchers have found that improvement in efficiency of biomass cookstove does not always ensure reduction in emissions. There exists a certain range of power levels where the correlation between efficiency and emissions is positive, while elsewhere it will be negative [18].

Improved biomass cookstoves are known as fuel efficient cookstoves as they reduce fuel consumption by 20–50% as compared with the traditional biomass cookstoves [25]. Some of the examples of improved biomass cookstoves are: *Swosthee* cookstove [26], rocket cookstove [27], *Patsari* cookstove [28], Envirofit cookstove [29], *Jiko* cookstove [30] etc.

4. Advanced biomass cookstoves

To improve thermal and emission performance of metal biomass cookstoves, efforts have been made by the researchers which include: application of scientific principles for designing the cookstoves, insulating the combustion chamber, supplying correct amount of primary and secondary air at right place into the combustion chamber, use of fan to create draft, use of gasification techniques, use of high density pellets as fuel etc. [31–34]. Such efforts have helped in accelerating the process of design of advanced metal biomass cookstoves, both in natural and forced draft versions, across the globe.

According to method of combustion of biomass fuel into combustion chamber, advanced biomass cookstoves can be classified into two types *viz.* normal combustion cookstoves and gasifier cookstoves. The best example of the biomass cookstove which can efficiently operate in both these modes is Philips forced draft biomass cookstove [34].

During normal combustion mode, the biomass is fed in terms of small batches to the combustion chamber (oven). The pyrolysis products burn near the top of the combustion chamber using secondary air whereas the char combustion occurs using primary air at the bottom of the oven. During gasification mode, the whole combustion chamber is filled with the biomass fuel. The cookstove is lit at the top and the fire slowly passes to the bottom of the combustion chamber. Unlike the combustion mode, no fuel is added to the cookstove until the fire goes off. This gasification mode of operation of cookstove is also known as Top Lit Up Draft (TLUD) gasification, as the cookstove is lit at the top and the flow of both primary as well as secondary air goes in upward direction. **Figure 3** shows schematic diagram of Philips forced draft cookstove.

Jetter et al. [35] conducted experimental studies on 22 biomass cookstoves and reported that the efficiency of Philips forced draft cookstove was about 38% where as its CO and $PM_{2.5}$ emissions were very small. The authors also reported that cookstove operating on TLUD mode showed the lowest CO and $PM_{2.5}$ emissions. Some examples of TLUD gasifier cookstoves are rice husk gas cookstove [36], Oorja cookstove [37], pellet-fed gasifier cookstove [38] etc. The main advantages of using TLUD type of gasifier cookstoves are: highly efficient operation, clean combustion with negligibly small levels of emissions, use of densified pellets made up of crop residues and other biomass wastes for waste to energy conversion.

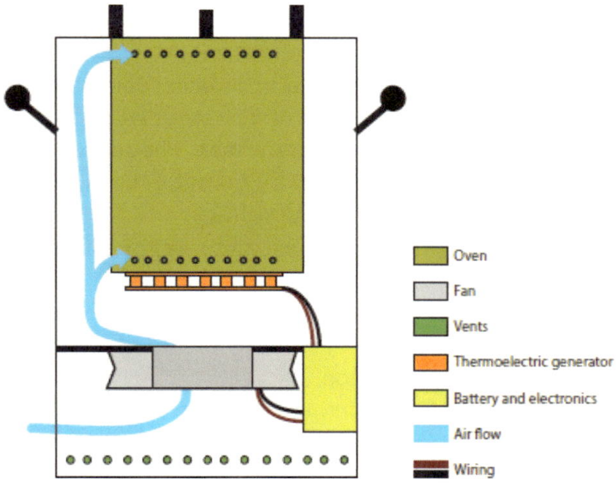

Figure 3.
Schematic diagram of Philips forced draft cookstove [34]. [© 2006, Royal Philips. Password, Philips Research innovation magazine, issue#28].

Tier	Thermal efficiency (%)	CO (g/MJ$_d$)	PM (mg/MJ$_d$)	Safety score	Durability score
5	≥ 50	≤ 3.0	≤ 5.0	≥ 95	< 10
4	≥ 40	≤ 4.4	≤ 62	≥ 86	< 15
3	≥ 30	≤ 7.2	≤ 218	≥ 77	< 20
2	≥ 20	≤ 11.5	≤ 481	≥ 68	< 25
1	≥ 10	≤ 18.3	≤ 1031	≥ 60	< 35
0	< 10	> 18.3	> 1031	< 60	> 35

MJ$_d$: Mega Joule of energy delivered.

Table 2.
Default values of voluntary performance targets for biomass cookstoves [40].

Research groups, non-government agencies and some government departments have developed protocols for testing the thermal and emission performance of biomass cookstoves. Comparative studies on testing protocols for biomass cookstoves are available in literature [9, 39]. An ISO technical committee comprising of experts from 45 countries and 8 international organizations published voluntary performance targets for biomass cookstoves in 2018 [40] in the form of a document called ISO Workshop Agreements (IWA) [41]. These targets cover five performance indicators *viz.* thermal efficiency, fine particulate matter emissions, carbon monoxide emissions, safety, and durability of biomass cookstoves. For each indicator, laboratory test results are rated along 6 tiers (0: for lowest performing cookstove to 5: for highest performing cookstove). **Table 2** reports default values of these voluntary performance targets. The three types of biomass cookstoves *viz.* traditional, improved and advanced can easily be categorized as per the tier rating. Most of the traditional cookstoves fall in tier 0 to 1 categories. The improved biomass cookstoves will have tier rating of 2–3 whereas the advanced biomass cookstoves shall be given the tier rating of 3–5. With increase in tier rating of the cookstove, its efficiency increases, CO and PM emissions decrease, its safety score increases i.e., it can be operated safely and its durability score decreases i.e., it becomes durable.

5. Adoption of biomass cookstoves

On one side, the design and development of biomass cookstoves is being done by the researchers in research laboratories but on the other side, dissemination of these cookstoves to the end users is a very important task. For successful dissemination and adoption of a cookstove, it must be locally manufactured, easy to operate, durable and it must result in clean combustion [42].

Bielecki & Wingenbach [43] reported that adoption of improved cookstoves depends on three factors *viz*. social, functional, and cultural. The social factor includes family size and meal occasion; functional factor includes ability of improved cookstove to provide space heat and ambient light, and the cultural factor includes local norms and traditional foods.

Adane et al. [44] categorized the factors affecting adoption of biomass cookstoves into four types: (i) household and setting related factors, (ii) cookstove technology related factors, (iii) cookstove users' knowledge and perception related factors, and (iv) financial and market development related factors. Household and setting related factors include: gender of the household head, educational level of the household head, family size of the household, house ownership, location of cooking quarter, and source of fuel. Cookstove technology related factors include: fuel processing requirement, durability of cookstove, fuel saving benefit of cookstove, health benefit of cookstove, time saving benefit of cookstove and safety benefit of improved cookstove. Cookstove users' knowledge and perception related factors include: optimistic previous social interaction, traditional suitability of cookstove and live demonstration experience. Financial and market development related factors include price and availability of the cookstove.

Nzengya et al. [45] reported that cost of cookstove, availability of cookstove, cost of fuel, availability of fuel, design of cookstove, time required for starting the cookstove, and time required for cooking the food are the factors affecting adoption of a biomass cookstove.

According to Jan [46], following factors act as key barriers to the adoption of improved cook stoves: lack of education of the women, non-participation of women in household decision making processes, low family income, lack of knowledge of health and environmental impacts associated with inefficient use of biomass, insufficient funds allocated by governments and NGOs for such programs, and poor monitoring system for the long-term cookstove use.

Jauland et al. [47] reported the evidence of saving in cooking time and fuel saving in the households which started using improved cookstoves. The authors did not find any evidence of health benefits in these households.

Jana and Bhattacharya [48] reported sustainable cooking energy options for rural people in Bargaon block of Odisha, India. Assessment of different cooking options such as traditional biomass cookstoves, improved cookstoves, gasifier cookstoves, biogas cookstove, LPG cookstove, electric cookstove and kerosene cookstove was conducted in terms of levelized cost of each cooking device per unit of useful cooking energy. While calculating the levelized cost of cookstove, the factors such as its capital cost, maintenance cost, estimated life, efficiency, cost of fuel, interest rate and energy equivalent per unit of energy source were considered. The levelized costs of different cooking devices per MJ without subsidy were: 1.3- traditional cookstove, 0.87-improved cookstove, 3.49-briquette gasifier cookstove, 2.72-kerosene cookstove (1.06 with subsidy), 1.88-LPG cookstove (1.33 with subsidy), 2.49-electric heater and 1.92-biogas cookstove (1.65 with subsidy). The authors found that the cookstoves using kerosene, LPG, briquettes, electricity and biogas were beyond reach of the poor people due to their high levelized costs though they could become cleaner cooking options for traditional and improved cookstoves.

DOI: http://dx.doi.org/10.5772/intechopen.101886

6. Recommendations for promotion of clean cooking energy options

Petroleum Conservation Research Association (PCRA) [49] has given some guidelines for about 30% saving in LPG and kerosene fuels. These guidelines will be very useful for all types of cookstoves. These guidelines include: plan before you start actual cooking; use pressure cooker of capacity corresponding to the family size; use optimum quantity of water; reduce the flame when boiling starts; soak the rice and pulses for about 15 minutes prior to their actual cooking; use shallow, wide vessels while cooking the food; put the lid to avoid heat losses; use small burner which saves fuel; use ISO/ISI marked cooking devices.

Council on Energy, Environment and Water (CEEW), India has published a report on roadmap for access to clean cooking energy in India [50]. Recommendations given by the authors in this report will be very useful for promotion of clean cooking energy options among the end users. These recommendations given for different cooking energy options along with author's own views are as reported here.

6.1 Biomass cookstoves

- Advanced biomass cookstoves are tier 4 and tier 5 cookstoves which are very expensive and are beyond reach of the common people. Most of the improved cookstoves found today are of tier 2 or tier 3 which are cheaper than the advanced biomass cookstoves and can be affordable to the poor people. In such a case the government must encourage use of tier 3 improved cookstoves equipped with chimneys for adequate ventilation. Also, to encourage use tier 4 and 5 cookstoves, subsidies must be provided to them.

- Labelling cookstoves with their efficiency and emissions rating will aid customer awareness and also will help them in taking decision on selection of cookstove for their family.

- Government shall provide subsidized training in pellet manufacturing and improved cookstove manufacturing, assembling, and marketing to local entrepreneurs and workers. This will help in enhancing local employment. It will reduce the cost of transportation and overhead charges. It will reduce the cost of pellets and initial cost of the cookstoves.

- It is observed that two third of the households using LPG cookstoves for cooking also use traditional cookstoves for heating of water and for space heating due to freely available biomass fuel and to ensure long lasting of LPG cylinder due to its high refill price. Hence, to fully eliminate household air pollution, it is important to address space heating and water heating for bathing using biomass cookstoves.

6.2 LPG

- Providing subsidy on the basis of socioeconomic characteristics will improve affordability of LPG among households.

- The thermal efficiency of the LPG cookstoves is about 55–57%. Research and development with a focused target of improving efficiency of LPG burners by about 10% must be undertaken. It can be done by modifying burner size, burner material, number of ports in a burner and also by improving burner pot interaction. The spacing between burner top and pot bottom can also be optimized. For a particular family size and for a given cooking process, pot sizes can be standardized.

- Sutar et al. [51] conducted preliminary experiments on domestic LPG stove to investigate the best combination of pot size for common cooking processes *viz.* heating of milk, making of tea and cooking of rice. During experimentation, a fixed amount of food item was used for each type of cooking process. The quantity of these food items was decided as per need of a family of four members based on the survey conducted in 80 LPG using households. The major findings were: LPG consumption was found to be more for larger pressure cookers as compared with smaller ones on account of higher thermal mass and higher convective and radiative losses through them; a pressure cooker of 1.5 liter capacity was found to be optimum for a family size of four members; round bottom pots caused reduction in LPG consumption than flat bottom pots of same sizes; it was estimated that for a four member family, use of optimum pot size, low flame setting and smaller burner will result into saving in about 14 kg of LPG annually.

- In India, only 41% of rural households received LPG cylinders at their doorstep in 2018. Permitting local institutions to stock LPG and to supply directly to households will reduce the distance traveled by users to procure LPG cylinders.

- The typical rural LPG distributor struggles for survival with low demand for LPG refills due to high cost of LPG and uncertainty of subsidies. In India, in 2016, over 80% of households that did not use LPG reported high recurring costs as a barrier [50].

6.3 Biogas

- The main hurdle in accelerating biogas use is the regular maintenance of biogas plants. If regular maintenance and servicing of a biogas plant is provided by an entrepreneur, households need not take on the hassle of operating, cleaning and maintaining the plant.

- It is important to train users to operate biogas plants in a manner that minimizes the need for operation and maintenance.

- A centralized toll-free helpline could be useful for people to lodge complaints regarding any issues with the biogas plants, which shall be immediately addressed by the local entrepreneur.

6.4 General guidelines on promotion of clean cooking energy options

- In rural areas, primary health centers and sub-centers are the closest access points to healthcare for the rural population, and thus could be effective venues for communication regarding the clean cooking technologies [50].

- Reliable information on consumers' willingness to pay for access to clean cooking energy will solve lot many issues. It will help the government agencies in providing the right cooking technology to right household [50].

- Greater focus on technology development, stricter quality standards and awareness drives to increase usage of new cooking technologies will be very important steps for the adoption of clean cooking technologies [9].

- Government must provide grants for the promotion of new technologies that are less effort intensive and/or more efficient e.g., advanced biomass cookstoves [9].

- Fabrication of cookstoves must involve a stringent quality control to keep the critical dimensions as per the requirement otherwise performance of the cookstoves will be affected adversely during its actual use.

- Performance of cookstove drastically affects due to changes in critical dimensions on account of lack of periodic maintenance [9].

- If trained personnel required for periodic maintenance of advanced cookstoves is available locally, then adoption rate of such cookstoves will enhance.

- Advanced biomass cookstoves generally require pellets or prepared fuels for their optimum performance. Availability and cost of the prepared fuel plays a very important role in the acceptance of the cookstove by the end user.

- The probability of acceptance of the new designed cookstove will be higher if the users are made familiar with the operation or if its operation is very similar to the cookstove they were previously using.

- While designing any clean cookstove, care must be taken that lower the capital as well as running cost of cookstove, higher is its acceptance among the users [50].

7. Conclusions

The present chapter reported four important aspects related to biomass cookstoves *viz.* energy efficiency, emissions, different designs of cookstoves and their adoption. Following are the conclusions drawn:

- The overall performance of biomass cookstoves is affected by two types of parameters *viz.* thermal and emission performance parameters. Thermal efficiency is the most important thermal performance parameter whereas emission factor of different pollutants is the most important emission performance parameter. In this regard, knowledge about safer limits of these pollutants in ambient air is very important. Also, knowledge of tier categories for biomass cookstoves is also important.

- Various ways to convert traditional biomass cookstoves into improved biomass cookstoves are: use of metals as cookstove materials, provision of grate for better air circulation, air preheating, provision of swirl and secondary air, provision of insulation, use of chimney and baffles

- Different techniques used by the researchers for development of advanced biomass cookstoves are: application of scientific principles for designing the cookstoves, insulating the combustion chamber, supplying correct amount of primary and secondary air at right place into the combustion chamber, use of fan to create draft, use of gasification techniques, use of high density pellets as fuel etc.

- The factors affecting adoption of biomass cookstoves, reported in literature are also discusses. These factors are: social, functional, cultural, affordability, women education, availability of fuel, availability of cookstove, timely servicing help, training, subsidies and grants etc.

- Recommendations on promotion of clean cooking energy options such as fuel saving guidelines, necessity of research and development in advanced cooking technologies, their promotion among the society, and financial support to the new clean cooking technologies etc. are given.

Conflict of interest

"The author declares no conflict of interest."

Author details

Kailasnath B. Sutar
Bharati Vidyapeeth (Deemed to be University) College of Engineering Pune, Pune, Maharashtra, India

*Address all correspondence to: kbsutar@bvucoep.edu.in

References

[1] World Energy Data. World Total Final Consumption [Internet]. 2020. Available from: https://www.worldenergydata.org/world-total-final-consumption/ [Accessed: June 20, 2021]

[2] Statista. Primary Energy Consumption Worldwide from 2000 to 2020 [Internet]. Available from: https://www.statista.com/statistics/265598/consumption-of-primary-energy-worldwide/ [Accessed: January 2, 2022]

[3] The Times of India. LPG Price has Doubled in 7 Years, and the Subsidies are gone. [Internet]. 2021. Available from: https://timesofindia.indiatimes.com/business/india-business/lpg-price-has-doubled-in-7-years-and-the-subsidies-are-gone-too/articleshow/81420847.cms [Accessed: June 20, 2021]

[4] Wrangham R. Catching Fire: How Cooking Made Us Human. New York: Basic Books; 2009

[5] Rupp R. A Brief History of Cooking With Fire. National Geographic: The Plate [Internet]. 2015. Available from: https://www.nationalgeographic.com/culture/article/a-brief-history-of-cooking-with-fire [Accessed: June 20, 2021]

[6] Did You Know Homes: History of Cooking [Internet]. Available from: https://didyouknowhomes.com/history-of-cooking/ [Accessed: June 20, 2021]

[7] Kumar M, Kumar S, Tyagi SK. Design, development and technological advancement in the biomass cookstoves: A review. Renewable and Sustainable Energy Reviews. 2013;**26**:265-285. DOI: 10.1016/j.rser.2013.05.010

[8] Kshirsagar MP, Kalamkar VR. A comprehensive review on biomass cookstoves and a systematic approach for modern cookstove design. Renewable and Sustainable Energy Reviews. 2014;**30**:580-603. DOI: 10.1016/j.rser.2013.10.039

[9] Sutar KB, Kohli S, Ravi MR, Ray A. Biomass cookstoves: A review of technical aspects. Renewable and Sustainable Energy Reviews. 2015;**41**: 1128-1166. DOI: 10.1016/j.rser.2014.09.003

[10] Mehetre SA, Panwar NL, Sharma D, Kumar H. Improved biomass cookstoves for sustainable development: A review. Renewable and Sustainable Energy Reviews. 2017;**73**:672-687. DOI: 10.1016/j.rser.2017.01.150

[11] Samal C, Mishra PC, Mukherjee S, Das D. Evolution of high performance and low emission biomass cookstoves—An overview. AIP Conference Proceedings. 2019;**2200**(1):020021. DOI: 10.1063/1.5141191

[12] International Energy Agency. Access to Clean Cooking [Internet]. 2020. Available from: https://www.iea.org/reports/sdg7-data-and-projections/access-to-clean-cooking [Accessed: July 15, 2021]

[13] World Health Organization. Household Air Pollution and Health [Internet]. 2018. Available from: https://www.who.int/news-room/fact-sheets/detail/household-air-pollution-and-health [Accessed: July 15, 2021

[14] Bailis PR, Ogle D, Maccarty N, From DS, Smith KR, Edwards R, Energy H. The Water Boiling Test (WBT) [Internet]. 2007. Available from: https://energypedia.info/images/3/38/Wbt_version_3.0_jan2007.pdf [Accessed: July 15, 2021]

[15] World Health Organization. WHO Guidelines for Indoor Air Quality: Household Fuel Combustion [Internet]. 2014. Available from: https://apps.who.int/iris/bitstream/handle/10665/

141496/9789241548885_eng.pdf [Accessed: July 15, 2021]

[16] United States Environmental Protection Agency. National Ambient Air Quality Standards (NAAQS) Table [Internet]. 2021. Available from: https://www.epa.gov/criteria-air-pollutants/naaqs-table [Accessed: July 27, 2021]

[17] Central Pollution Control Board, India. National Ambient Air Quality Standards [Internet]. 2009. Available from: https://cpcb.nic.in/uploads/National_Ambient_Air_Quality_Standards.pdf [Accessed: July 27, 2021]

[18] Bhattacharya SC, Salam PA. Low greenhouse gas biomass options for cooking in the developing countries. Biomass and Bioenergy. 2002;**22**(4): 305-317. DOI: 10.1016/S0961-9534(02)00008-9

[19] De Lepeleire G, Krishna Prasad K, Verhaart P, Visser P. A Woodstove Compendium. Eindhoven, Netherlands: Woodburning Stove Group, Technical University; 1981

[20] Krishna Prasad K, Verhaart P. Wood Heat for Cooking. Bangalore: Indian Academy of Science; 1983

[21] United Nations Educational, Scientific and Cultural Organization. Consolidation of Information, Cooking Stoves Handbook. Paris: Tata Energy Research Institute Pbs.; 1982

[22] Smith KR. Biofuels, Air Pollution and Health: A Global Review. New York: Plenum Press; 1987

[23] Stroad Ament K, Joseph S. Wood Conserving Cookstoves: A Design Guide. USA: Volunteers in Technical Assistance; 1980

[24] Verhaart P. On designing woodstoves. Proceedings of the Indian Academy of Sciences Section C: Engineering Sciences. 1982;**5**(4):287-326

[25] Ekouevi K, Freeman KK, Soni R. Understanding the Differences between Cookstoves [Internet]. 2014. Available from: https://openknowledge.worldbank.org/bitstream/handle/10986/18411/880580BRI0REPL00Box385214B00PUBLIC0.pdf?sequence=6&isAllowed=y [Accessed: July 15, 2021]

[26] Mukunda HS, Shrinivasa U, Dasappa S. Portable single-pan wood stoves of high efficiency for domestic use. Sadhana. 1988;**13**(4):237-270. DOI: 10.1007/BF02759888

[27] Still D, Kness J. Capturing Heat: Five Earth-Friendly Cooking Technologies and How to Build Them [Internet]. 2001. Available from: http://aprovecho.org/publications-3/ [Accessed: July 15, 2021]

[28] Pine K, Edwards R, Masera O, Schilmann A, Marrón-Mares A, Riojas-Rodríguez H. Adoption and use of improved biomass stoves in Rural Mexico. Energy for Sustainable Development. 2011;**15**(2):176-183. DOI: 10.1016/j.esd.2011.04.001

[29] Envirofit. Smarter Living [Internet]. 2017. Available from: https://envirofit.org/products/cookstoves/ [Accessed: September 9, 2021]

[30] Climate Technology Centre & Network (CTCN). Improved Cook Stoves [Internet]. Available from: https://www.ctc-n.org/technologies/improved-cook-stoves [Accessed: September 10, 2021]

[31] Reed TB, Larson R. A wood-gas stove for developing countries. In: Developments in Thermo-Chemical Biomass Conversion. Dordrecht: Springer; 1997. pp. 985-993. DOI: 10.1007/978-94-009-1559-6_79

[32] Reed TB, Anselmo E, Kircher K. Testing and modeling the wood-gas Turbo stove. In: Progress in Thermochemical Biomass Conversion

Conference. Tyrol, Austria; 2000. pp. 17-22. DOI: 10.1002/9780470694954.ch55

[33] Bhattacharya SC, Leon MA. Prospects for biomass gasifiers for cooking applications in Asia. In: World Renewable Energy Regional Conference. Jakarta, Indonesia; 2005

[34] Hegarty D. Satisfying a burning need. Phillips Research Magazine. 2006; **28**:28-31

[35] Jetter J, Zhao Y, Smith KR, Khan B, Yelverton T, DeCarlo P, et al. Pollutant emissions and energy efficiency under controlled conditions for household biomass cookstoves and implications for metrics useful in setting international test standards. Environmental Science & Technology. 2012;**46**(19): 10827-10834. DOI: 10.1021/es301693f

[36] Belonio AT. Rice Husk Gas Stove Handbook. Iloilo City, Philippines: Central Philippine University; 2005

[37] Mukunda HS, Dasappa S, Paul PJ, Rajan NK, Yagnaraman M, Kumar DR, et al. Gasifier stoves—Science, technology and field outreach. Current Science. 2010;**98**(5):627-638

[38] Champion WM, Grieshop AP. Pellet-fed gasifier stoves approach gas-stove like performance during in-home use in Rwanda. Environmental Science & Technology. 2019;**53**(11):6570-6579. DOI: 10.1021/acs.est.9b00009

[39] Lombardi F, Riva F, Bonamini G, Barbieri J, Colombo E. Laboratory protocols for testing of Improved Cooking Stoves (ICSs): A review of state-of-the-art and further developments. Biomass and Bioenergy. 2017;**98**:321-335. DOI: 10.1016/j.biombioe.2017.02.005

[40] Voluntary Performance Targets. Clean Cooking Alliance [Internet]. 2018. Available from: https://www.cleancookingalliance.org/technology-and-fuels/standards/iwa-tiers-of-performance.html [Accessed: July 27, 2021]

[41] ISO Workshop Agreements (IWA). Clean Cooking Alliance [Internet]. 2018. Available from: https://www.cleancookingalliance.org/binary-data/DOCUMENT/file/000/000/6-1.pdf [Accessed: July 27, 2021]

[42] Barnes DF, Openshaw K, Smith KR, van der Plas R. What Makes People Cook with Improved Biomass Stoves? World Bank Technical Paper No. 242 [Internet]. 1994. Available from: https://documents1.worldbank.org/curated/en/738011468766789505/text/multi-page.txt [Accessed: October 9, 2021]

[43] Bielecki C, Wingenbach G. Rethinking improved cookstove diffusion programs: A case study of social perceptions and cooking choices in rural Guatemala. Energy Policy. 2014;**66**: 350-358. DOI: 10.1016/j.enpol.2013.10.082

[44] Adane MM, Alene GD, Mereta ST, Wanyonyi KL. Facilitators and barriers to improved cookstove adoption: A community-based cross-sectional study in Northwest Ethiopia. Environmental Health and Preventive Medicine. 2020;**25**:1-2. DOI: 10.1186/s12199-020-00851-y

[45] Nzengya DM, Maina Mwari P, Njeru C. Barriers to the adoption of improved cooking stoves for rural resilience and climate change adaptation and mitigation in Kenya. African Handbook of Climate Change Adaptation. 2020:1-8

[46] Jan I. What makes people adopt improved cookstoves? Empirical evidence from rural northwest Pakistan. Renewable and Sustainable Energy Reviews. 2012;**16**(5):3200-3205. DOI: 10.1016/j.rser.2012.02.038

[47] Jeuland MA, Pattanayak SK, Samaddar S, Shah R, Vora M. Adoption and impacts of improved biomass cookstoves in rural Rajasthan. Energy for Sustainable Development. 2020;**57**: 149-159. DOI: 10.1016/j.esd.2020.06.005

[48] Jana C, Bhattacharya SC. Sustainable cooking energy options for rural poor people in India: An empirical study. Environment, Development and Sustainability. 2017;**19**(3):921-937. DOI: 10.1007/s10668-016-9774-y

[49] Ministry of Petroleum & Natural Gas, Government of India. Make Gas and Kerosene Last Longer. Petroleum Conservation Research Association (PCRA) [Internet]. Available from: http://www.pcra.org/pages/display/41-Make-Gas-and-Kerosene-last-longer/20 [Accessed: July 27, 2021]

[50] Patnaik S, Tripathi S, Jain A. Roadmap for Access to Clean Cooking Energy in India. Council on Energy, Environment and Water, NITI Aayog, Government of India [Internet]. 2019. Available from: http://www.niti.gov.in/sites/default/files/2019-11/CEEW-Roadmap_for_Access_to_Clean_Cooking_Energy_in_India-Report.pdf [Accessed: July 27, 2021]

[51] Sutar KB, Kumar M, Patel MK, Kumar A, Mokashi SR. Experimental investigation on pot design and efficiency of LPG utilization for some domestic cooking processes. Energy for Sustainable Development. 2020;**56**: 67-72. DOI: 10.1016/j.esd.2020.04.006

Chapter 9

Energy Efficiency: The Overlooked Energy Resource

Ali Al-Qahtani, Zeeshan Farooq and Sami Almutairi

Abstract

The objective of this chapter is to draw the attention of government policy makers internationally to a strategy for alleviating global warming through proven cost-effective energy efficiency measures. The Saudi Arabian government has embraced the approach with demonstrable success over the past 20 years, with rates of return on investments averaging more than 25%. Even though Saudi Aramco is the National Oil and Gas company, the company takes the threat of climate change to the world's economies very seriously and initiated programs for, systematically and responsibly, transition to less-polluting energy sources. Primarily, the chapter will define the supply chain components of Saudi Arabia's energy sector and explain the existing conditions and efficiencies of each of its components. It analyzes the existing energy management framework and its achievements, as well as its current and forthcoming commitments, status, and updates. It will also explain the vital equipment, systems, and processes in the supply chain, with possible energy efficiency improvement gaps based on existing literature/Energy Assessment Reports conducted by Saudi Aramco professionals in numerous industrial facilities. The chapter will pinpoint the highest achievable efficiencies areas in major systems, processes, or equipment and discusses its impact on the primary energy fuels and green house gas (GHG) emission reduction.

Keywords: energy efficiency, industiral energy power generation, energy transmission, energy oil and gas

1. Introduction

Oil and gas are the world's most used energy sources based on the share of each source of global energy consumption. More than half of the global energy demand is fulfilled by oil and gas, as shown in **Figure 1** [1]. The production of these primary energy fuels (oil and gas) involves them going through various stages of processing before it is used directly in the vehicle or converted to electricity in the power plant for other end-users. It can be characterized as a typical supply chain, which is defined as a complex structure of supply facilities linked together in order to serve end customers, collectively called the "supply chain" network. In the present context, it can be referred to as the energy supply chain. The main objective of the energy supply chain is to deliver crude oil, natural gas, and refined products safely and economically to customers. These energy supply chain networks are subject to various losses in primary energy (oil & gas) as well as secondary energy (electricity), some of which are unavoidable while some are not. Improvements in the

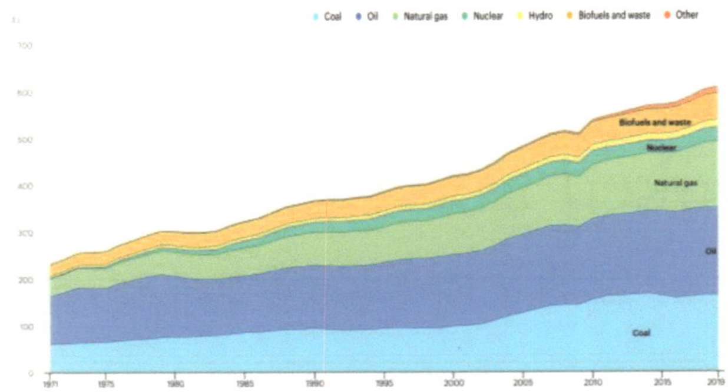

Figure 1.
World total energy supply by source [1].

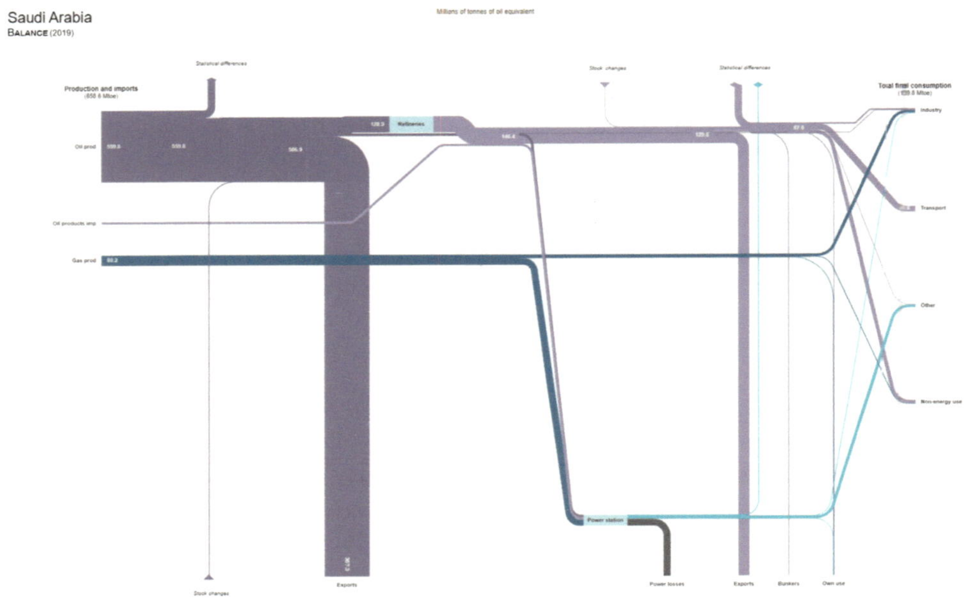

Figure 2.
Sankey diagram of Saudi Arabian energy flows 2019 [2].

overall efficiency of the energy supply chain certainly result in enhanced profit margins and mitigated environmental impacts. Consequently, a comprehensive strategy to develop efficient energy supply chain network is inevitable.

The Kingdom of Saudi Arabia (KSA) is among the top crude oil producing and exporting countries in the world as well as one of the major producers of natural gas. KSA has invested heavily to improve the overall efficiency of its energy supply chain and demonstrated an approach that is driving the business towards excellence with a noticeable improvement in the preservation of the livable environment. To identify the most significant opportunities for increasing energy efficiency and reducing energy losses, it is vital to determine where and how energy is used—how much is used, where are the losses—how much is lost, where energy losses could potentially be recovered or reduced, and to what extent. **Figure 2** shows an overall picture of KSA's Energy flow, as a Sankey diagram, represents KSA's total production, consumption, and exports [2]. It will aid the understanding of overall energy usage that occurs from source to end-user in KSA's energy supply chain and consequently provides insights to identify areas of improvement and overall efficiency enhancement. **Figure 2** [2] clearly establishes KSA as the leading oil exporter, however, the main emphasis here is about understanding the energy flow and the overall efficiency of the energy supply chains. There are certainly energy consumptions as the energy flows from sources to end-users, but a reduction in the amount of energy used to process and deliver it to the final users have greater implications as it will improve the performance as well as save the product i.e., energy, which will eventually be added to the product. Even though some losses occur at every stage in the energy supply chain, it is evident from **Figure 3** that a significant energy loss occurs at power generating stations.

From an overall efficiency improvement standpoint, primary energy is the best place to look at for energy use as well as for losses. As illustrated in **Figure 3**, a typical industrial pumping system utilizes only 10% of the primary energy resources, if pumping is considered to be the end-use of the energy from the primary energy sources and typical losses for all components in the supply chain is considered. Energy "footprints" could be created for all users, illustrating energy flows along the utility supply chain from energy sources to an industrial end-user based on which energy use, loss, and opportunities analysis shall be conducted to prioritize efforts to improve the overall efficiency of the energy supply chain.

To establish the effectiveness of the policy framework of overall efficiency enhancements for the energy supply chain, it is essential to evaluate macroeconomic benefits from such an approach. It is a very common and well-established causality relationship between energy consumption and gross domestic product (GDP). The ratio of energy use to GDP indicator is referred to here as "aggregate energy intensity" or "economy-wide energy intensity". Economic-wide energy intensity is measured by dividing the cumulative energy consumption requirement of a particular region by its GDP. Its trend indicates the general relationship of energy consumption to economic development and provides a rough basis for projecting energy consumption and its environmental impacts on economic growth. It estimates the absolute amount of energy needed to generate a single unit of gross domestic product. GDP is represented at a consistent exchange rate and an increasing parity of power to exclude inflation, which influences and indicates the diversity of energy consumption and general energy price levels in the real economic scenario. The economic-wide energy intensity and GDP of some major countries are shown in **Figures 4** and **5** respectively [3, 4]. The trend for the USA shows that even though the GDP is growing, energy consumption is declining. Most of this is due to a shift away from low-margin energy-intensive manufacturing to more profitable financial and IT services, not due to better energy efficiency. A similar profile can be observed for Germany. In KSA however, it appears that energy consumption is rising faster than GDP till the year 2010 which reflects energy inefficiency in the energy supply chain including end-users inefficiency.

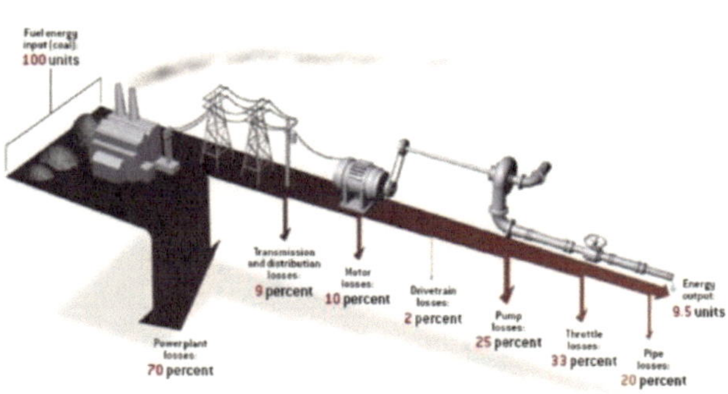

Figure 3.
A typical industrial pumping system.

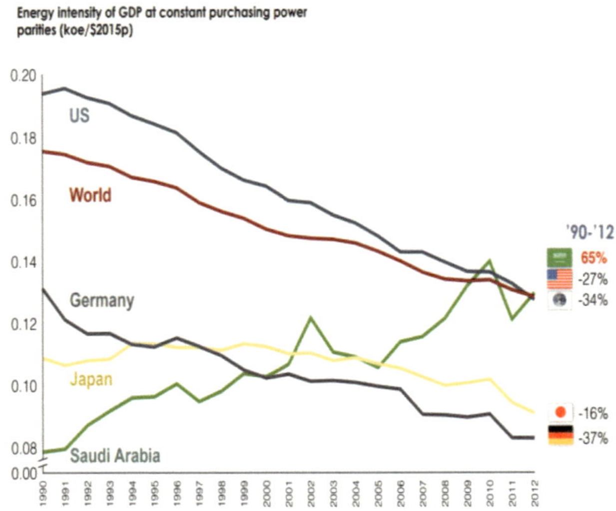

Figure 4.
Economic-wide energy intensity of countries [3].

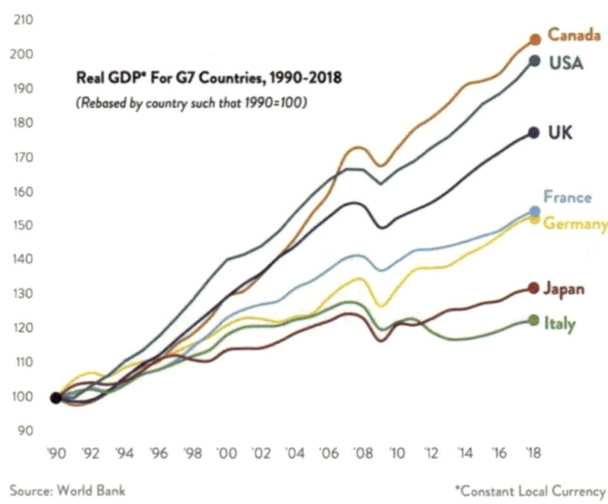

Figure 5.
World Bank provided GDP of countries [4].

It is important to have a look at the energy consumption in different sectors to improve the energy scenario and provide recommendations to meet the country's goal of rational energy consumption patterns. According to the Saudi Energy Efficiency Center (SEEC), 90% of domestic energy consumption in Saudi Arabia is consumed by the construction, transport, and industry sectors [5]. The industrial sector consumed 47%, the construction sector consumed 29%, while the transportation sector's consumption was about 14% of the country's primary energy in 2017 [5, 6].

Figure 6 shows trends of energy consumption in different sectors from the year 1990 to 2014 [7], indicating a sharp rise in industrial and building sectors. The trends with inference from GDP (**Figure 7**) movement suggest that energy consumption is increasing as a result of economic activity without any improvement in the consumption patterns. Consequently, energy efficiency policies need to be developed with an emphasis on the three most energy-intensive sectors i.e., industrial, transport, and building sectors.

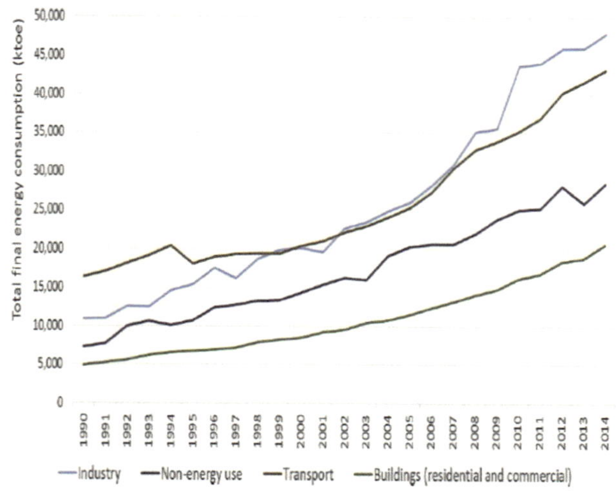

Figure 6.
Total energy consumption in different sectors for the Kingdom of Saudi Arabia [6].

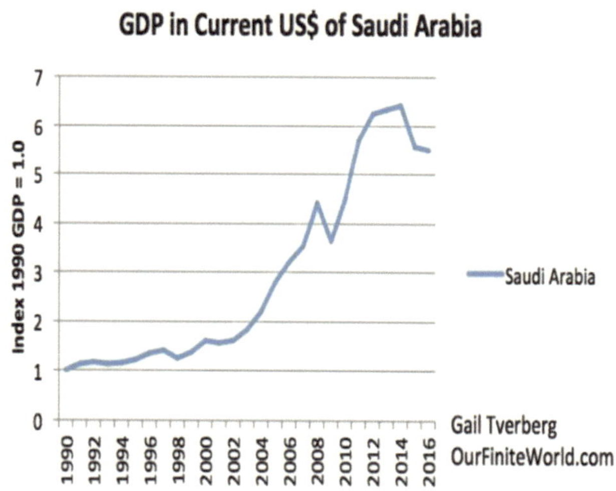

Figure 7.
GDP of the Kingdom of Saudi Arabia [4].

The best way to improve energy productivity as a way forward for the Kingdom's strategy would build on the competitive advantages by enabling a strong and energy-efficient industrial sector. As for the other two sectors i.e., transport and building sectors, they need more regulatory and behavioral improvements. For example, given the low energy prices in the Kingdom, it is difficult to invest in energy-efficient home appliances (AC units, refrigerators, or efficient lightings) to improve buildings' energy performance. Similarly, for transport, fuel-efficient vehicles will not be preferred by the masses if the gasoline prices are very low. It is obvious that there will be very little to no incentive for owners to invest in energy efficiency. Consequently, this will likely remain an issue, till the energy price regulation/reforms are fully implemented. However, when we see the supply side of these sectors, it is part of the energy supply chain i.e., part of the industrial sector, thus these sectors have a unique feature, where its boundaries are not completely dictated by its sector but by other sectors too. If the benefits from avoided energy consumption or improved efficiency in the supply side (power plant) which resulted in the avoidance of the new electricity generation facility, are considered, energy efficiency investments seem highly cost-effective.

In common with other parts of the energy sector transformation, it is important for actions to be based on an integrated strategy with clear goals. Energy efficiency and other demand-reduction measures will need to be analyzed together with supply expansions to find the best balance in terms of both service delivery and costs. It is critical to ensure that the opportunities offered by new digital technologies are fully utilized to enhance the efficient interaction of ever-more integrated energy system supply and demand elements. The system is first modified to use energy in a more effective manner, more energy efficiency opportunities are readily available to meet the emissions targets, within the given time frame. Moreover, it will have positive effects on energy transition as it will minimize demand and result in a lesser number of needed renewable/green energy installations. However, energy efficiency measures often need policy support to be implemented and strategies must address the main barriers to the adoption of energy efficiency measures and promote structural and behavioral changes. Furthermore, they must be considered across different sectors and areas, for instance, buildings, transport, and industrial sectors.

To address the global agenda of enhancing energy productivity, KSA's vision 2030 program has identified and addressed many areas in which energy efficiency can be improved significantly and cost-effectively. One of the outcomes of KSA's vision 2030 program is the Saudi Energy Efficiency Center (SEEC), which has taken wider initiatives to address national energy efficiency improvement and carbon emissions. It is functional from the inception of the year 2010. In 2012, SEEC launched a national program to raise energy efficiency in the Kingdom, using initiatives designed according to local market potential, by involving all stakeholders (government, companies, and the public). The program focuses on three key sectors (buildings, transport, and industry), which consume about 90% of the total energy in the Kingdom [6]. The program developed the factors and possible supporting mechanisms to boost its activities.

The program was launched as a dedicated system for energy efficiency improvements, to ensure the implementation and enforcement, including a mechanism to update when necessary, with an executive committee that holds all the power necessary to manage the program through an organizational structure. Since its formation to date, the impact of the programs on the overall national-level energy efficiency index is visible as shown in **Figure 4** (from the year 2012 onwards). It is important to note that other agencies and their initiatives contributed to this energy productivity improvement.

The Kingdom is implementing many other initiatives as well, including renewable energy (wind, solar), safe nuclear power, cost-effective energy efficiency, and minimization of needless fuel emissions through flare management. Best of all, these technologies are mostly well-established and proven for all commercial applications. It has been clearly observed that from 2012 to 2018 the overall supply chain energy efficiency improved significantly because of major efficiency improvement in utility plants, one of the most significant components of the supply chain. Overall national level utility plants efficiency has improved from 31.8–38% and is targeted to reach 45% by 2030 [8] through the incorporation of combined cycles and integration between power generation and seawater desalination at the same site. Towards this end, the formerly separate Ministries of Water and Electricity were merged by a Royal Decree into a single entity—MOWE. The combination of such strategic decisions justifies the reasons for such significant efficiency improvement in Saudi Arabia's Industrial and Public Utility sectors.

Energy efficiency in Saudi Arabia has included the establishment of a framework for an energy efficiency market involving energy service companies and a range of regulatory measures to drive the market. These were focused on the building, transport, and industry sectors which covered around 90% of energy consumption in the Kingdom [5, 6]. The approach adapted is to develop a baseline for setting policies, establish performance relative to international benchmarks, prioritize initiatives based on potential impact, achieve consensus and coordination among implementation agencies, and establish execution teams and empowering policy environment. Then, finally, to monitor and evaluate progress, with a view of registering feedback into the design of the overall approach.

Energy supply chain is an essential part of the kingdom, as it fulfills energy requirements for all sectors and also provides products to export. As the Kingdom is one of the largest exporters of crude oil, the industry sector, which is the largest energy consumer in the kingdom, is predominant with the components of the energy supply chain. Accordingly, any improvement in the energy supply chain will result in a greater effect on the energy productivity of the whole kingdom. There are two key drivers to improve the energy productivity of the Kingdom, firstly, structural change in the economy by moving away from energy-intensive to a high margin value manufacturing and, secondly, energy efficiency in energy-intensive manufacturing. Both aspects of energy productivity are important for the Kingdom but the energy efficiency improvements provide a quick win for the kingdom. Moreover, the solutions to improve the energy efficiency of the energy supply chains are applicable to other sub-sectors of the industrial sector and could be leveraged across all industrial sectors. Improvements in the energy supply chain will have great implications on the abundant natural resources of the Kingdom i.e., the counts of barrels saved in the processing will be added to the export/usage.

2. Energy supply chain networks

Energy efficiency of the energy supply chain network is aligned with the Kingdom's Vision 2030 which is aimed at delivering more sustainable, socially inclusive, and prosperous economic development. The energy supply chain in KSA mainly constitutes of Hydrocarbon Supply Chain (HCSC) i.e., crude oil and natural gas supply chains, along with utility plants. HCSC chain is wholly owned by the state-owned oil giant Saudi Aramco while utility plants which are mainly for power generation plants are owned by Saudi Electric Company (SEC). Although SEC owned power generation plants and distribution of electricity, Saudi Aramco also has power generation capacity through cogeneration and generates power which is excess of its total consumption and supplied to SEC for distribution. **Figure 8**, provides an overview of the energy supply chain network in KSA which constitutes of different stages/components and could be defined in many ways but usually constitute of the following main components:

1. **Reservoir and facility management:** This term in the present context refers to the exploration and production of oil and natural gas. Geologic surveys and any information gathering used to locate specific areas where oil and gas are likely to be found, commonly called "exploration". The term also includes the steps involved in the actual drilling and bringing oil and natural gas resources to the surface, referred to as "production". In short, from an operation perspective, it is associated with the extraction of hydrocarbons from the "wells" that bring the crude oil or/and natural gas to the surface as well as the reinjection of produced water.

2. **Upstream facilities:** It is referred here to include all facilities which do the separation of wellhead fluids into constituent vapor (gas) and liquid (oil and

Figure 8.
Energy supply chain losses diagram.

produced water) components. It includes Gas and Oil Separation Plants (GOSPs), crude stabilization facilities, and gas compression plants along with utilities to supply energy to these plants.

3. **Downstream facilities:** These are the facilities involved in the processes of converting oil and gas into the finished product. These include refining crude oil into gasoline, natural gas liquids, diesel, and a variety of other energy sources. The closer an oil and gas company is to the process of providing consumers with finished petroleum products, the more it adds value to the hydrocarbons and generally results in higher energy intensity.

4. **Terminals & distribution:** These facilities deliver gas, crude oil, and refined hydrocarbon products in a safe, reliable, cost-effective, and environmentally friendly manner to the customers. It includes a network of pipelines and storage facilities to deliver the wellhead fluids to upstream facilities, upstream products i.e., oil and gas to downstream facilities i.e., refineries, fractionation plants, and also finished products from downstream facilities for other consumers including end-users.

5. **Utilities plants:** It is referred to all facilities that provide energy to the end-users in the form that can be directly usable by them and mainly, transform primary energy in the final form to be used by end-users. It supports energy requirement for both upstream and downstream activities and also includes power plants (one of the major contributors to energy inefficiency) that supplies electricity to end-users.

6. **End-users:** Transportation, building, industry, agriculture, etc. are the end-users. All the energy produced is consumed by end-users. It is the energy that reaches the final consumer's door and excludes that which is used by the energy sector itself. Final energy consumption excludes energy used by the energy sector, including for deliveries, and transformation. Final energy consumption in "households, services, etc." covers quantities consumed by private households, commerce, public administration, services, agriculture, fisheries etc.

Although, there are not very sharp and distinct boundaries among all components of the energy supply chain but it is needed to unify efforts to improve energy efficiency. The KSA's energy supply chain, which is mainly HCSC consisted of two parallel supply chains of crude oil and natural gas, which are divided into three echelons: production areas (starting at the well head), processing plants, and distribution terminals. An overlap exists between the two networks because crude oil is strictly a two-phase fluid that contains both oil and dissolved associated gas. **Figure 9** [9] depicts a schematic representation of the HCSC network, it clearly indicates that Echelon 1 is mainly a reservoir facility management component, Echelon 2 covers upstream, downstream, distribution as well as utilities facilities while Echelon 3 belongs to end-users.

The reservoir facility management is very different from the above surface facilities. It includes sub-surfaces, highly dependent on the reservoir characteristic, and emphasis on reliability and not conversed in the chapter. Most of the equipment and systems, such as steam, compressors, furnaces and boilers, rotary equipment (air compressors, pump, and fan motors), combined heat and power, cogeneration is common for Echelon 2. The discussed solutions might be applicable to different components of the energy chain but considering the frequent applicability, they could be leveraged across an Echelon or beyond.

2.1 General strategies for energy efficiency improvement

There are several well-established and proven strategies for improving energy efficiency in industrial processes. Some of them pertain to equipment, and others to systems. For example, a heat exchanger is an item of equipment, but a Heat Exchanger Network (HEN) that preheats the feed to a distillation column is a system. Both categories must be addressed, keeping in mind that improving system efficiencies has multiple potentials for savings than component/equipment efficiency upgrades.

This chapter will list some of the more important techniques generally applicable to all processes, without attempting to describe any of them in detail. It highlights some of the findings of the energy assessments conducted by Saudi Aramco and based on the case studies;

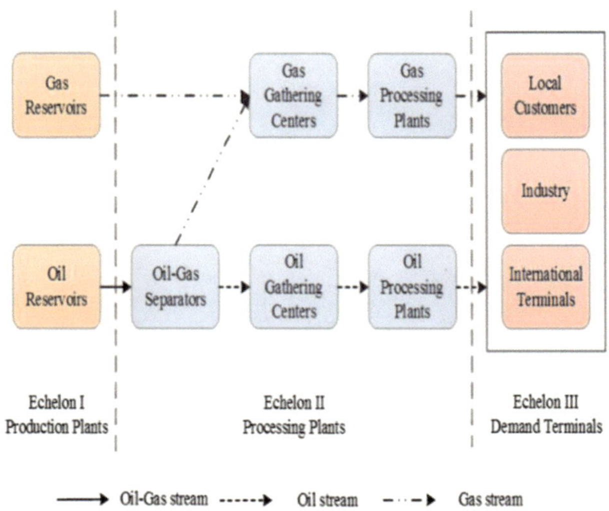

Figure 9.
Details of upstream hydrocarbon supply chain network [9].

- Heat recovery optimization using pinch analysis.
- Variable speed drives—VFDs for motors, steam turbines—for pumps and compressors.
- Load management of series/parallel networks of boilers, furnaces, pumps, compressors, etc.
- Integrated optimization of utilities (combined heat, power, and cooling systems) with the process needs.

Rather, we will give examples of the application of these methods for specific projects under-taken or under consideration at Aramco facilities.

The following subsections will explain each energy supply chain network component. The energy consumption and production in upstream, downstream, utilities including power plants and end-users—is very complex and linked.

3. Upstream facilities

Several processing steps are required to separate the underground well fluids—a multiphase mixture of gas, oil, water, and solids (both dissolved and suspended) as shown in **Figures 10** and **11**. These separation processes not only include the separation of "produced water" from the oil/gas mixture, but also removal of dissolved gases, acid gases and extraction of light-end distilled products from the crude oil, and the separation of acid gases, water, condensate, and NGL from the associated gases.

The hydrocarbons from wells are transported to gas-oil separation plants (GOSPs) which is generally equipped with a three-phase separator and separate the associated gas and water from oil. The hydrocarbon stream from the three-phase separator might undergo water removal in desalters depending on the crude oil's characteristics. The dry crude is stabilized in the stabilizer column and transferred to terminals either for exports or further processing in downstream facilities.

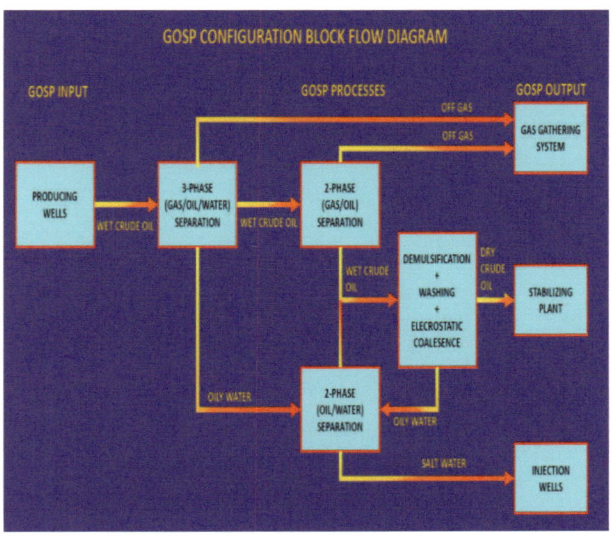

Figure 10.
GOSP process block diagram.

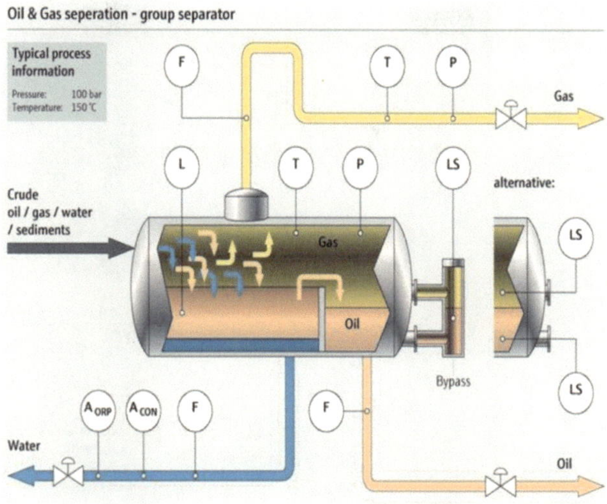

Figure 11.
Typical oil & gas separation process.

Natural gas from dry-gas reservoirs (i.e., non-associated gas) and associated gas from GOSPs are collected at the gas gathering centers and fed to gas processing plants.

At Saudi Aramco, the gas processing plants are classified as part of the "Upstream" sector, whereas in most other parts of the world they are classified as midstream. At our gas processing plants, H_2S (used for sulfur production) and CO_2 (vented to atmosphere) are removed by *absorption and stripping* processes, while methane and NGL are produced as separate products by low-temperature fractional distillation. After that, NGL is fractionated further in a separate plant into its components (viz., ethane, propane, butane, and natural gasoline (C5+). Finally, sweetened crude oil and gas are distributed to different storage terminals (local, industrial, and international) for shipment to customers. All of the above processes involve several levels of heating, cooling, pumping, compression, expansion, refrigeration, and other supporting operations.

In **Figures 12** and **13** [9], reservoir and natural gas networks in Saudi Arabia are shown.

3.1 Applications in upstream

The first process after the well-head typically involves a high-pressure test trap, a low-pressure degassing tank, electrostatic dehydrator, electrostatic de-salter, water-oil separator, several types of pumps (booster, shipping, disposal, condensate, wash, stabilizer bottom, transfer, and chemical injection), air compressors, gas compressors, refrigeration compressors, shell and tube heat exchangers, air coolers, a crude stabilizer column, etc. All this equipment and instruments involve multiple utility systems: process water, fire water, instrument/plant air, flare, gravity and pressure sewer, chemical injection, power generation, hot oil, boiler, and steam, etc.

Imbalance and inefficiencies in these systems and equipment have a significant impact on the GOSP, and thus on the overall energy supply chain, so must be carefully assessed and optimized. Typically, in GOSP configuration, essential energy consumers are water injection pumps (39%); gas compressors (44%); and pumps or internal liquids transport and deliver the crude oil via pipeline from the GOSP to the oil terminal or refinery (17%).

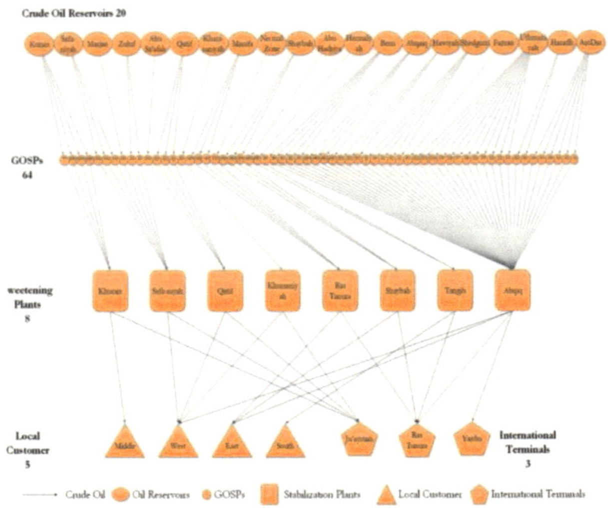

Figure 12.
The upstream crude oil supply chain network in Saudi Arabia [9].

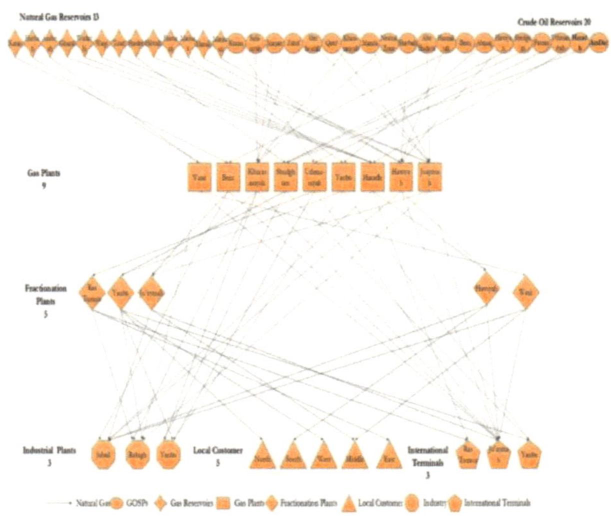

Figure 13.
The natural gas supply chain network in Saudi Arabia [9].

Reliability is the main focus in the crude extraction process especially from the underground well to the GOSP. So, while working to improve the efficiency of that stage of the cycle, it is very important to improve other stages from the GOSP to transportation. The following few common measures are widely discussed in the upstream process energy improvement.

3.1.1 Compressors load management and process optimization

Overall energy efficiency of a compressor is the ratio of absorbed energy by the process gas to the energy consumed by the driver. Compressor load management means ideally turning only the minimum number of units in the network, and optimizing the load on each one according to the demand. This needs to be controlled and monitored with enough time delay function to reduce frequent

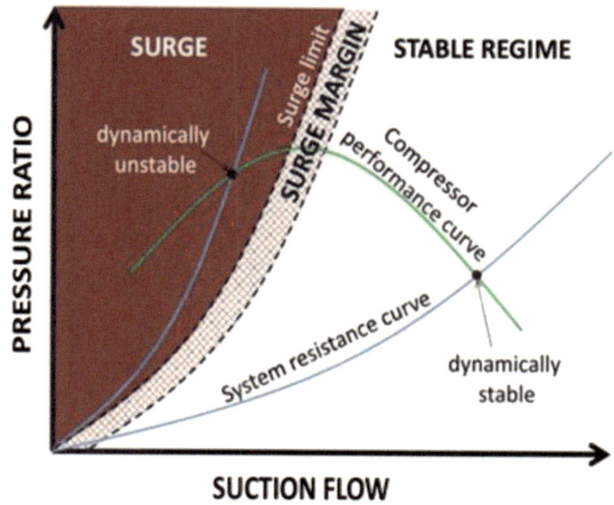

Figure 14.
Performance curve and system resistance curve for a typical compressor system.

activation/inactivation. The use of variable speed drives may or may not be the best option, since it is difficult to control flow rates under turn-down conditions where the system curve may be nearly parallel to the compressor curve (see **Figure 12**). Properly controlled compressor load management can bring 3–5% savings from total compression.

The methodology for estimating savings potential from load management of compressors is similar to that for pumps, except that the fluid is gas instead of liquid in the pump. In general, the head vs. capacity curve (also called the "performance" curve, **Figure 14**) for a centrifugal compressor operating at a fixed speed is relatively flat, with the total head at the minimum throughput (the surge point) typically being only 105–115% of the head at design throughput. The system curve is also relatively flat because the static head usually dominates the frictional (dynamic) head. The operating point occurs at the intersection of the compressor performance curve and the system curve.

3.1.2 Reduction and recycling through VFD and impeller trimming

The compressor operation can be controlled, within limits, by installing a variable frequency drive (VFD). For extreme cases where the machine was grossly oversized for the duty and if the desired flow reduction is permanent, impeller trimming may be preferable compared to a VFD. The savings can be in the order of 5–10% of existing power and energy.

3.1.3 Fired heater efficiency improvement by controlling the excess O_2

The main operating parameters affecting combustion process efficiency are the flue gas temperature and the excess air ratio. The target FG exhaust temp should be about 50 F above the dew point, and the % of O_2 in the flue gas should be 2–3%). Approximately 3–5% savings in fuel consumption can be expected.

3.1.4 Design modification on the fired heater convective/economizers

Fired heaters transfer the combustion heat from the fuel to process streams. Most of the heat transfer occurs by radiation in the radiant zone, while some of the

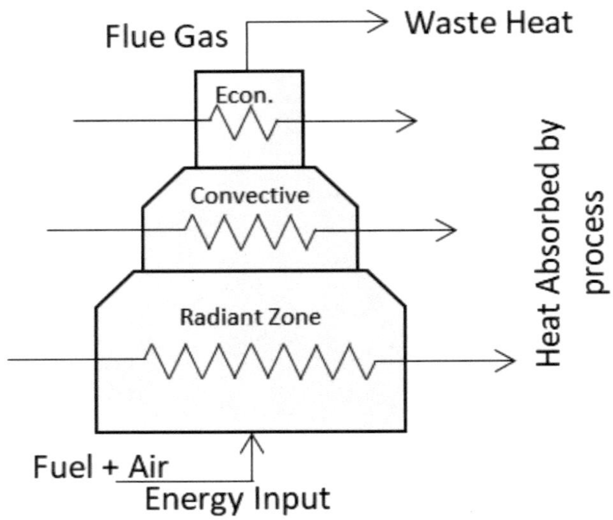

Figure 15.
Fired heater zones.

heat in the flue gas is absorbed by convection in the convective section (common) and the economizer (not as common) zones (**Figure 15**). The remaining heat in the flue gas leaves the fired heaters through the stack and is lost.

The heater efficiency is defined as the ratio of heat absorbed by the process and the total heat released by the combustion of fuel. Even small efficiency improvements in fired heaters can yield considerable energy savings and green house gas (GHG) emissions reduction due to their large energy consumption. Usually, the fluid being vaporized is preheated in the convective section, while the economizer is used to recover heat into colder process or utility stream such as Boiler Feed Water (BFW). Since most furnaces come with a built-in convective section, to start with, the only remaining heat recovery retrofit opportunity is usually to add an Economizer. Unfortunately, for piping layout and structural support reasons, this is often not economically feasible but could be feasible with a separate economizer at the surface (i.e., not mounted on the top of convection section) connected through the duct.

3.1.5 Well-head to GOSP gas turbo expander generator

A turboexpander (**Figure 16**), also referred to as an expansion turbine, is a centrifugal or axial-flow turbine, through which high-pressure gas is expanded to produce work, and can often be used to directly drive a compressor or generator. Most gas from the wells is produced at high pressure, and after the first knock-out drum in a gas processing plant is usually let down to lower pressures across a valve. Instead of destroying the pressure energy, it could be let down through a turboexpander to recover and could be utilized energy is utilize as shaft power or to generate electricity in a generator.

3.1.6 Hydraulic turbine (liquid)

Similar in concept to a turbo expander, the hydraulic turbine can be used to recover some power from high-pressure liquids being let down to lower pressure. They are effective pumps operating in reverse but are not commonly used in oil and gas plants.

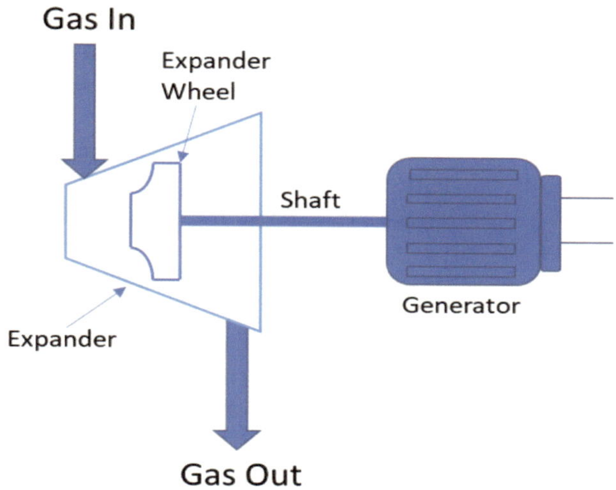

Figure 16.
Turboexpander schematic.

There are many other energy efficiency techniques available that can be applied in GOSPs, gas plants, stabilization units, and related supporting facilities, which have the potential for significant energy savings. It is estimated that 5–10% of total supply chain energy is used in upstream activities, out of which 15–20% energy savings are possible.

4. Downstream facilities

The oil refining sector is vast and diverse with multiple processes that are tailored to the type of crude oil feed (light, medium, heavy, etc.). Everything starts with the CDU (crude distillation unit) which fractionates the desalted feed into various cuts—light-end gases, naphtha (gasoline), kerosene (jet fuel), diesel, Light Gas Oil (LGO), Heavy Gas Oil (HGO), and tower Bottoms. The bottoms stream may or may not to send to a vacuum distillation unit (VDU) for further fractionalization into more valuable cuts, depending upon the flow rate and composition. Other process units include typically include hydro-treaters for sulfur removal, catalytic reformers for raising the octane value, hydro-crackers, delayed cokers, visbreakers, alkylation units, hydrogen production, sulfur recovery units, etc. An oil refinery requires a multitude of supporting facilities and utilities, such as fuel storage and distribution, steam generation and distribution, power distribution, on-site power generation (whether as mechanical shaft work or electricity), compressed air, cooling water, refrigeration, freshwater supply, wastewater treatment and disposal, chemicals storage and distribution (e.g., caustic soda, nitrogen, etc.), firewater system, chillers, boilers, steam turbines, and gas turbines. Although all are important for the smooth operation of the processes, they are not equally important from an energy consumption perspective. The seven main ones—steam, fuel gas, fuel oil, electric power, air cooling, water cooling, and refrigeration are energy-intensive and important from the energy conservation point of view. All of these systems could involve significant energy losses.

4.1 Applications in the downstream

To identify the gaps, it is important to consider the energy assessment of the entire unit instead of just equipment by equipment. It starts with drawing the boundary around the processing unit, calculating energy losses, and comparing them by equipment component first, followed by comparing with the unit's "target" energy consumption to identify the efficiency gaps for the whole integrated unit and subsequently for the whole facility.

The CDU + VDU system is generally the largest single energy-consuming unit in a refinery, the majority of them utilize fuel for the fired heaters to provide heat for reboiler in the columns, followed by steam used in the feed preheat train, for side strippers, and for vacuum jets. The goal of the preheat trains is to deliver the feed streams to the fired heater at the highest achievable inlet temperature on a sustained basis. The efficiency of the preheat train (PHT) can be easily measured as the actual heat recovery divided by the target heat recovery.

PHT efficiency is governed by two principal factors: (a) design the heat exchanger network (HEN) so as to follow the temperature profile of the hot and cold streams (using Pinch Analysis), and (b) the rate of asphaltene fouling in critical heat exchangers, generally at crude temperatures above 200°C (390°F). Fouling, especially in the processing of heavier high-boiling crudes, is a major problem. It can shorten the run length between maintenance shutdowns by as much as 1 year, with a significant negative impact on profitability. So, fouling control is a critically important energy efficiency improvement measure. However, it is not easy to manage.

Ebert and Panchal (1995) first identified and modeled the fouling rate as a competition between deposition and removal mechanisms. Deposition rate depends mostly on the surface temperature while removal depends on the mechanical shear rate (flow velocity). Fouling will be negligible if the removal rate surpasses the formation rate. Significant progress has been made in the 25 years since 1995, and there is commercial software available that is able to predict fouling rates throughout the PHT, and therefrom the optimum HX cleaning strategy (**Figure 17**), while the refinery is running, which can extend run times back to near non-fouled conditions [10].

An important perspective is that there is no such thing as waste heat; there is only WASTED Heat, whether deliberate (due to reluctance to invest in recovery) or accidental wasted due to ignorance). Because of historical anti-investment bias in the industry, there will almost always be opportunities to recover such wasted heat

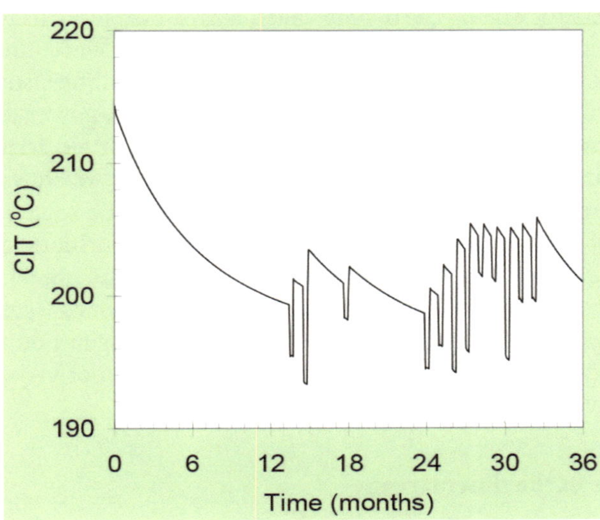

Figure 17.
Furnace coil inlet temp was maintained within target range via optimized HX cleaning strategy.

at a payback that far exceeds what one can hope to get by other legally permitted investments.

One major area of opportunity is where systems are oversized/overdesigned. These are the consequence of engineers habitually putting extra safety margins into the original design. So, by clever modifications, the excess equipment capacity can be converted into better energy efficiency.

Traditional refinery energy efficiency "optimization" initiatives normally focus only on the obvious equipment performance improvements: furnace stack temperature, furnace excess air, adding extra HX surface to the PHT in an effort to raise column feed temperature, steam usage for stripping, and process heating, recovering obviously wasted heat, reducing flow rates recycle streams, cleaning of poorly-performing heat exchanger, power recovery, rotating equipment and motors/machinery efficiency improvements, adding insulation, reducing steam leaks and improving condensate return. While these all are good things to do, but the benefits of optimal system integration are far greater, yet they are seldom addressed.

4.1.1 Optimum energy retrofit of a catalytic reformer

A project was undertaken to convert an existing Platformer unit to a Continuous Catalytic Reformer (CCR). The newly established Energy Systems unit of the company (Saudi Aramco) was tasked to assess the energy efficiency of the licensor's heat recovery design (**Figure 18**).

An optimized HEN (**Figure 19**) for the process was developed by Aramco, with estimated utility cost savings of 16%. A patent was applied for the same and deemed

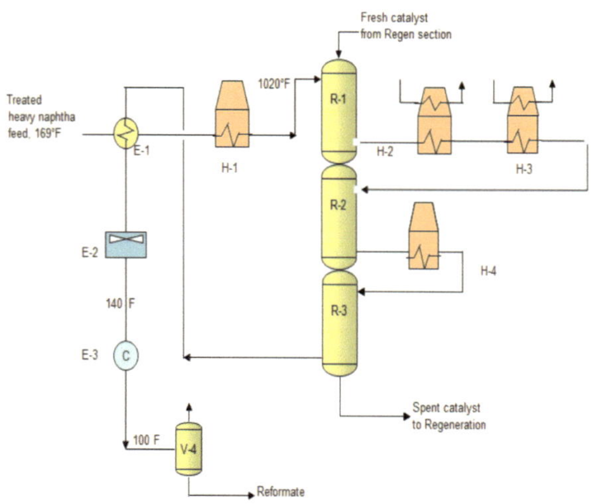

Feed rate =	49.2 MBD			
Operating rate =	8400 hr/yr			
			$/unit	Cost, K$/yr
Fuel consumption	458.9 MMBtu/h		1.25	4818
Steam consumption	0.0 Klb/h		2.67	0
Air cooling duty	87.3 MMBtu/h		0.09	66
Seawater cooling duty	21.5 MMBtu/h		0.27	49
				4933
Exothermic temp rise in R-1 =	221	F		
Furnace Fuel consumption =	223.6	Kbtu/bbl		

Figure 18.
Licensor's proposed design and costs.

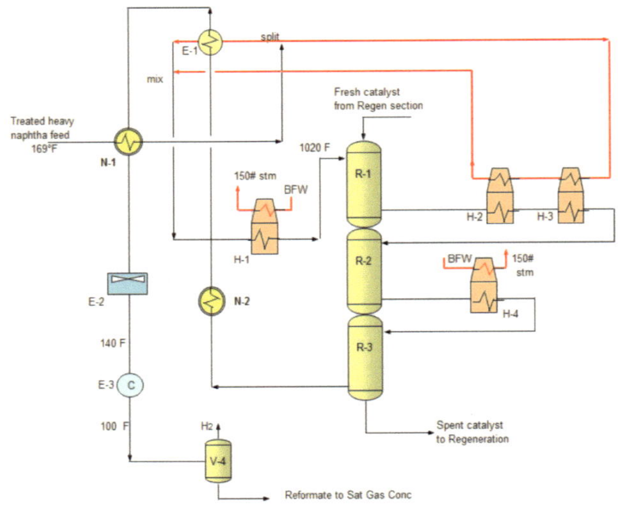

Figure 19.
Optimized HEN design.

to be qualified by the patent attorneys, but later with-drawn by the company for certain commercial and legal considerations, it was never implemented.

4.1.2 Power savings via pump load management at a refinery product loading terminal

The example refinery had multiple storage tanks for its various liquid products which had to be loaded onto ships at the dock about 5 km away. A schematic diagram is shown in **Figure 20**. All pumps were equal-sized with 3550 rpm fixed-speed motors and no flow control valves.

The original design basis assumptions were:

- All ships want to load up at the fastest possible rate, to minimize time at the dock.

- The maximum loading rate is determined by pumping capacity.

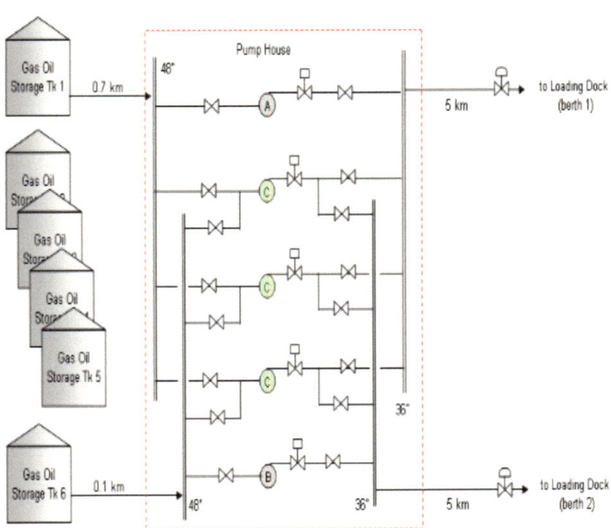

Figure 20.
Flowsheet for product loading station at a refinery.

- The operating policy was to run each pump at its maximum flow capacity (the "run-out" point).

- No flow control valve needed.

During the optimization study, the reality was found to be quite different. Each ship has its own max loading rate (to prevent capsizing), significantly less than the pump run-out flow rate (a case of oversizing). The system curve had low ΔP, which meant that the valve at the loading dock had to be severely throttled. The configuration resulted in a system which has an opportunity for significant power saving. The best solution turned out to be replacement of the existing 3550 rpm motors on 4 of the 5 pumps with dual-speed 3550/1770 rpm motors, and then apply load management techniques to match the pump network flowrate to the required ship loading rate. Total savings for a typical year worked out to be 76 kWh/1000 bbl of product shipped, about 68% savings compared to the previous operating policy. Until the 2-speed motor replacement project was approved, the no-cost savings from load management of the high full speed motors was estimated to be 30 kWh/1000 bbl, or 26% savings compared to the base case.

4.1.3 Improved catalysts for better conversion and lower energy consumption

The use of newer catalysts for enhanced yield and higher selectivity will reduce the recycling and separation of undesired species/streams. This can provide up to 5% efficiency gain (Btu/lb of desired product) compared to older catalysts, in many refinery processes.

4.1.4 Process improvement for maximum product valuation

Aramco uses the PIMS linear programming (LP) software for planning refinery and olefins operations, enabling optimization of feedstock selection, product slate, plant design, and operational execution. It includes assay management, making it easier to add, modify, and re-cut assays, and helps refinery planners develop more accurate plans that deliver greater profitability. The estimated benefits from energy efficiency are in the range of 3–5% of total energy consumption.

4.1.5 New design standards for energy efficiency

It is always more cost-effective in the long term to build energy efficiency and operating reliability into the original design of a new plant than to retrofit a poorly designed facility. Yet there appears an ingrained bias among corporate decision-makers towards the latter approach, following short-term policies such as minimizing construction time at the cost of lower safety, reliability, and efficiency.

Recognizing this reality, Saudi Aramco revised its published corporate standards and procedures in the mid-2000s to include an energy efficiency study between the preliminary process design stage and the detail engineering stage. In order to provide flexibility, however, these standards were made "advisory" not "mandatory", to be followed at the discretion of the Project Manager. It is not hard to imagine the outcome.

To be effective, such standards must necessarily be mandatory, but with the flexibility to waive them by a senior corporate executive, and not at the sole discretion of the Project Manager.

Furthermore, the energy efficiency audit of the preliminary process design must be done by a qualified independent company reporting to the operating company, not to the engineering contractor that produced the preliminary design in the first place. Without such terms, the standards will be toothless and ineffective in achieving the stated corporate goals.

Additionally, the incentive bonuses paid to the company's Project Manager and to the EPC contractor should be based not on beating the construction schedule, but on (a) trouble-free startup and commissioning and (b) meeting long term (e.g., 2–3 years) performance metrics for the facility.

An optimized new design will improve all performance metrics—yields, production rates, energy efficiency, reduced maintenance cost, etc. Even though the initial investment may be slightly higher by about 5%the long-term operating cost savings can be expected to be in the range of 15–20%.

Aramco, like most oil and gas companies worldwide, purchases process technology from international licensors. The scope of supply subject to the performance guarantee typically includes not only the reactors and catalysts but also the separations (distillation columns) and the heat exchange network. There are two issues that must be addressed. The first issue is that most licensors typically guarantee the performance of their units only if their design is followed exactly in all respects, with no deviations unless approved. Now it makes sense to purchase the guarantee for reactors and separations but makes no sense to accept voiding the guarantee if the operator modifies the HEN, or the pumping and compression network, or the utilities mix. Since 2001, the Energy Systems Unit (now a Division with over 50 staff) of Aramco has developed considerable expertise in optimizing HEN design as well as pumping, compression, and process-utility integration. All we need is to license the reactor/separator technology, and absolve the licensor for the performance of the HEN, pumps, and utilities. This is the direction in which the rest of the industry also needs to move to remain competitive.

The second issue is that as a matter of policy Aramco usually prefers to enter into joint venture agreements with international oil, gas, and petrochemical companies for new facilities. Usually, we have to compromise on which company's standards to follow and are often forced to sacrifice energy efficiency in return for some other concession.

5. Terminals and distribution facilities

These facilities are spread across all components and connect all components together by supplying either as feed to one sector which is the finished product of the other or the final product to the end-users. It includes a network of pipelines and storage facilities to deliver the wellhead fluids to upstream facilities, upstream products i.e., oil and gas to downstream facilities i.e., refineries, fractionation plants, and also finished products from downstream facilities for other consumers including end-users. As it includes mainly a network of pipelines, pumping systems, compression systems with various pressure requirements, consequently, energy losses are resulted mainly due to friction, pressure let-downs, types of drivers, etc.

5.1 Applications in terminals and distribution

Swing/transfer pipelines are the principal way of providing flexibility in transferring oil and gas to allow optimum dispatching of production among existing facilities in the most efficient manner. Aramco uses a Dammam-7, a new supercomputer among the top 10 most powerful in the world, to manage its facilities. These swing line connections are utilized to effectively optimize the value chain from GOSP to product shipment, and save energy whenever the opportunity arises.

This requires close coordination between operations foremen for various facilities. For lower production, fewer gas compressors are needed to handle the associated gas. Similarly, for higher loads, the gas compressor and pump operation need attention.

The corporate energy conservation team helped develop such operating practices and incorporated them into the company's Standards and Procedures manuals. In addition, the team undertook a company-wide communications initiative to make sure production engineers, operations, and shift coordinators in the field were made aware of these revised documents.

5.1.1 Turboexpanders in high-pressure pipelines

Turbo-expanders are well-proven technologies for pressure-energy recovery (discussed previously under Upstream Sector applications) and are widely used in the processing industry to produce work or generate electricity. In distribution, high pressure (700–100 psig) pipelines are typically used to transport "wet gas" from upstream gas processing plants to centralized fractionation plants where high-value C2–C5 components are separated from methane by cryogenic distillation. Instead of letting down the inlet pressure across throttling valves, they can be passed through a turboexpander for power recovery. This is essentially free power that easily justifies capital investment. Up to 20–25% additional power can be generated by preheating the inlet gas prior to expansion using low-grade heat that has no other beneficial use (e.g., necessary cooling of HP steam condensate return, if available nearby, to avoid water hammer).

5.1.2 Pump load management

One of the biggest sources of essentially free power savings is to minimize the number of pump trains being operated in parallel. Two important considerations must be kept in mind: net positive suction head (NPSH), which generally becomes an issue when the flow falls below 60% of flow at the best efficiency point, and the flow achievable by using N pumps in parallel will be less than N times the flow through a single pump. The methodology used is to develop the composite performance curves for the pump network, and match them to the system pressure drop curve.

5.1.3 Drag reducing agents to reduce pipeline frictional losses

Pipeline drag reducing additives have proven to be an extremely powerful tool in fluid transportation. High molecular weight polymers are used to reduce the frictional pressure loss ratio in crude oil pipelines, refined fuel, and aqueous pipelines. The drag reducer used in the Mostorod to Tanta crude oil pipeline in Egypt reportedly achieved a 35.4% reduction in pressure drop and a 23.2% flow increase. The experimental application of DRA by Arab Petroleum Pipelines Company (SUMED) in a pipeline from Suez to Alexandria in Egypt achieved a flow increase ranging from 9 to 32% [11].

Aramco is using DRA technology (**Figure 21**) for the Riyadh-Qassim pipeline capacity expansion project and for future projects.

5.1.4 Proper piping networks design

For proper design of piping networks, the key optimization parameters that should be considered include pipe sizing and design code, materials, piping

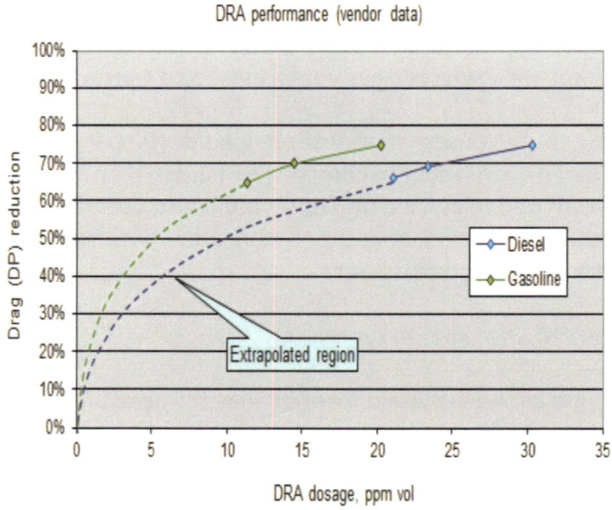

Figure 21.
ΔP reduction vs. DRA dosage.

connection, connection to the header, maintenance, etc. Saudi Aramco has engineering standards and best practices (SAES & SABP) which are continuously updated to improve the energy efficiency of the upcoming as well as existing facilities.

5.1.5 Utilization of high-temperature heat from exhaust of gas turbine-driven rotating machinery

While pumps and compressors are usually driven by electric motors, when the units are large enough and run mostly at a steady state, it can be advantageous to use gas turbine drives, with the hot exhaust (which must be above the process pinch temperature) being used for process thermal heating or even local steam generation linked to the site Utility System. Aramco has successfully implemented several such projects.

5.1.6 Let-down of imported HP fuel gas pressure using turbo-expander

Power recovery turbines (**Figure 22**) are one of the largest sources of clean power generation in many industrial facilities and high-pressure distribution pipelines in particular. The high-pressure fuels gas is let down to lower psig to supply the HP fuel gas header, and a part is further let down to supply the LP fuel gas header. An expander/generator can be installed between the HP sales gas and the HP fuel gas header to recover some power. Turbo-expander generators offer great promise from an energy efficiency perspective in that they have the potential to provide power at very high isentropic efficiencies over 90% and are extremely reliable, with availability factors approaching 99%. Furthermore, the inlet HP gas should be preheated if possible, using low-grade heat (defined as less than 350°F) such as steam condensate return from the refinery). Up to 20–25% additional power generation is possible at zero emission of CO_2.

Although the reported data so far are limited, potential energy efficiency improvement in the T&D sector is estimated to be 10–15% of baseline consumption.

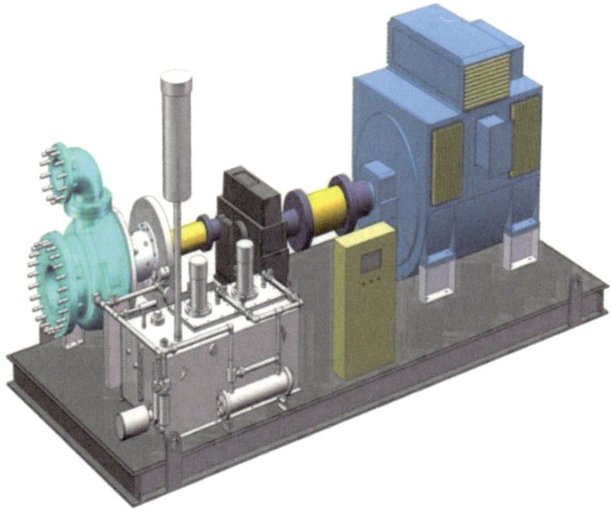

Figure 22.
Model of typical turbo-expander generator equipment.

6. Utilities and power plant

It is the components of the energy supply chain which generally provide energy in usable form to the end-users and also provide energy to the other components of the energy supply chain. Electricity is the most used secondary form of energy and power plants are the part of utilities that covert primary form of energy supplied by HCSC to electricity. Also, it is the component where high energy inefficiencies could have resulted. In 2017, the total power generation capacity of Saudi Arabia was 84 GW (100%), mainly from thermal sources (oil and gas). By 2030, it is projected to be 153–118 GW (65.5%) from fossil fuels, 58.7 GW (32.5%) from wind and solar, and 2.8 GW (2%) from nuclear [12]. This includes dedicated central power plants as well as industrial cogeneration. Other components of the energy supply chain need energy either as power or as heat, the conversion of primary energy to heat and power results in many configurations with differs a lot in terms of its efficiency.

6.1 Simple cycle power plant

The vast majority of fossil fuel power generation in the world is based on a simple Rankine cycle with steam as the working fluid. Water is heated to generate high-pressure steam, driving a steam turbine to either electric energy through a generator. The steam turbine cycle is highly versatile, adaptable to any fuel that provides enough energy by combustion to vaporize water. The main drawback is that the power generation efficiency is limited to 30 to 40%, versus overall energy efficiencies of 55–80% for modern alternative cycles.

6.2 Combined cycle power plant

The combined cycle for power generation offers efficiencies of about 58–60% (electricity only) in theory. It combines the gas-turbine (GT) based Brayton cycle (which has simple cycle efficiency of 18–27%) as the front end followed by a Rankine cycle at the back end. The GT fuel has to necessarily be in the vapor phase—natural gas, synthesis gas from coal, vaporized diesel, etc. The hot GT exhaust containing about 15% O2 issued as combustion air for a downstream heat recovery steam generator (HRSG) which uses the HP steam to make additional power by the Rankine cycle.

6.3 Power generation integrated with brackish water desalination

This concept applies to both simple Rankine cycle and combined cycle power plants. Instead of condensing the steam turbine exhaust against a cooling utility, the steam is exhausted at a back-pressure of about 15–25 psig, which is then used to drive a multiple-effect evaporator (MEE) desalination plant, increasing the overall site energy efficiency to about 75–80%. The MEE makes potable water from seawater or municipal and industrial wastewater treatment facilities effluent, thereby augmenting increasingly scarce freshwater sources. This is the direction KSA has taken.

To facilitate this transition, the formerly independent Ministries of Electricity and Water were merged at the direction of then King Abdullah. To the knowledge of the author, no other country in the world has taken such an enlightened policy initiative to promote national energy efficiency while simultaneously conserving the planet's precious limited freshwater resources. It can serve as a model for the rest of the world.

The energy efficiency initiatives undertaken in the Kingdom's Utility sector are estimated to improve the overall 2012 efficiency of 31.5% is to reach 45% by 2030 [8] as illustrated in **Figures 23** and **24**.

Concurrently, the power, fuel, and water utility supplies must be increased, and distribution infrastructure expanded to accommodate the rapidly growing population of the Kingdom.

6.4. Applications in utilities and power plant

Saudi Aramco plays a critical role in supporting the Kingdom's evolving energy infrastructure. The overall average thermal efficiency of Aramco's interconnected on-site power generation facilities (CHP and CC as well) is 71.9%, while the national power grid efficiency in 2018 was around 38.7%. The savings from the

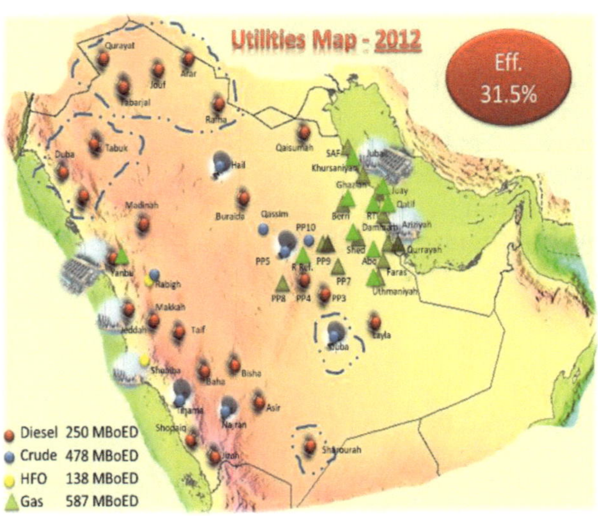

Figure 23.
Utilities map: 2012.

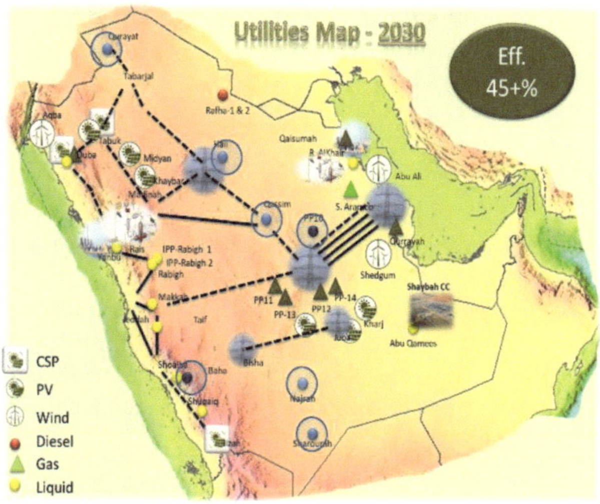

Figure 24.
Utilities map: 2020.

corporate energy program have reduced our CO_2 emissions footprint since 2001 by about 50% per unit of production compared to "Business as Usual" prior to the year 2000. No other company worldwide has been able to match this documented metric to our knowledge. It gives the lie to companies that claim energy efficiency is not economically viable in order to conceal their hidden agendas.

The optimal energy strategy from the Kingdom's perspective would be for Aramco to operate its process-integrated cogeneration plants at full capacity, and to export any surplus power into the national electricity grid, to minimize fuel burning for electricity generation. As a result of the high thermal efficiency in 2018, Saudi Aramco saved approximately 263.8 MMSCFD of natural gas, compared to the average national energy efficiency. To improve power plant efficiencies, some best practices are as follow:

6.4.1 GT compressor inlet air cooling

The lower the inlet air temperature to the compressor of a GT, the higher the capacity as well as the energy efficiency. The air can be cooled in two ways by injection and evaporation of de-ionized water if the air is dry *(as in arid regions), or by indirect cooling using absorption chillers driven by low-grade heat from the process, or by LP steam. The overall efficiency of the GT increases 10–15% from capacity gain. Both systems have their pros and cons.

The direct injection approach, also known as a "fogging system" is simple and cheap, and became very popular worldwide. Aramco jumped on the bandwagon too. However, it was soon noticed that our facilities began to report erosion of the initial efficiency gains over time. A long-term 4-year study of their efficacy conducted with the help of SRI international (San Antonio, USA) found that dissolved salts and fine particulate deposits on the compressor blades were the cause. On balance, the fogging systems did not deliver a net benefit, at least in Aramco GT installations.

Absorption chiller cooling technology offers the advantage of not introducing water which tends to favor adhesion of solids to the compressor blades. However, it is much more capital intensive and incurs higher maintenance costs.

The verdict on these options seems theoretically attractive, is neutral when practical contemplations are considered.

6.4.2 Maximizing cogeneration units operation for power plants

Provided the high incentive for exporting power from a facility, it's always recommended to maximize the operation of cogeneration units. The overall efficiency improvement from a combined cycle with a cogeneration unit vs. a simple Rankine cycle is about 30–35%.

6.4.3 Conversion of simple cycle to combined cycle power plant with cogeneration

Energy efficiency in a simple cycle power plant is in the range of 30–35% and that in a combined cycle power plant is 50–55%, and a combined cycle with cogeneration option can go up to 75–80%. So, it is strongly recommended that the system moves from a simple cycle to a combined cycle with the cogeneration option.

6.4.4 Steam and water conservation by using steam traps and management

Industry data show that the average steam trap has a service life of 4 years. This means that on average 25% of traps will fail every year, usually in the leaking position. Therefore, is imperative that every plant should have a permanent ongoing steam trap monitoring and maintenance program This can give around 3–5% boiler fuel savings. A good steam system management program can improve plant efficiency significantly. It should include self-regulating electrical tracing, condensate recovery, piping insulation, minimizing or eliminating the use of steam-reducing stations and vents.

A simulation model of the Combined Heat and Power (CHP) system for the plant is the most effective tool to check the steam balance and to estimate losses. Although CHP simulation modeling is commercially available, they all lack an important feature which is data reconciliation. Aramco is currently developing in-house software to rectify this deficiency.

6.4.5 Supplementary firing

Supplementary firing may be used in combined cycles (in the HRSG) raising exhaust temperatures from 600°C (GT exhaust) to 800 or even 1000°C. Using supplemental firing will not raise the combined cycle efficiency but is used instead to increase peak power production of the unit, or to enable higher steam production in an emergency. Supplementary firing can raise the temperature of the GT exhaust gas from 800 to 900°C., enabling higher steam generation flows, pressures and temperatures.

6.4.6 Boiler load management

The performance curve for each boiler is basically a relation between fuel consumption and the steam production of the boiler. Boiler efficiencies can vary, even for nominally identical units, by 2–3%. By maximizing base load on the more efficient boilers in a set of parallel units and using the next lower efficiency boiler for swing production, overall steam gen efficiency can be increased 2–3%. Installing economizers in boilers, heat recovery from utility system blowdown, boiler minimum load reduction, minimum steam reserve reduction, and excess O_2 minimization to 2%, are a few other initiatives to improve the boiler system performance. Air preheating is usually not economic, with simple payback typically exceeding 10 years.

6.4.7 Minimize excess low-pressure steam

There should never be excess LP steam at any plant site which includes fuel-fired boilers. It is a symptom of gross steam system mismanagement and can be corrected using a CHP system model to pinpoint the causes of needless waste.

6.4.8 Minimize PRV letdown of high or medium pressure steam

Excess HP or MP steam is a symptom of CHP system mismatch between demand and supply. In such cases, the CHP system operating and control practices should be modified to generate less steam. It can be easily done at a near-zero capital cost, so there is no excuse for plant management to allow such a situation to fester.

6.4.9 Recovery of water from humid boiler or furnace flue gases (condensing economizers)

This approach was touted by academics without industrial experience back in the early 1970s in the wake of the first "energy crisis". Many got US government funding for pilot plants to demonstrate the concept. In practice, operating below the acid dew point accelerates corrosion inside the condenser. Teflon-coated internals was tried, but the coating quickly peeled off, and this approach was largely abandoned.

7. Endusers

End-user is Saudi Arabia mainly belongs to two main energy sectors i.e., transportation and buildings, the transport sector is utilizing the majority of its usage from hydrocarbon (from refineries) while building sector depends mainly on electricity (from power plants).

7.1 Energy efficiency gap in end-user

The majority of demand from buildings is from HVAC, consequently, a closer look into the air conditioning is needed from an energy efficiency perspective. Data is not available on existing plant efficiencies in Saudi Arabia. But worldwide, it is common that without any efficiency improvement measure, air conditioning efficiency usually lies between 1 and 2 kW/ton. With that assumption, and based on the world's best efficiency practices, it is estimated that 40–70% efficiency improvement is possible across the air conditioning. Accordingly, more than 50% in energy savings is possible, by improving the existing old T8 and T12 lighting to LED lightings. A significant portion of the energy is lost in buildings, due to users' habits, and mainly results from a lack of knowledge about energy efficiency and improvement.

Energy efficiency in transport can be enhanced in three ways. First, by reducing transport demand through urban planning and information technology; then by shifting transport of passenger and goods away from more energy-intensive modes, such as road, to fewer intensive means, such as public transportation for passengers and rail and sea for goods; and finally, by improving the fuel economy of the vehicles used, be they road vehicles, aircraft, trains or ships.

8. Building components and their efficiency

According to the International Energy Agency (IEA) (2013), the residential sector accounted for more than a quarter of global electricity consumption in 2011.

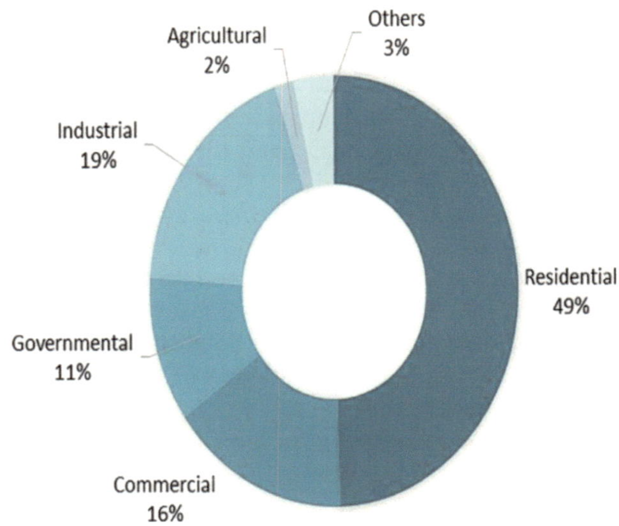

Figure 25.
Existing energy supply from utilities plant to different sector [13].

In Saudi Arabia, this share is almost double at around 50%, largely because of very high average ambient temperatures, and the use of power-hungry air coolers for HVAC systems (**Figure 25**).

All buildings, including residential, in Saudi Arabia, use 40–70% (average 59.4%) of their energy for air conditioning. The combination of heating and ventilation consumption with air-conditioning systems i.e., HVAC, systems consumes about 70% of the total energy (**Figures 26** and **27**) [13]. In addition, lighting electrical energy represents around 15% of the total used electrical energy in buildings all over the Kingdom (19% globally). The remaining 11% of energy is consumed by other appliances/equipment items.

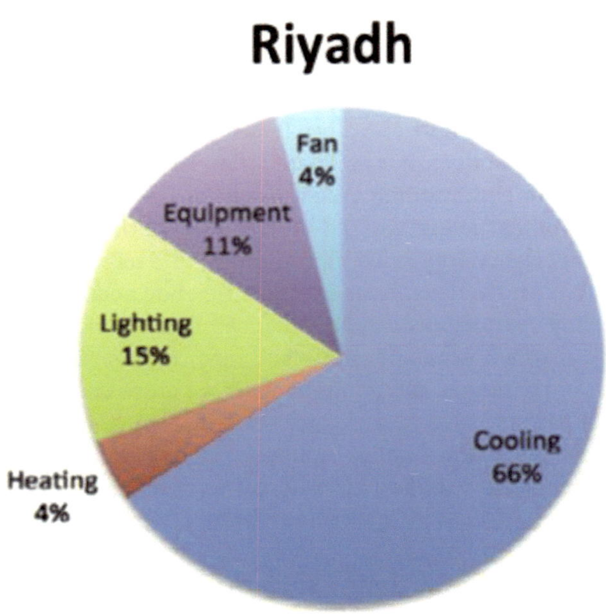

Figure 26.
Energy flow for residential buildings in Riyadh, source [13].

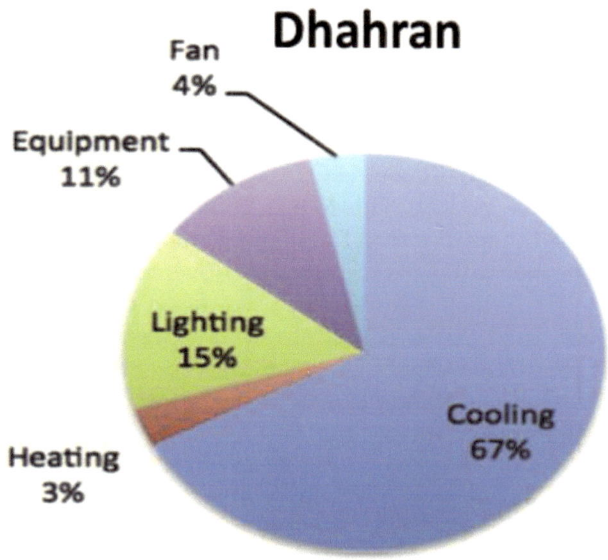

Figure 27.
Energy flow for residential buildings in Dhahran, source [13].

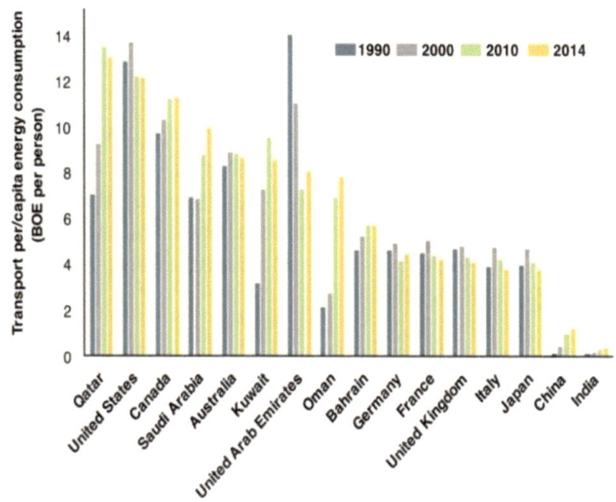

Figure 28.
Transport energy consumption per person [6].

To provide an indicative benchmark of energy efficiency in the sector, transport energy consumption per capita for road transport can be compared across a range of countries (**Figure 28**) [6]. It will be noted that while per capita transport energy consumption is relatively high in Saudi Arabia, it is still lower than in Canada and the U.S. However, the major difference is that per capita transport energy consumption is stable or declining in most OECD countries, whereas in Saudi Arabia it has been growing strongly.

8.1 Applications in end-users

There are several key areas, improvement of which can improve the end-user energy efficiency significantly. A few major energy savings opportunities are listed.

8.1.1 Holistic system design approach

Analysis revealed that changing the renovation design to a whole-systems approach could dramatically improve comfort, quadruple energy efficiency, and cost about the same as normal renovations. Super windows, deep daylighting, and efficient lights, and office equipment could reduce the cooling load (except that caused by the occupants) by 85%. This in turn could make the replacement cooling equipment three-fourths smaller than the original system, four times as efficient, and significantly cheaper to pay for the other improvements. The annual energy bill would then fall by 75%, which will result in a significant reduction in rent per square foot per year, and be at least 10 times the competitive rent difference in the local market. The fourfold energy efficiency improvement would cost essentially the same as the standard renovation that was about to be done anyway, with far better amenity, esthetics, and rentability.

8.1.2 Optimizing the design by providing wider piping and laying it first

Research shows that significant energy losses occur due to piping. In the study, it is found that the fatter pipes and cleaner layout yielded not only 92% lower pumping energy at a lower total capital cost, but also simpler and faster construction, less use of floor space, more reliable operation, easier maintenance, and better performance. As an added bonus, easier thermal insulation of the pipe.

8.1.3 Measurement of existing air conditioning plant efficiency

It is important that measurement of air conditioning plant efficiency including small residential split systems to large chiller plants should be conducted throughout the Saudi Arabian buildings and plants. Understanding the existing systems efficiency condition will help the building owner/operator to identify the savings potential and will raise interest to act. Proper understanding and retrofitting will make the system 15–25% more efficient.

8.1.4 Building data analytics

Buildings are loaded with a very large volume of data and information. Overall, there is a need for an automated solution to process this information. Building analytics can help turn data into action and savings. The overall building analytics services include: data monitoring and analyzing, automation and controlling, sustainability reporting and strategy development, metering and billing, asset management, measurement and verification, comfort management, carbon and waste reporting, energy and water reporting, demand analysis, etc. Utilizing the Advance Building Data Analytics can bring 5–10% savings from the existing operation.

8.1.5 Implementation of low hanging fruits (e.g., lighting replacement/refurbishment)

Buildings can easily save 5–10% of the consumption by installing easy solutions. There is technology, such as LED lighting that can be easily installed and bring more than 50% savings from its existing baseline. Simple set point adjustment will also help savings from air conditioning and other system operation savings.

In addition to that, as electric motors account for more than 30% of all electricity consumed in commercial buildings, more emphasis should be placed on motor efficiency. Finally, the system loss in the pumping and ducting, especially in all the throttling mechanisms, should be carefully reviewed. All of these are possible if the user and the building owner/operator work sincerely towards a sustainable building, living and working environment.

8.1.6 Transportation sector improvements

In Saudi Arabia, all three transportation options are being pursued, with high-density urban areas being planned, a metro system being constructed in Riyadh, and nationwide railway infrastructure also under development. In terms of energy efficiency regulations, SEEC issued its fuel economy standard for passenger cars in November 2014, using the U.S. NHTSA CAFE standards as a reference, and has established a fuel economy testing lab for monitoring and evaluating actual performance against this standard. SEEC passenger car regulations also include a requirement for vehicle fuel efficiency labels and a low rolling resistance tire standard. In addition, fuel economy standards for heavy-duty vehicles and a "cash for clunkers" vehicle scrapping scheme are under development. For example, Egypt's Greater Cairo Region Old Vehicles Scrapping and Recycling Program is one example that involved replacing old taxis with compressed natural gas (CNG) fueled vehicles.

8.2 Energy efficiency impact in end-user

The overall end-user loss in the supply cycle is about 10%. Based on current new technology and building upgrade along with transportation sector reforms, more than 20–30% of those losses can be recovered.

8.3 Impact of energy efficiency on primary energy and fuel

Energy supply chain efficiency is a global concern. It is a common misconception that the demand side primarily impacts the energy supply chain. With a detailed breakdown, as summarized in the chapter, it can be clearly observed that each component of the energy supply chain has a significant impact on the overall energy supply chain efficiency. It is also noted that the most critical two components that have significant losses and are impacted highly are the power plant and the end-user equipment/system. We propose to have an improvement in all components. A little improvement in the end-user system or in the utility power plant creates compounded impact throughout the overall chain.

9. Sensitivity analysis

Figure 29 shows the impact of end-user energy savings on overall primary fuel in 2030. With each portion of the improvement in end-user savings, the overall primary fuel will be saved significantly.

10. Overall summary and recommendation

Following section will summarize the overall losses, challenges, and recommendations for improvement.

10.1 Summary

The possible energy performance improvement percentage from existing conditions for different sectors is represented in **Figure 30** (GOSPs are shown as a sector

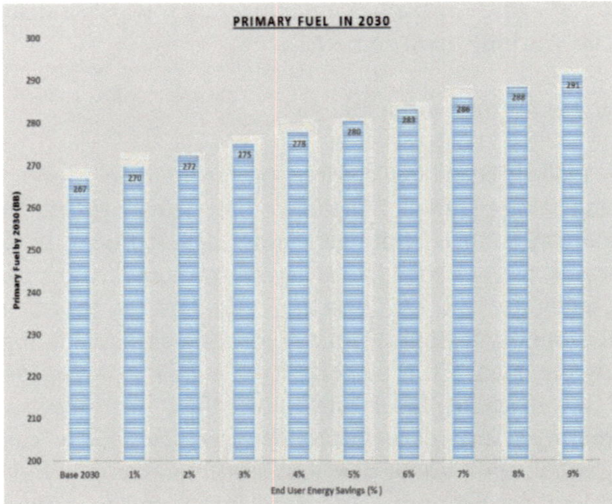

Figure 29.
Impact of 2030 primary energy fuel for different end-user efficiency improvement (Keeping all other parts of the network constant).

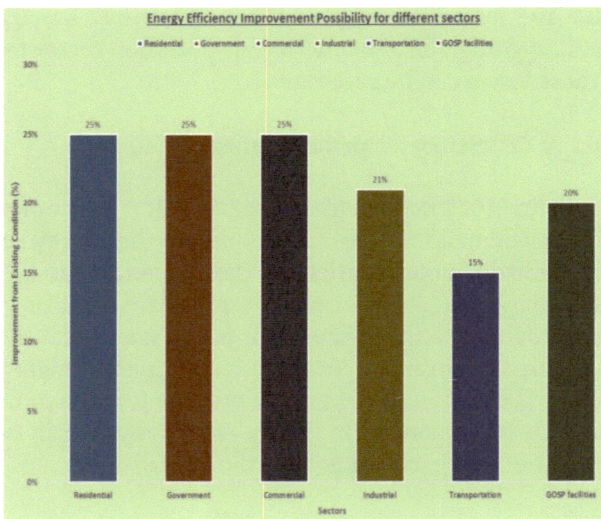

Figure 30.
Energy performance improvement (%) possibility.

as it is part of the industrial sector but due to its impact it is mentioned separately). Even buildings are segregated in commercial, residential, and government due to different ways to tackle energy efficiencies in these sub-sectors. On average, approximately 20–25% energy conservation and efficiency improvement are possible from all sectors. These figures are conservative and can go much higher in all three major sectors (i.e., industrial, transportation, and buildings).

There is a significant technological improvement for building sectors and there is huge hidden wastage due to consumer attitude. The sensitivity analysis represented earlier only considers up to 10% improvement from its existing wastage; however, the efficiency improvement can go up to 50% if holistic approaches are applied.

The analysis suggests that there is significant scope for energy efficiency improvement in the transportation sector. Improved urban planning, public transport, and the implementation of energy efficiency vehicle regulations will play a key role. As awareness is a key factor to this sector, policies like incentives for carpooling, privilege parking for hybrid cars, have the popular social appeal that could have a significant impact in improving energy efficiency in this sector.

The industrial sector is the most energy-intensive sector in Saudi Arabia and it is principally made up of HCSC and the utility sector. The discussed improvements are predominantly covering the entire industrial sector and the energy enhancements are enormous creating opportunities for reducing GHG emissions along with monetary benefits. Improvements in the energy supply chain will have greater implications, the counts of barrels saved in the processing will be added to the export/usage, preserving natural resources and environment for the future generations. It is likely that by following the approach, manufacturing industry of the future will become energy efficient and fully embrace the best practices to optimize resources utilization while consuming less energy. All case studies are summarized to enrich and illustrate the subject and demonstrate the methodology appropriately to put them on track to achieve international climate and energy goals. The adoption of cogeneration technology, pressure energy recovery, heat integration, load management, etc. helps in promoting energy efficiency, lowering the energy intensity of operating plants, adding value to hydrocarbon resources, and protecting the environment. The analysis clearly indicates that there is an enormous potential for improvements in the complete energy supply chain and well-capitalized by the Kingdom's policy makers and can be leveraged to other countries.

10.2 Challenges

Improvement in energy efficiency has multi-dimensional challenges and needs some attention from the policy makers and strategists. Even though the technological knowledge gap could be minimized, yet there is a huge shortage of knowledgeable professionals in this specialty, (i.e., energy efficiency/energy conservation), worldwide. The unavailability of commercially available tools to conduct energy efficiency analysis is also an area of attention for the policy makers, to develop and promote the expertise and tools by providing investment for the required infrastructure. Technological development without strategies to realize the life-cycle basis where energy efficiency is an element to explore sustainability concepts along with the progression. End-users lack understanding/awareness about the significance of energy wastage, which in turn impacts global issues like global warming, making it very difficult to tackle. Upgrading from old inefficient systems and technologies to newer efficient ones demands a clear perspective on investment and a relative understanding of the impact of alternative investment.

10.3 Conclusions and recommendations

The outcome is that energy efficiency is essential for further and vigorous growth in all three major sectors. It is worth stating that using energy efficiency as a framework for environmental strategy and energy goals enabled the Kingdom to implement a plan that is feasible and robust. It provided the Kingdom a competitive advantage by enabling a strong and energy-efficient economy and reaping the benefits of selling avoided energy consumption. Other policymakers can leverage these findings and adapt a similar approach to alleviate global warming through proven cost-effective energy efficiency measures. Although, all sectors have similar savings potential percentage but the industrial sector is of greater importance as it is very energy-intensive and of larger magnitude. Kingdom's industrial sector is dominated by hydrocarbon supply chain, owned by Saudi Aramco, which is leading the effort and playing a key role in positioning the strategy. Saudi Aramco has a vital role in the formulation of the strategy and also provided its support in deployment and further improvement.

Some of the key takeaways from the adapted approach of improving a livable environment through the implementation of energy efficiency measures are summarized below as quick wins;

- Implementation of more cogeneration systems for all simple cycle and combined cycle power plants.
- Emphasize process improvement through heat/process integration and evaluate each improvement based on life-cycle cost.
- Emphasize on innovation in utilities and other end-user technologies via providing incentives to the energy management professionals/institutions.
- Demonstrate key energy-efficient retrofit projects to share the experience and confidence to execute projects and adopt new technologies.
- Emphasize on energy efficiency monitoring by displaying a dashboard for end-user engagement.
- Promote green energy production utilizing turbo-expander, hydraulic turbine, etc.
- Improvement of refrigeration cycles (including air-conditioning) efficiency across all sectors (residential, commercial, industrial, utility plants, etc.).
- Implementation of building codes and advanced data analytics for end-user applications.

Nomenclature

BB	billion barrels
BBL	barrels
BCM	billion cubic meter
CHP	combined heat and power
CC	combined cycle
DRA	drag reducing agents
GOSP	gas oil separation process
GW	giga watts
HCSC	hydro carbon supply chain
HVAC	heating, ventilation and air-conditioning
IEA	International Energy Agency
kW/ton	the ratio of input electrical power to out thermal power. Ton of refrigeration is approximately 3.5 times higher than kW of thermal power.
LED	light emitting diode
MBD	millions barrel per day
OECD	Organization for Economic Co-operation and Development (36 countries)
ORC	organic Rankine cycle

SABP	Saudi Aramco best practice
SAES	Saudi Aramco engineering standards
SEEC	Saudi Energy Efficiency Centre
VFD	variable frequency drive

Author details

Ali Al-Qahtani, Zeeshan Farooq and Sami Almutairi*
Saudi Aramco, Dhahran, Saudi Arabia

*Address all correspondence to: sami.almutairi.1@aramco.com

References

[1] World Total Energy Supply by Source. Paris: International Energy Agency (IEA); 1971–2019. Available from: https://www.iea.org/data-and-statistics/charts/world-total-energy-supply-by-source-1971-2019

[2] Sankey Diagrams for Saudi Arabia. International Energy Agency (IEA); 2018. Available from: https://www.iea.org/sankey/#?c=Saudi Arabia&s=Balance

[3] Fawkes S, Galeotti M, Howarth N, Krarti M, Lanza A, Padmanabhan P. Growth through Diversification and Energy Efficiency: Energy Productivity in Saudi Arabia. KAPSARC; 2017. Available from: https://www.kapsarc.org/research/publications/growth-through-diversification-and-energy-efficiency-energy-productivity-in-saudi-arabia/

[4] Al Ghamdi A. Saudi Arabia Energy Report. 2020. Available from: https://www.kapsarc.org/research/publications/saudi-arabia-energy-report/#

[5] World Bank National Accounts Data. The World Bank, 1960–2020. World Bank Group. Available from: https://data.worldbank.org/

[6] Global Energy Statistical Yearbook 2021. Enerdata; 2021. Available from: https://yearbook.enerdata.net/total-energy/world-energy-intensity-gdp-data.html

[7] Ahmad T, Zhang D. A critical review of comparative global historical energy consumption and future demand: The story told so far. Energy Reports. 2020; **6**:1973-1991

[8] Attia AM, Ghaithan AM, Duffuaa SO. A multi-objective optimization model for tactical planning of upstream oil & gas supply chains. Computers & Chemical Engineering. 2019;**128**

[9] Abdou HAM. Case study in economic impact of using drag reducing agent on cost of pumping crude oil in western desert, Egypt. In: Pipeline Technology Conference. 2014

[10] Krarti M, Dubey K, Howarth N. Evaluation of building energy efficiency investment options for the Kingdom of Saudi Arabia. Energy. 2017;**134**:595-610

[11] Edward I, Jimmy K, Graham P, Pugh S. Improved energy efficiency in CDUs through fouling control. In: AIChE Annual Meeting. 2010

[12] Amran YHA, Amran YHM, Alyousef R, Alabduljabbar H. Renewable and sustainable energy production in Saudi Arabia according to Saudi Vision 2030: Current status and future prospects. Journal of Cleaner Production. 2019;**247**:119602

[13] Soummane S, Ghersi F. Projecting Saudi Sectoral Electricity Demand in 2030 Using a Computable General Equilibrium Model. KAPSARC; 2021. Available from: https://www.kapsarc.org/research/publications/projecting-saudi-sectoral-electricity-demand-in-2030-using-a-computable-general-equilibrium-model/#

Chapter 10

The Use of Computational Fluid Dynamics in the Analysis of Gas-Liquid-Liquid Reactors

Godfrey Kabungo Gakingo and Tobias Muller Louw

Abstract

Gas–liquid–liquid reactors are typically found in bioprocess setups such as those used in alkane biocatalysis and biological gas stripping. The departure of such reactors from traditional gas–liquid setups is by the introduction of a secondary (dispersed) liquid phase. The introduction of the latter results in complicated hydrodynamics as observed through measurements of velocity fields, turbulence levels and mixing times. Similarly, changes in mass transfer occur as observed through measurements of gas hold up, bubble diameters and the volumetric mass transfer coefficients. The design and analysis of such reactors thus requires the adoption of an approach that can comprehensively account for the various observed changes. This chapter proposes Computational Fluid Dynamics as an approach fit for this purpose. Key considerations, successes and challenges of this approach are highlighted and discussed based on a review of previously published case studies.

Keywords: Gas–liquid–liquid reactors, stirred tanks, hydrodynamics, mass transfer, Computational Fluid Dynamics, predictive modelling

1. Introduction

Multiphase systems comprising of more than two phases are a common occurrence in the fields of chemical and bioprocess engineering. Such multiphase systems may be comprised of a gas phase, a liquid phase and a solid phase as is the case in froth flotation processes in the minerals sector [1, 2]. Alternatively, such multiphase systems may be comprised of a gas phase and two immiscible liquid phases as is often found in biological gas stripping [3] or biocatalysis [4]. Irrespective of the application field, a common expectation among such multiphase systems is that they are characterised by more complex hydrodynamics than two phase systems which are reasonably well understood [5, 6]. Similarly, mixing and mass transfer effects are expected to be more complex in such systems.

Given the above considerations, the design and analysis of multiphase systems requires the use of comprehensive frameworks that are capable of taking into account the various mechanisms of action that are at play. Computational Fluid Dynamics (CFD) has been proposed as one such framework since it is able to describe the hydrodynamics of multiphase systems based on fundamental equations of flows [7]. Furthermore, coupling of CFD simulations to sub-models of mass transfer, mixing or flotation can enable the description of these effects at finer resolutions than can be obtained based on empirical modelling. Thus, significant effort has been recently directed towards the development and application of CFD techniques to simulate multiphase systems comprising of two phase reactors [8–10] as well as those with more than two phases [2, 11–13].

This chapter builds upon recent work by presenting a discussion on the use of CFD in the design and analysis of gas–liquid–liquid reactors within the context of mass transfer. Key considerations informing the modelling approach have been discussed with their implementation illustrated by the review of recently modelled case studies [12, 13]. Furthermore, the successes and challenges attending the CFD-based modelling of gas–liquid–liquid reactors have been highlighted and on the basis of these, recommendations have been given on areas requiring further investigation. The chapter thus addresses itself to graduate students, academics and industrial practitioners interested in a comprehensive modelling framework for the design and analysis of gas–liquid–liquid reactors.

2. Considerations in gas–liquid–liquid systems

Gas–liquid–liquid reactors are a common occurrence in the bioprocess field (see **Table 1**). Such reactors tend to be comprised of one continuous liquid phase (primary liquid phase) and two dispersed phases (gas phase and secondary liquid phase) as illustrated in **Figure 1**. The departure of such reactors from traditional gas–liquid reactors is through the introduction of a secondary liquid phase within

Primary liquid phase	Secondary liquid phase (volume fraction)	Gas phase	Reactor type (objective)	Reference
Water	n-C_{11-18} alkane cut (0–100%)	Pure oxygen	STR[1] (oxygen transfer)	[14]
	Oleic acid (0–100%)	Pure argon	STR (argon mass transfer)	
Water	n-C_{12-13} alkane cut (0–20%)	Air	STR (oxygen transfer)	[15]
Water	n-Dodecane (0–100%)	Pure oxygen	STR (oxygen transfer)	[16]
	n-Heptane (0–100%)			
	n-Hexadecane (0–100%)			
Water	Silicone oil (0–10%)	Air	BCR[2] (oxygen transfer)	[17]
		Air dosed with styrene	BCR (biological gas stripping[3])	
Water	Anisole (0–10%)	Air	BCR (oxygen transfer)	[18]
	2-ethyl-1-hexanol (0–10%)			
	Decyl alcohol (0–10%)			
	Toluene (0–10%)			
	n-Heptane (0–10%)			
	n-Decane (0–10%)			
	Dodecane (0–10%)			

[1]*STR – stirred tank reactor.*
[2]*BCR – bubble column reactor.*
[3]*For more on biological gas stripping, see [3].*

Table 1.
Selection of experimental work on gas–liquid–liquid reactors.

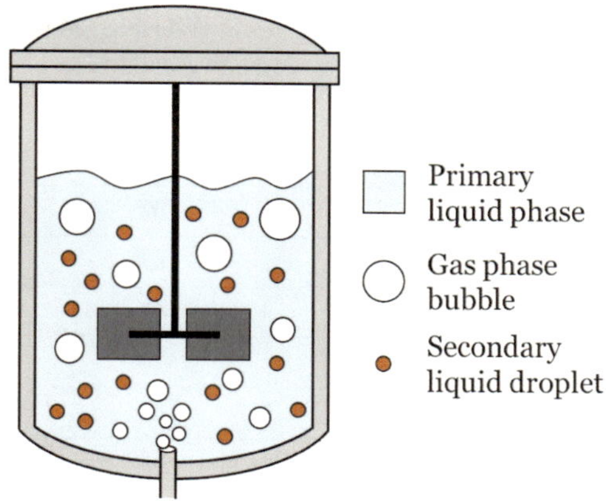

Figure 1.
Illustration of gas–liquid–liquid stirred tank reactor.

the system. The introduction of the latter modifies the system in a manner that impacts mass transfer and this can be seen by changes in the gas hold up, the bubble diameters and the overall volumetric mass transfer coefficients [4]. Underpinning these observable changes, however, are modifications to the fluid properties, mass transfer properties and pathways as well as the reactor hydrodynamics. A consideration of these modifications is necessary for appropriate modelling and as such a brief review on the same is presented herein. In-depth reviews are available elsewhere [4, 19, 20].

2.1 Changes in fluid properties

The introduction of a secondary liquid phase into a traditional gas–liquid reactor can result in a change of fluid properties. For example, the surface tension between the primary liquid phase and the gas phase can change depending on the degree of solubility of the secondary liquid phase in the primary liquid phase. To characterise this, surface tension values based on mutually saturated liquids have been reported in literature [16, 18]. Such values have generally been obtained by mixing the primary liquid phase with the additional liquid phases for long periods followed by a separation of the phases and surface tension measurements using a tensiometer [18]. **Table 2** illustrates selected results from literature.

The results in **Table 2** generally point to a decrease in the saturated surface tension (σ_{sat}) with addition of the secondary liquid phase. Furthermore, this change in surface tension has been observed to be greater when the secondary liquid phase is more soluble in the primary liquid phase [18]. As a decrease in the surface tension results in smaller gas bubbles, an enhancement in mass transfer can be expected. However, this is not always the case. For example, Kundu et al. [18] observed a negative impact on mass transfer upon the addition of toluene, anisole, decyl alcohol and 2-ethyl-1-hexanol despite a decrease in σ_{sat} (refer to **Table 2**). This points to the presence of additional factors that need to be taken into account for a proper description of mass transfer in such systems.

Two additional points need to be highlighted with regard to surface tension. First, discrepancies in the reported values of σ_{sat} exist as seen in **Table 2**. For example, Ngo & Schumpe [16] measured an insignificant change in σ_{sat} (71.8 mN/m) upon the addition of n-Heptane whereas Kundu et al. [18] measured a significant change (σ_{sat} = 65 mN/m). Such discrepancies point to a need for additional experimental measurements of σ_{sat}.

Primary liquid phase	Secondary liquid phase	Saturated surface tension, σ_{sat} (mN/m)	Reference
Water	—	72.8	[18]
	Dodecane	71	
	n-Decane	67	
	n-Heptane	65	
	Anisole	65	
	Toulene	44	
	2-ethyl-1-hexanol	43	
	Decyl-alcohol	38	
Water	—	72	[16]
	n-Heptane	71.8	
	n-Dodecane	68.2	
	n-Hexadecane	71.2	
Deionised water	—	71.69 ±0.14	Author's laboratory
	n-C_{14}-C_{20} alkane cut	60.85 ± 0.50	

Table 2.
Saturated surface tension for various liquid–liquid combinations.

A second point to note is that researchers have also attempted to report on dynamic values of the surface tension [21]. This was achieved by preparation of a "stable" liquid–liquid dispersion followed by measurement of the surface tension [21]. The values obtained were in the range of 17–26 mN/m for an n-C_{10}-C_{13} alkane cut mixed with water [21]. These values were lower than those reported in **Table 2** and tended towards the surface tension values of pure alkanes (23.9 mN/m for n-Decane, 25.41 mN/m for n-Dodecane [18]). Consequently, it may be suggested that a degree of separation of the liquid–liquid dispersion occurred during surface tension measurements despite the best efforts of the researchers. In this case, the settled-out and less dense alkane phase would form an intervening layer between the gas phase and the dispersion.

Besides the above changes to the surface tension, the presence of a secondary liquid phase can lead to changes in *effective* fluid properties such as the effective density, the effective viscosity as well as the effective solubility. The pre-qualifying term, "*effective*", is used in this case since such properties are defined based on a view of the liquid–liquid dispersion as a single pseudo-homogenous liquid. This view permits for a simplification of the 3-phase gas–liquid–liquid reactor to an effective 2-phase reactor. Consequently, correlations derived for traditional 2-phase reactors, such as Eq. (1) with an empirical basis and Eq. (2) with a theoretical basis [5], can be used as a starting point for the design of 3-phase reactors. In these equations, K_La represents the overall volumetric mass transfer coefficients in a 2-phase reactor whereas x_g and d_g respectively represent the gas hold up and the bubble diameters (see Eq. (3) and (4)). Other variables are as defined in the Nomenclature.

$$K_La = \Gamma \cdot \left(\frac{P_g}{V}\right)^x \nu_s^y \tag{1}$$

$$K_La = \Lambda \cdot \sqrt{D_c}\left(\frac{P_g/V}{\mu_c}\right)^{0.25} \cdot \left(\frac{6x_g}{d_g}\right) \tag{2}$$

$$d_g = 0.7\left(\frac{\sigma_{gc}^{0.6}}{\rho_c{}^{0.2}\left(\frac{P_g}{V}\right)^{0.4}}\right)\left(\frac{\mu_c}{\mu_g}\right)^{0.1} \tag{3}$$

$$\frac{x_g}{1-x_g} = 0.819\frac{\nu_s{}^{0.67}N^{0.4}T^{0.267}}{g^{0.33}}\left(\frac{\rho_c}{\sigma}\right)^{0.2}\left(\frac{\rho_c}{\rho_c-\rho_g}\right)\left(\frac{\rho_c}{\rho_g}\right)^{-0.067} \tag{4}$$

Effective fluid properties can be specified on the basis of simple mixing rules such as Eq. (5) for the effective density (ρ_{eff}). Such equations consider the contributions of the individual liquid phases based on their volumetric proportions (x_i) but neglect non-ideal effects that may arise from molecular interactions, commonly found in homogeneous mixtures, which may lead to excess volumes. This notwithstanding, experimental evidence suggests that such equations are sufficient for the effective density [22] and the effective solubility (see next section). This may be due to the poor mutual solubility of liquid phases typically tested for gas–liquid–liquid reactor applications: the properties of the individual liquid phases remain unchanged by the presence of a second, immiscible phase and the bulk properties can be estimated by simple volume averaging.

$$\rho_{\text{eff}} = \sum_{i=1}^{n} x_i\rho_i \tag{5}$$

With regard to effective viscosity (μ_{eff}), models of a more complicated form than that given in Eq. (5) are required. This is due to the need to account for the perturbation of fluid flow in the presence of droplets of the secondary liquid phase. Models varying in complexity have been proposed. For example, the Taylor model (see Eq. (6)) has been proposed for dilute dispersions of spherical droplets [23]. Additionally, models such as those given in Eq. (7) and (8) have been proposed for non-dilute dispersions where the hydrodynamic interactions among droplets need to be accounted for [23, 24]. Besides these, however, models have also been proposed to account for non-zero shear rates in the fluid which introduce effects such as droplet deformation [23, 25].

$$\frac{\mu_{eff}}{\mu_c} = 1 + 2.5x_d\left[\frac{0.4+\mu_r}{1+\mu_r}\right]; x_d \to 0 \tag{6}$$

$$\frac{\mu_{eff}}{\mu_c}\left[\frac{2\frac{\mu_{eff}}{\mu_c}+5\mu_r}{2+5\mu_r}\right]^{1.5} = \exp\left(\frac{2.5x_d}{1-x_d/x_m}\right) \tag{7}$$

$$\frac{\mu_{eff}}{\mu_c}\left[\frac{2\frac{\mu_{eff}}{\mu_c}+5\mu_r}{2+5\mu_r}\right]^{1.5} = \left(1-\frac{x_d}{x_m}\right)^{-2.5x_m} \tag{8}$$

In the equations above, μ_r refers to the ratio of viscosity of the secondary liquid phase (μ_d) to that of the primary liquid phase (μ_c). Variables x_d and x_m, on the other hand, refer to the volume fraction of the secondary liquid phase and the maximum packing limit of its droplets respectively.

2.2 Changes in mass transfer properties and pathways

Properties of interest affecting mass transfer include the solubility and diffusivity of a species within the liquid–liquid dispersion. Taking the liquid–liquid dispersion as a pseudo-homogenous liquid, effective properties have been defined. For example, it has been observed that the effective solubility of a species (C^*_{eff}) can be specified according to the volumetric proportions of the liquids involved, once again recognising that the presence of immiscible phases has a negligible effect on the solubilities associated with individual phases [22, 26]. This is illustrated in Eq. (9) below, with C^*_i representing the solubility in liquid phase i and n representing the total number of liquid phases.

$$C^*_{eff} = \sum_{i=1}^{n} x_i C^*_i \tag{9}$$

Given the application fields of gas–liquid–liquid reactors (such as biological gas stripping), the effective solubility is usually higher than the solubility of a species in the primary liquid phase ($C^*_{eff} > C^*_c$). This implies that a longer duration is required to saturate the liquid–liquid dispersion with a given species in comparison to the time required to saturate the pure primary liquid phase [27, 28]. Consequently, for a fixed mass transfer rate, lower values of the overall volumetric mass transfer coefficients can be obtained upon the addition of a secondary liquid phase [28, 29].

With regard to the molecular diffusion of a species, the effective diffusivity (D_{eff}) has been defined as illustrated in Eq. (10) for dilute liquid–liquid dispersions [30]. This equation, being of a general nature, has also been used to describe the effective diffusivity of solid–liquid suspensions [31]. The variable D_r in this equation represents the diffusivity ratio whereas D_d and D_c represent the diffusivity of the species in the secondary and primary liquid phases respectively.

$$D_{eff} = D_c \left[\frac{D_r(1 + 2x_d) + 2(1 - x_d)}{D_r(1 - x_d) + (2 + x_d)} \right]; D_r = \frac{D_d}{D_c} \tag{10}$$

It should be noted that the effective properties defined by Eqs. (9) and (10) can be, in a general sense, regarded as bulk properties. Appropriate as it may be to define them, a full description of mass transfer in a liquid–liquid dispersion also requires an examination of changes introduced by the secondary liquid phase at the mass-transfer interface. Tied to this are the questions whether the secondary liquid phase will be present at the interface and whether it is involved in active uptake of dissolving species at the interface. Indeed, the concept of a pseudo-homogenous liquid implies that the secondary liquid phase will be present, not only in the bulk of the primary liquid, but also at the interface. Furthermore, homogeneity implies that a similar distribution of the secondary liquid phase is found at the interface as is found in the bulk of the primary liquid phase. However, researchers have considered different possible configurations of the interface as illustrated in **Figure 2**. In this way, the requirement for homogeneity has been relaxed at the interface while being maintained in the bulk.

Assuming active uptake by the secondary liquid phase, different possible pathways of mass transfer have arisen. These have included, for example, parallel mass transfer and series mass transfer with or without the shuttling of the droplets of the secondary liquid phase [14, 30, 32, 33]. These various pathways have been associated with an enhancement in mass transfer besides that occurring due to changes discussed earlier

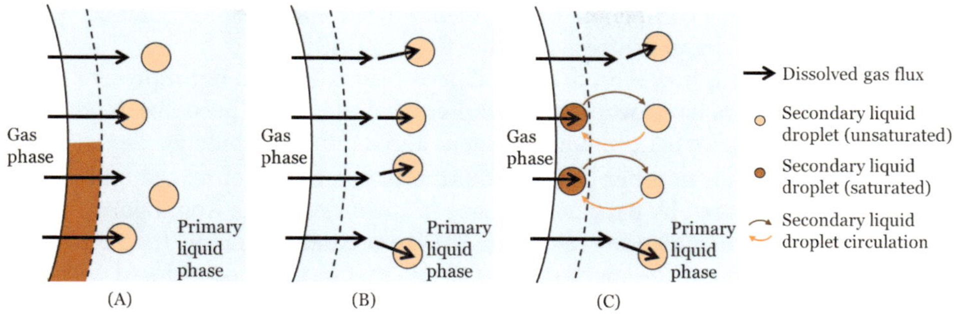

Figure 2.
Mass transfer interface illustrating possible configurations – (A) parallel mass transfer, (B) series mass transfer without shuttling and (C) series mass transfer with shuttling.

(refer to Eq. (9) and (10)). A review of these pathways as well as their associated enhancement factors can be found in the work by Dumont & Delmas [19].

2.3 Changes in reactor hydrodynamics

The hydrodynamics of a reactor are taken to refer to the mean velocity field and the turbulence field within a reactor. These affect mass transfer in various ways. For example, the transport of gas bubbles within a reactor (and hence overall gas hold up) is dependent on the magnitude and orientation of the mean velocity field. Additionally, mass transfer at the interface between the gas bubbles and the liquid phase will depend on the prevailing local turbulence. Consequently, a change in the reactor hydrodynamics will lead to a change in the mass transfer.

The addition of a secondary liquid phase has been observed to change both the mean velocity field and the turbulence levels in stirred tank reactors [34–36]. Direct measurements of the velocity through techniques such as Particle Image Velocimetry [34, 36] and Laser Doppler Anemometry [35] have revealed that a secondary liquid phase can dampen the mean velocities [34, 35] while either increasing or decreasing the turbulence levels [34–36]. Further evidence in literature for a change in the hydrodynamics has been largely indirect – inferred from examining the change in, for example, mixing time upon addition of a secondary liquid phase [37, 38].

Two schools of thought have been postulated to explain the interaction of the secondary liquid phase with the hydrodynamics of a reactor. The first has been focused on the change in effective fluid properties. In this line of thinking, it has been suggested that a decrease in the effective density should lead to an increase in the velocities [34]. On the other hand, it has been suggested that an increase in the effective viscosity should have a dampening effect on both the mean velocities and the turbulence levels [34, 35].

The second school of thought, on the other hand, has been focused on the augmentation or dampening of turbulence by the droplets of the secondary liquid phase [34, 35]. Various mechanisms have been suggested in this regard although these are still the subject of active research [39–42]. For example, whether a particle (solid, liquid or gas) augments or dampens turbulence has been traditionally associated with the size of the particle (d_p) in relation to the integral (or large) scales of turbulence (l) [43]. Large particles $(d_p/l > 0.1)$ with characteristically large particle's Reynolds numbers $(Re_p > 400)$ have generally been associated with turbulence augmentation through mechanisms such as vortex shedding in the wakes behind the particles [43, 44]. On the other hand, small particles $(d_p/l < 0.1)$ have generally been associated with a dampening of turbulent kinetic energy (TKE) that occurs as the particles are accelerated/dragged by the flow [39, 43].

It should be noted, however, that the above observations do not represent a fixed rule; exceptions have been observed. For example, it has been observed that particles of a size $d_p/l < 0.1$ can both augment and dampen turbulence depending on the flow's Reynolds number [45]. Furthermore, it has been observed that turbulence can be augmented by particles of a size in the order of the Kolmogorov (smallest) scales of turbulence [46–50]. In the latter cases, however, the non-uniform modification of the spectrum of TKE by particles was considered with the augmentation of TKE observed to occur at the small scales of turbulence [46–50]. Mechanisms that were proposed for the turbulence augmentation included flow forcing due to the inertia [50] or buoyancy [49] of small particles that were well correlated with the fluctuating fluid flow.

3. Computational fluid dynamics for gas–liquid–liquid reactors

As illustrated in the previous sections, the introduction of a secondary liquid phase into a traditional gas–liquid reactor can result in a variety of changes. Consequently, the modelling of mass transfer in gas–liquid–liquid reactors requires a comprehensive modelling framework that can account for the different changes. An empirical approach, such as illustrated in Eq. (11) [51], may not suffice as the numerous effects introduced by the secondary liquid phase are reduced into a single term with an adjustable exponent requiring optimisation (compare to Eq. (1)). This is not an easily generalizable approach.

$$K_L a' = \Gamma \cdot \left(\frac{P_g}{V}\right)^x v_s^y (1 - x_d)^z \tag{11}$$

Computational Fluid Dynamics (CFD), on the other hand, offers a fundamental framework that can be built upon to incorporate as much level of detail (or physics) as necessary/desired. This is the case since CFD offers an approximate/numerical solution to the fundamental equations of flow governing a system/reactor [7, 52]. Consequently, the hydrodynamics of a reactor are resolved in space and time and such hydrodynamic data can be coupled to fundamental models of mass transfer so as to predict parameters of interest such as the overall volumetric mass transfer coefficient, $K_L a'$.

As CFD-based approaches are inherently computationally intensive, a major constraint to such approaches is the computational resources available [7, 52]. Thus, a compromise has to be made between the level of physics to be captured and computational resources available [7]. For example, prior consideration must be given as to the level of resolution to be employed for the flow field. Similarly, an appropriate continuum description of the phases involved must be chosen *a priori*. Such compromises notwithstanding, a higher level of detail, accuracy and generality is still maintained with a CFD-based approach in comparison to empirical approaches. This will be illustrated in Section 4.

3.1 Modelling frameworks for hydrodynamics

As noted in the introduction to CFD above, several prior considerations have to be made with regard to the modelling approach. These considerations tend to give rise to different modelling frameworks or techniques. For example, as relates to the resolution of flow fields, it is possible to simulate the usually turbulent flow in a reactor at all length- and time-scales using Direct Numerical Simulations but the associated computational expense is prohibitive [53]. Therefore, filtering or averaging of the flow field is usually done [53]. Filtering techniques such as those used in Large Eddy Simulations offer an enhanced resolution of the flow field as compared to averaging techniques [53]. However, they are still considered computationally expensive and their use has been largely limited to single-phase reactors [53]. Averaging techniques such as those used to generate the Reynolds-Averaged Navier–Stokes (RANS) equations, on the other hand, lead to tractable simulations [53] and are a practical choice for the modelling of multiphase reactors.

With regard to the description of the phases, an Eulerian framework is typically used for the continuous phase (primary liquid phase) and it involves describing the flow based on a fixed observer position [54]. For the dispersed phases (gas and secondary liquid phase), on the other hand, either an Eulerian framework or a Lagrangian framework can be employed [54]. The Lagrangian framework involves the tracking of individual particles of the dispersed phase within the flow field of the continuous phase [54]. The particles either follow the flow field without interaction (one-way coupling) or interact with it thus modifying it (two-way coupling) [55].

The Lagrangian framework provides a greater degree of detail and as can be expected, it is costly to implement [54]. Consequently, an Eulerian framework for both the continuous and dispersed phases tends to provide a practical choice for modelling. An Eulerian description of both phases assumes that the phases involved can be treated as a continuum [56]. In this case, a phase indicator function is introduced into the governing equations of flow to account for the possible realisation of a phase i at a given position and a given time [56]. Averaging of the governing equations after decomposing the instantaneous flow field into its mean and fluctuating components results in Eq. (12) and (13) [57].

$$\frac{D}{\mathrm{Dt}}(\alpha_i\rho_i) = 0 \tag{12}$$

$$\frac{D}{Dt}\left(\alpha_i\rho_i\overline{V}_i\right) = -\alpha_i\nabla p + \nabla\cdot\bar{\bar{\tau}}_i + \alpha_i\rho_i\overline{g} + \sum_{j=1}^{n}\overline{R}_{ji} \tag{13}$$

In Eqs. (12) and (13), subscripts i and j represent individual phases whereas α represents the respective phase volume fraction. This volume fraction is based on the total volume of all phases as opposed to the total liquid volume (see variable x in earlier equations). The respective phases can be either the gas phase, the primary liquid phase or the secondary liquid phase. Alternatively, if the liquid–liquid dispersion is treated as a pseudo-homogenous liquid, then the subscripts would refer to either the gas phase or the pseudo-homogenous liquid.

Other variables in Eqs. (12) and (13) such as ρ and $\overline{V}$ represent the density and the mean velocity of the respective phases whereas p represents the shared pressure field. Additionally, $\overline{g}$ represents the gravitational acceleration while $\overline{R}$ represents the interphase momentum exchange terms. These terms include the lift force, the drag force, the turbulent dispersion force and the added mass force among others [56].

There are two key points to note regarding the interphase momentum exchange terms. First, their significance varies with the set up being considered. For example, the drag force has been observed to be the most significant interphase exchange term in the bulk of a stirred tank reactor [58, 59]. On the other hand, terms such as the lift force have been observed to significantly affect the flow in a bubble column reactor [60]. Consequently, modelling can be simplified by only accounting for significant terms.

The second point to note touches on the various models that have been proposed to specify the interphase momentum exchange terms. Such models chiefly consider 2-phase interactions, that is, gas–liquid or liquid–liquid interactions [61–63]. Though such models have been improved upon to consider effects such as particle-particle interaction at high volume fractions of a single dispersed phase [61–63], there is still a need for appropriate models that specify the interphase exchange terms when more than two phases are present. To this end, work such as that by Baltussen et al. [64, 65] investigating the effective gas–liquid drag in the presence of an additional solid phase may provide direction.

Finally, $\bar{\bar{\tau}}$ in Eq. (13) represents the stress tensor accounting for both viscous and turbulent (Reynolds') stresses. The specification of the stress tensor is non-trivial due to the Reynolds' stresses that need to be solved directly or modelled. A direct solution of the Reynolds' stresses provides a greater amount of detail and is able to resolve complex features of the turbulence field such as anisotropy [53, 54]. However, the computational costs associated with this approach as well as reported solution difficulties favour the specification of the stress tensor using alternative simplified approaches [53, 54]. Modelling of the stress tensor using the Boussinesq hypothesis is one such alternative approach [53, 54]. In the latter, the Reynolds' stresses are related to the gradients of the mean velocity with the eddy/turbulent viscosity (μ_t) arising as a proportionality constant (see Eq. (14)) [57].

$$\bar{\bar{\tau}}_i = \alpha_i(\mu_i + \mu_{t,i})\left(\nabla\overline{V_i} + \nabla\overline{V_i}^T\right) - \frac{2}{3}\alpha_i\left((\mu_i + \mu_{t,i})\nabla\cdot\overline{V_i} + \rho_i k_i\right)\bar{\bar{I}} \tag{14}$$

$$\mu_{t,i} = C_\mu \rho_i \frac{k_i^2}{\epsilon_i} \tag{15}$$

The eddy viscosity can be modelled in various ways such as that illustrated in Eq. (15) [53, 57]. In this case, it is related to the turbulent kinetic energy (k) and its dissipation rate (ϵ), both of which need to be solved for. Various models have been proposed for these latter parameters (k and ϵ) such as the dispersed k–ϵ turbulence model, the mixture k–ϵ turbulence model and the per-phase k–ϵ turbulence model [57]. These models represent an extension of the single phase k–ϵ turbulence model to multiphase situations and their applicability depends on the expected/prevailing type of flow [57]. For example, the mixture k–ϵ model is recommended for stratified flow whereas the dispersed k–ϵ model is recommended where there is one clear continuous phase and the other phases are dispersed within it [57].

It should be noted that irrespective of the choice of turbulence model, one key point to consider is the interaction of the dispersed phases with the prevailing turbulence. As noted in Section 2.3, the dispersed phase can modify the prevailing turbulence and this needs to be captured. To this end, various models have been proposed in literature and these vary in complexity depending on the level of detail captured. Various authors have recently examined the sufficiency of these models and their work is recommended [66–71].

3.2 Modelling frameworks for mass transfer

Similar to the modelling of hydrodynamics, the modelling of mass transfer involves the selection of appropriate frameworks prior to the actual simulation. The choice of a particular framework involves a compromise between the level of detail captured and computational cost involved. Furthermore, the choice of a modelling framework for mass transfer tends to be influenced by choices made during the modelling of the hydrodynamics. For example, the use of an Eulerian description of the phases together with averaged equations of flow offers a practical choice for the modelling of a reactor hydrodynamics. However, this approach involves a loss of specifics on the mass transfer interface thus necessitating the prescription of models to approximate the expected mass transfer behaviour. On the other hand, a Lagrangian tracking of particles coupled with interface tracking algorithms better resolves the mass transfer interface but is unfeasible for reactors with a large number of dispersed particles (droplets or bubbles). Consequently, the discussion below focuses on the modelling of mass transfer within the context of an Eulerian description of the phases and the modelling of hydrodynamics using averaged equations of flow.

Mass transfer in a gas–liquid–liquid reactor can be solved for by tracking the concentration of a species in each phase within the reactor. This results in three equations as given in Eqs. (16)–(18). In these equations, C represents the concentration of the species in the respective phase (subscripts g, c, d) whereas $\bar{J}$ and S represent the diffusive flux and the interphase mass transfer source terms respectively. Closure models are required for the interphase mass transfer source terms depending on the expected mass transfer behaviour.

$$\frac{D}{Dt}\left(\alpha_g C_g\right) = -\nabla \cdot \left(\alpha_g \bar{J}_g\right) - S_{gc} - S_{gd} \tag{16}$$

$$\frac{D}{Dt}\left(\alpha_c C_c\right) = -\nabla \cdot \left(\alpha_c \bar{J}_c\right) + S_{gc} - S_{cd} \tag{17}$$

$$\frac{D}{Dt}\left(\alpha_d C_d\right) = -\nabla \cdot \left(\alpha_d \bar{J}_d\right) + S_{gd} + S_{cd} \tag{18}$$

In the most general case, mass transfer can be assumed to occur between the gas phase and both the primary and the secondary liquid phases ($S_{gc} \neq 0, S_{gd} \neq 0$). This would correspond to the case of parallel mass transfer as illustrated in **Figure 2(A)**. Evidence for this mass transfer pathway has been provided based on observations of the formation by oil films on gas bubbles during "static" experiments [72]. The question arises, however, as to whether sufficient time for film formation occurs in a dynamic/agitated reactor [72, 73].

Series mass transfer is often taken to be the more probable mass transfer pathway in an agitated reactor (cases (B) or (C) in **Figure 2** with $S_{gd} = 0$) [34]. In addition, the formation of small droplets of the secondary liquid phase with a large interfacial area implies that mass transfer between the respective liquid phases is usually faster than that occurring between the gas phase and the primary liquid phase [72]. Thus, based on the timescale of mass transfer between the gas phase and the primary liquid phase, it may be assumed that equilibrium conditions exist between the respective liquid phases. Furthermore, one need only track the concentration of the species in two phases – the gas phase (Eq. (16)) and the primary liquid phase (Eq. (17) with $S_{cd} = 0$). With these simplifying assumptions, the only unknown left is the interphase source term between the gas phase and the primary liquid phase (S_{gc}). Gakingo et al. [13, 28] proposed definitions for S_{gc} that account for possible mass transfer pathways at the gas–liquid interface plus an apparent decrease in the overall volumetric mass transfer coefficients that occurs due to a larger total solubility of the species in the liquid–liquid dispersion (refer to Section 2.2). This illustrated in Eqs. (19)–(21) below [13, 28].

$$S_{gc} = K_L a' \left(C_c^* - C_c\right) \tag{19}$$

$$K_La' \approx E' \cdot \Lambda \cdot \sqrt{D_c}\left(\frac{\rho_c \epsilon_c}{\mu_c}\right)^{0.25} \cdot \left(\frac{6\alpha_g}{d_g}\right) \tag{20}$$

$$E' = \frac{1}{(1 - \alpha_d + \alpha_d m)^{\varphi}} \tag{21}$$

In the equations above, K_La' represents the overall volumetric mass transfer coefficient in the presence of the secondary liquid phase. The latter has been expanded in Eq. (20) based on an eddy cell model [74] with Λ and E' representing a constant and the enhancement factor respectively. The enhancement factor is given in Eq. (21) where m represents the solubility ratio ($m = C_d^* / C_c^*$) and the exponent φ varies between 0.5 and 1. A value of $\varphi = 1$ represents the series mass transfer pathway without shuttling whereas a value of $\varphi = 0.5$ represents the series mass transfer pathway with shuttling [13, 28]. Other variables are as previously defined.

4. Case studies

Few studies have reported on the CFD-based modelling of gas–liquid–liquid reactors. Two such studies are reviewed in this section with the one having focussed on the modelling of mixing time [12] while the other focussed on the modelling of mass transfer [13].

In the case studies of interest, stirred tank reactors were considered as illustrated in **Table 3**. In addition, the modelling in both cases was done based on an Eulerian description of the phases involved (refer to Eqs. (12) and (13)). There were certain similarities in the modelling and these included, for example, a consideration of drag force as the only interphase momentum exchange term. The latter was specified according to Eq. (22) [13] with different models for the drag coefficient (C_D) employed as illustrated in **Table 4**. Constant sizes were also assumed for both the gas phase bubbles and the droplets of the secondary liquid phase and these were predicted by Eqs. (3) and (23) respectively [12, 13]. Finally, turbulence was modelled based on the dispersed k–ϵ model (see Eqs. (24) and (25)) though the use of the Reynolds stress model was additionally considered in one study [12].

Experimental details		Cheng et al. [12]	Gakingo et al. [13]
Tank dimensions	Tank diameter	0.24 m	0.177 m
	Liquid height	0.24 m	0.22 m
	Impeller diameter	0.08 m	0.059 m
	Number of impellers	1	2
Aeration system	Sparger diameter	0.08 m	0.05 m
	Sparger holes (diameter)	16 (0.0015 m)	7 (0.001 m)
Operating conditions	Agitation rates (rpm)	170, 220, 300, 400, 425, 500	600, 800
	Aeration rates (L/min)	0.16, 0.24, 0.32, 0.4, 0.48, 0.64	4
	Primary liquid phase	Water	Water
	Secondary liquid phase (volume fractions)	Kerosene (0%, 3%, 5%, 7%, 10%, 12%, 15%, 20%)	n-C_{10}-C_{13} alkane cut (0%, 2.5%, 5%, 10%, 20%)

Table 3.
Details of experimental setups used in the studies of Cheng et al. [12] and Gakingo et al. [13].

Parameter	Cheng et al. [12]	Gakingo et al. [13]
Gas phase–primary liquid phase drag coefficient (with correction for turbulence effects)	$C_{D,\infty} = max\left\{\begin{array}{l}\frac{2.667Eo}{Eo+4}, \\ \frac{24}{Re_p}\left(1+0.15Re_p^{0.687}\right)\end{array}\right\}$	$C_{D,\infty}^{sph} = \frac{24}{Re_p}\left(1+0.1Re_p^{0.75}\right);$ $C_{D,\infty}^{ell} = \frac{2}{3}d_g\left(\frac{g\Delta\rho}{\sigma_{gc}}\right)^{0.5}\left(\frac{1+17.67\left(1-\alpha_g\right)^{1.29}}{18.67\left(1-\alpha_g\right)^{1.5}}\right)^2;$ $C_{D,\infty}^{cap} = \frac{8}{3}\left(1-\alpha_g\right)^2;$
	$\frac{C_D}{C_{D,\infty}} - 1 = 6.5\times 10^{-6}\left(\frac{d_g}{\lambda}\right)^3$	$\frac{C_D}{C_{D,\infty}} = \Theta \times \left[1-1.4St^{0.7}\exp\left(-0.6\,St\right)\right]^{-2}$
Secondary liquid phase–primary liquid phase drag coefficient	$C_D = C_{D,\infty} = \left(1+\alpha_d^{1/3}\right)\left(0.63+\frac{4.8}{\sqrt{Re_p}}\right)^2$	$C_D = C_{D,\infty} = \begin{cases}\frac{24}{Re_p}\left(1+0.15Re_p^{0.687}\right); & Re_p \le 1000 \\ 0.44; & Re_p > 1000\end{cases}$
Turbulence modulation by dispersed phases	$\Pi_{k_c} = 0.02\times\left\|\overline{R}_{ci}\right\|\sqrt{\left(\overline{V}_i-\overline{V}_c\right)^2}$	$\Pi_{k_c} = \frac{\rho_i}{\rho_i+C_{AM}\rho_c}\times\frac{K_{ic}}{\alpha_c\rho_c}\left[\zeta_{ic}-2k_c+\overline{V}_{dr}\bullet\left(\overline{V}_i-\overline{V}_c\right)\right]$
	$\Pi_{\epsilon_c} = C_{1\epsilon}\frac{\epsilon_c}{k_c}\Pi_{k_c}$	$\Pi_{\epsilon_c} = C_{3\epsilon}\frac{\epsilon_c}{k_c}\Pi_{k_c}$

Table 4.
Sub-models used in the modelling work of Cheng et al. [12] and Gakingo et al. [13].

$$\overline{R}_{ji} = -\overline{R}_{ij} = K_{ji}\left(\overline{V}_j - \overline{V}_i\right); K_{ji} = \frac{3}{4}\alpha_i\alpha_j\rho_i\frac{C_D}{d_j}\left|\overline{V}_j - \overline{V}_i\right| \tag{22}$$

$$\frac{d_d}{T} = \Omega\cdot\left(1+\gamma\cdot\alpha_{o,ave}\right)\left(\frac{\sigma_{cd}}{\rho_c N^2 T^3}\right)^{0.6} \tag{23}$$

$$\frac{D}{Dt}\left(\alpha_c\rho_c k_c\right) = \nabla\cdot\left(\alpha_c\left(\mu_c+\frac{\mu_{t,c}}{\sigma_k}\right)\nabla k_c\right) + \alpha_c G_{k,c} - \alpha_c\rho_c\epsilon_c + \alpha_c\rho_c\Pi_{k_c} \tag{24}$$

$$\frac{D}{Dt}\left(\alpha_c\rho_c\epsilon_c\right) = \nabla\cdot\left(\alpha_c\left(\mu_c+\frac{\mu_{t,c}}{\sigma_\epsilon}\right)\nabla\epsilon_c\right) + \alpha_c\frac{\epsilon_c}{k_c}\left(C_{1\epsilon}G_{k,c} - C_{2\epsilon}\rho_c\epsilon_c\right) + \alpha_C\rho_c\Pi_{\epsilon_c} \tag{25}$$

The above similarities, notwithstanding, there were some notable differences between the two studies. For example, Gakingo et al. [13] tested two approaches in the modelling of the liquid–liquid dispersion. The first approach involved the treatment of the dispersion as a pseudo-homogenous liquid (herein referred to as the P-HOM approach) and the mixture properties were obtained from Eq. (5) for effective density or through experimental measurements for effective viscosity. The second approach, also used by Cheng et al. [12], involved a consideration of the heterogeneous nature of the dispersion and a modelling of each individual liquid phase (herein referred to as the HET approach). In this second approach, there was a need to specify models for turbulence modulation by the dispersed phases (Π_{k_c} in Eq. (24)) and different models were employed as illustrated in **Table 4**. Last but not least, mass transfer was considered in only one study [13] where the authors employed the previously reviewed frameworks, that is, Eqs. (16) and (17) with $S_{gd} = 0$, $S_{cd} = 0$ and Eqs. (19)–(21). Further specifics on the implementation of the modelling approaches (meshing, boundary conditions and solver settings) can be found in the respective studies [12, 13].

Several key observations can be made from the reported modelling works [12, 13]. First, it was observed that the hydrodynamics of a gas–liquid–liquid stirred tank reactor were better captured based on the HET modelling approach as opposed to the P-HOM modelling approach [13]. The failure by the latter approach was attributed to the fact that the hydrodynamics in a turbulent stirred tank reactor are dominated by turbulence as opposed to the effective (mixture) viscosity. To this end, it was reported that the ratio of turbulence viscosity to mixture viscosity was in the order of $O(10 - 100)$ based on the P-HOM modelling approach [13]. Consequently, a minimal change in the hydrodynamics was observed despite an almost 2-fold increase in the values of mixture viscosity [13]. This is illustrated through the minimal change in gas hold up trends shown in **Figure 3**.

As pertains to the HET modelling approach, the better performance at capturing the hydrodynamics and hence gas hold up trends (see **Figure 3**) was attributed to the capture of turbulence modulation by droplets of the secondary liquid phase [13]. In particular, it was reported that there was an increase in the turbulence viscosity which served to dampen the mean velocity field (see **Figure 4**) thus reducing the effective drag and dispersion experienced by the gas bubbles [13]. Cheng et al. [12], on the other hand, did not make an explicit mention of changes in the turbulence viscosity. Rather, they hypothesised that their mixing time was reduced at low volume fractions of the secondary liquid phase due to an increase in turbulence caused by the droplets of the latter [12]. Furthermore, they hypothesised that an increase in mixing time at high volume fractions of the secondary liquid phase was due to a dampening of the turbulence arising from an increase in the effective viscosity [12]. This reference to notable changes due to the effective viscosity stands in contrast with the findings on the P-HOM modelling approach [13].

A recent study on a gas–liquid–solid reactor has reported similar gas hold up trends as those reported for the HET approach [75]. Furthermore, the modelling of such a reactor based on non-Newtonian models for the liquid–solid slurry has been observed to result in the formation of regions of localised fluid motion near the impeller (caverns) with stagnant fluid elsewhere in the tank [75, 76]. This corresponds to the reports of a dampened mean velocity field according to the HET approach as seen in **Figure 4**. Observations such as these suggest that the dampening of the mean velocity field may be a common feature in three-phase reactors. However, questions arise as to the appropriate manner of describing such effects. For example, the question may be posed as to whether liquid–liquid dispersions in stirred tanks should be treated based on non-Newtonian models with a yield stress and shear-thinning behaviour rather than equations of the type given in Section 2.1 (Eqs. (6)–(8)). It is to be noted that Eqs. (6)–(8) predict the effective viscosity of the dispersion assuming no non-Newtonian behaviour. Furthermore, the

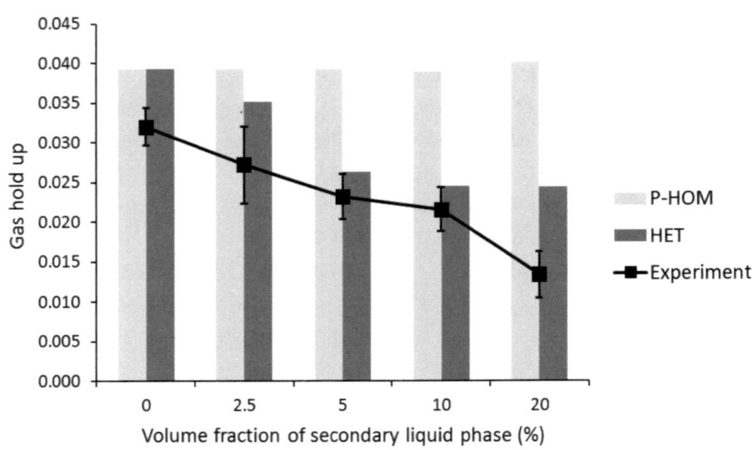

Figure 3.
Gas hold up versus volume fraction of secondary liquid phase at 600 rpm. P-HOM refers to the pseudo-homogenous modelling approach whereas HET refers to the individual treatment of the liquid phases. Data from Gakingo et al. [13].

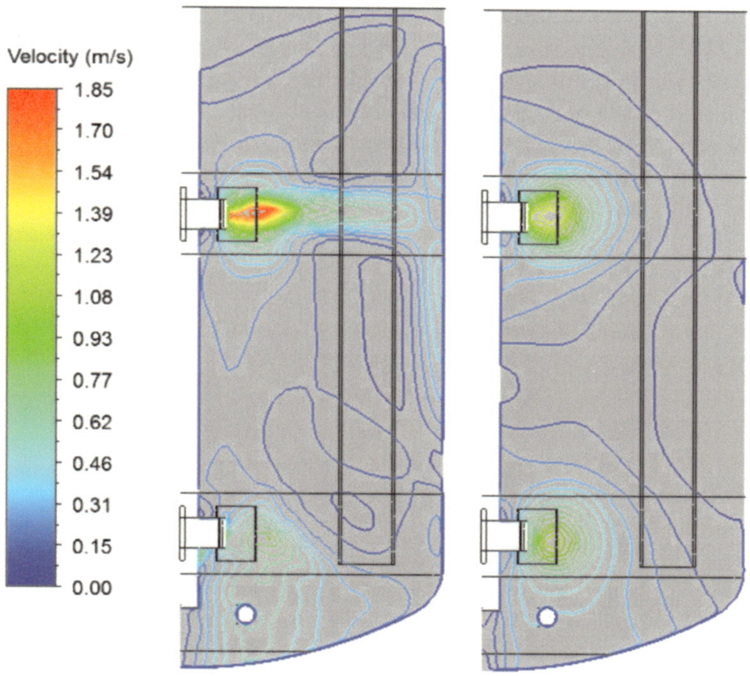

Figure 4.
Velocity contours based on the HET modelling approach on a mid-baffle plane at 600 rpm. Left – 0% alkane volume fraction (2-phase). Right –10% alkane volume fraction.

experimental effective viscosity values that were used for the P-HOM modelling approach were in close agreement with those predicted by these equations [13].

Further observations made in the case studies of interest touched on the trends of pumping capacity and power drawn by the impellers. It was observed that the pumping capacity of the impellers decreased at high volume fractions of the secondary liquid phase [12, 13]. This was attributed to a decrease in the mean velocity field as dissipated by an increasing turbulence viscosity [13]. As for the power drawn, both an increase and a decrease in the power was reported. The decrease in power, as noted by Cheng et al. [12], was attributed to a decrease in the effective density upon addition of the secondary liquid phase. On the other hand, Gakingo et al. [13] attributed a noted increase in power drawn to increased energy dissipation by the droplets of the secondary liquid phase. This latter effect was not reported by Cheng et al. [12] and this could have been due to the use of different models to capture turbulence modulation by the dispersed phases (see **Table 4**). Cheng et al. [12] used a model by Katoaoka et al. [77] which considers only the mean velocity differences $\left(\overline{V_i} - \overline{V_c}\right)$ in computing the work done by interfacial drag. Gakingo et al. [13], on the other hand, employed the model by Simonin et al. [57, 78–80] which considers not only the effects arising from mean velocity differences but also the effects arising from the fluctuating velocities of the dispersed phase particles. These latter effects were expected to become more significant than the former as the size of the dispersed phase particles decreased ($\overline{V_i} - \overline{V_c} \rightarrow 0$ as $d_p \rightarrow 0$) [68]. Consequently, it may be suggested that the use of the model by Simonin et al. [57, 78–80] was more appropriate as it is more comprehensive.

With regard to mass transfer, it was illustrated that potentially accurate predictions can be obtained by using the HET modelling approach for hydrodynamics and Eqs. (16), (17) and (19)–(21) for mass transfer with $S_{gd} = S_{cd} = 0$ (see **Figure 5**) [13]. These equations correspond to a case where the concentrations of a species in the two liquid phases are at equilibrium and no direct uptake by the secondary liquid phase occurs at the gas–liquid interface. Though this scenario implied the occurrence of either series mass transfer or series mass transfer with shuttling, minimal differences were reported based on a consideration of these two

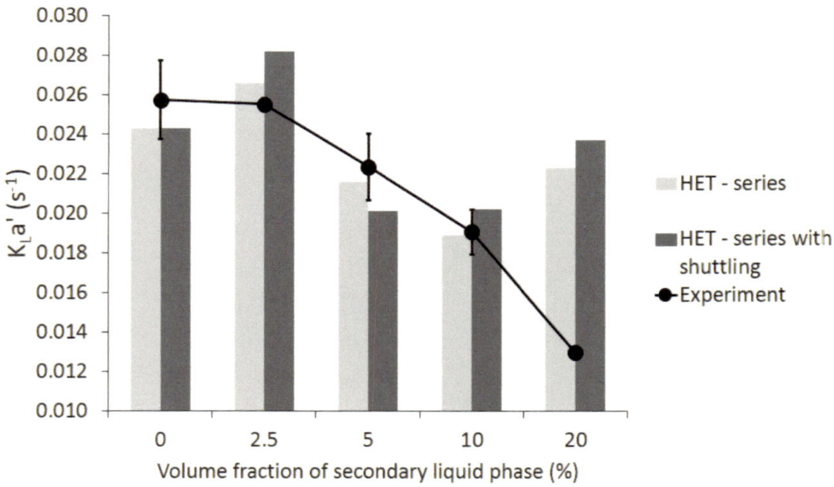

Figure 5.
Comparison of predicted versus experimental overall volumetric mass transfer coefficients at 600 rpm and varying alkane concentrations. Experimental measurements have been obtained by the pressure step method. Data from Gakingo et al. [13].

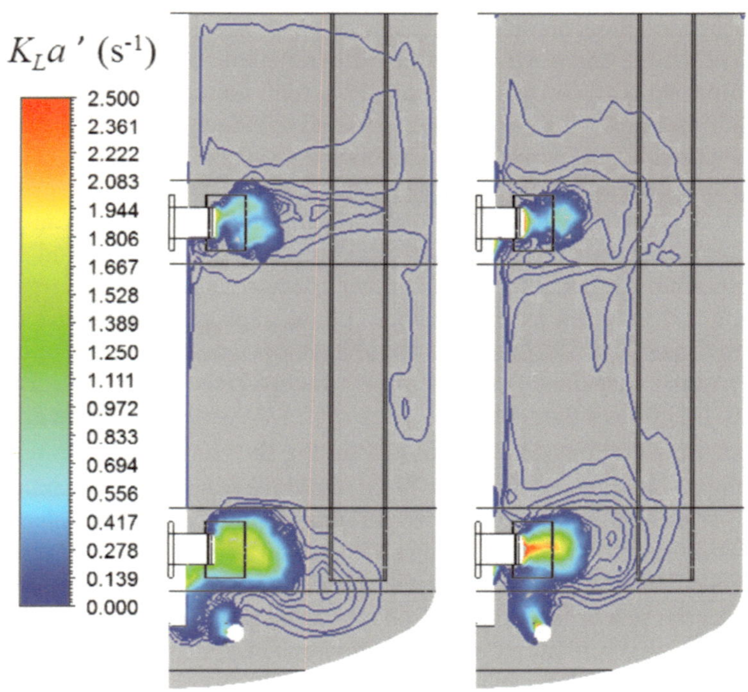

Figure 6.
Contours of local K_La' values predicted based on the HET modelling approach on a mid-baffle plane at 600 rpm. Left – 0% alkane volume fraction (2-phase). Right –10% alkane volume fraction.

alternatives [13]. Thus, it may be suggested that mass transfer in gas–liquid–liquid reactors is less sensitive to the assumed configuration at the gas–liquid interface and more sensitive to other varying parameters such as the energy dissipation rate and the gas hold up. More evidence for or against this suggestion should, however, be provided based on a re-examination of mass transfer using a different set of modelling assumptions.

In conclusion, the usefulness of a CFD-based mass transfer model may be glimpsed from the amount of detail that it can generate. For example, spatial resolutions of variables of interest can be quickly generated as illustrated in **Figure 6** to support the visualisation of changes made during *in situ* reactor design. This is indeed the chief advantage of a CFD-based approach over an empirical approach. However, before the full potential of the CFD-based approach can be realised, a number of areas will need improvement or further investigation. More on this is presented in the subsequent section.

5. Areas requiring further investigation

The results reviewed in the previous section illustrate that the addition of a secondary liquid phase can have a great impact on both the hydrodynamics and the mass transfer in a reactor. In particular, the results suggest that turbulence modulation by the secondary liquid phase is the key mechanism of action through which changes in a reactor occur. A CFD-based framework that is able to capture this mechanism of action has also been proposed and its potential has been illustrated. This notwithstanding, care ought to be taken in generalising the results obtained and the implications arising from them. This is because there still are a number of issues that need further investigation. Two such pertinent issues are highlighted below.

The first pertinent issue concerns the applicability of the obtained results to different reactors. It is the view of the authors that the results reported above should be taken as being particular to stirred tank reactors operating in the turbulent regime. This would be in line with existing experimental evidence for turbulence modulation by secondary liquid phases in turbulent stirred tank reactors [34–37]. Different reactors, on the other hand, may have alternative mechanisms of action. For example, recent studies in a gas–liquid–liquid–solid bubble column reactor have illustrated that the addition of a secondary liquid phase did not significantly impact the hydrodynamics of the reactor [81, 82]. Rather, a greater impact on the hydrodynamics was observed for the solid phase and this depended on the type of the solid phase employed [81, 82]. Consequently, it may be suggested that a pseudo-homogenous treatment of the liquid–liquid dispersion would suffice for such a bubble column reactor contrary to what has been established in the case study above.

The second point worth investigation is the sensitivity of CFD-based results to the sub-models and simplifying assumptions employed. As noted in sections 3.1 and 3.2, quite a number of decisions have to be made prior to the actual simulations. Though these decisions are necessitated by a need to keep the simulations tractable, the quality of results obtained may be impacted. It is the view of the authors that particular attention should be given to the sub-models used to capture turbulence modulation. This is because of the seemingly large effects that turbulence modulation had on the results presented in the case study. To this end, it is recommended that comprehensive sets of experiments should be conducted on gas–liquid–liquid reactors and these should involve a concurrent measurement of the mass transfer and the hydrodynamics. Current literature is fragmented with authors who have investigated mass transfer not having measured the potential changes in the hydrodynamics and vice versa.

Acknowledgements

The authors would like to acknowledge Ayman A. Abufalgha for performing the surface tension measurements.

The first author would also like to acknowledge funding received from the DST-NRF Centre of Excellence in Catalysis (c*change) and the DAAD-AIMS In Region Scholarship Programme.

Nomenclature

d	Diameter (m).
$\bar{g}$,g	Gravitational acceleration vector, gravitational constant (m/s^2).
k	Turbulent kinetic energy (m^2/s^2).
l	Integral length scale of turbulence.
m	ratio of concentration of a species dissolved in secondary liquid phase to that dissolved in primary liquid phase at equilibrium.
n	Total number of phases.
v_s	Superficial gas velocity (m/s).
x_g	Volume fraction of gas phase based on total volume in a gas–liquid reactor.
x_d	Volume fraction of secondary liquid phase based on total volume of un-gassed reactor.
x_m	Maximum packing limit.
C	Dissolved concentration of a species (mol/m^3).
C^*	Dissolved saturation concentration of a species (mol/m^3).
C_D	Drag coefficient.
$C_{D,\infty}$	Drag coefficient in the absence of turbulence.
D	Diffusivity of a species in a liquid phase (m^2/s).
D_r	Diffusivity ratio
E'	Enhancement factor.
Eo	Eotvos number.
$\bar{\bar{I}}$	Identity tensor.
$\bar{J}$	Total diffusive flux of a species.
$G_{k,c}$	Production of turbulent kinetic energy from mean velocity gradients.
K_La	Overall volumetric mass transfer coefficient in the absence of a secondary liquid phase (s^{-1}).
K_La'	Overall volumetric mass transfer coefficient in the presence of a secondary liquid phase (s^{-1}).
K_{ji}	Interphase momentum exchange coefficient between phases i and j.
N	Impeller speed (rps).
P_g	Gassed power (W).
Re	Reynolds number.
St	Stokes number.
T	Impeller diameter (m).

V	Volume (m^3).
$\overline{V}$	Velocity vector (m/s).
$\overline{V}_{dr}$	Drift velocity (m/s).
α	Volume fraction based on total volume in a gassed reactor.
ϵ	Turbulent kinetic energy dissipation rate (m^2/s^3).
ρ	Density (kg/m^3).
σ	Surface/interfacial tension (N/m).
$\overline{\overline{\tau}}$	Stress tensor (Pa).
μ	Dynamic viscosity (Pa s).
μ_t	Turbulent viscosity (Pa s).
μ_r	Viscosity ratio.
λ	Kolmogorov length scale.
ζ_{ic}	Covariance of fluctuating velocities between phase i and the continuous liquid phase (subscript c).
Π_{k_c}	Production (destruction) of turbulent kinetic energy by motion of dispersed particles.
Π_{ϵ_c}	Production (destruction) of turbulent kinetic energy dissipation rate by motion of dispersed particles.
x, y, z, φ	Exponents.
$\Gamma, \Lambda, \gamma, \Omega, \Theta, C_\mu$ $\sigma_k, \sigma_\epsilon, C_{1\epsilon}, C_{2\epsilon}$ $C_{3\epsilon}, C_{AM}$	Constants.

Subscripts/superscripts

c	Continuous/primary liquid phase.
d	Secondary liquid phase.
g	Gas phase.
p	Dispersed particle (gas, liquid or solid).
i,j	Phases i,j.
cap	Spherical cap.
ell	Elliptical shape.
eff	Effective.
sph	Spherical shape.
sat	Saturated.

Author details

Godfrey Kabungo Gakingo and Tobias Muller Louw*
Department of Process Engineering, Stellenbosch University, Stellenbosch, South Africa

*Address all correspondence to: tmlouw@sun.ac.za

References

[1] Kasat GR, Pandit AB. Review on mixing characteristics in solid-liquid and solid-liquid-gas reactor vessels. Canadian Journal of Chemical Engineering. 2005;83:618–43. DOI: 10.1002/cjce.5450830403.

[2] Karimi M. CFD analysis of solid-liquid-gas interactions in flotation vessels [PhD thesis]. Stellenbosch University, 2014.

[3] Muñoz R, Daugulis AJ, Hernández M, Quijano G. Recent advances in two-phase partitioning bioreactors for the treatment of volatile organic compounds. Biotechnology Advances. 2012;30:1707–20. DOI: 10.1016/j.biotechadv.2012.08.009.

[4] Clarke KG, Correia LDC. Oxygen transfer in hydrocarbon-aqueous dispersions and its applicability to alkane bioprocesses: A review. Biochemical Engineering Journal. 2008; 39:405–29. DOI:10.1016/j.bej.2007.11.020.

[5] Garcia-Ochoa F, Gomez E. Bioreactor scale-up and oxygen transfer rate in microbial processes: An overview. Biotechnology Advances. 2009;27:153–76. DOI:10.1016/j.biotechadv.2008.10.006.

[6] Besagni G, Inzoli F, Ziegenhein T. Two-Phase Bubble Columns: A Comprehensive Review. ChemEngineering. 2018;2:13. DOI: 10.3390/chemengineering2020013.

[7] Joshi JB, Nandakumar K. Computational modelling of multiphase reactors. Annual Review of Chemical and Biomolecular Engineering. 2015;6: 347–78. DOI:10.1146/annurev-chembioeng-061114-123229.

[8] Sajjadi B, Raman AAA, Ibrahim S, Shah RSSRE. Review on gas-liquid mixing analysis in multiscale stirred vessel using CFD. Reviews in Chemical Engineering. 2012;28:171–89. DOI: 10.1515/revce-2012-0003.

[9] Sajjadi B, Raman AAA, Shah RSSRE, Ibrahim S. Review on applicable breakup/coalescence models in turbulent liquid-liquid flows. Reviews in Chemical Engineering. 2013;29:131–58. DOI:10.1515/revce-2012-0014.

[10] Shah RSSRE, Sajjadi B, Raman AAA, Ibrahim S. Solid-liquid mixing analysis in stirred vessels. Reviews in Chemical Engineering. 2015;31:119–47. DOI: 10.1515/revce-2014-0028.

[11] Wang G, Ge L, Mitra S, Evans GM, Joshi JB, Chen S. A review of CFD modelling studies on the flotation process. Minerals Engineering. 2018;127: 153–77. DOI:10.1016/j.mineng.2018.08.019.

[12] Cheng D, Wang S, Yang C, Mao ZS. Numerical Simulation of Turbulent Flow and Mixing in Gas-Liquid-Liquid Stirred Tanks. Industrial and Engineering Chemistry Research. 2017; 56:13050–63. DOI:10.1021/acs.iecr.7b01327.

[13] Gakingo GK, Clarke KG, Louw TM. A numerical investigation of the hydrodynamics and mass transfer in a three-phase gas-liquid-liquid stirred tank reactor. Biochemical Engineering Journal. 2020;157. DOI:10.1016/j.bej.2020.107522.

[14] Linek V, Beneš P. A study of the mechanism of gas absorption into oil-water emulsions. Chemical Engineering Science. 1976;31:1037–46. DOI:10.1016/0009-2509(76)87024-8.

[15] Clarke KG, Williams PC, Smit MS, Harrison STL. Enhancement and repression of the volumetric oxygen transfer coefficient through hydrocarbon addition and its influence

on oxygen transfer rate in stirred tank bioreactors. Biochemical Engineering Journal. 2006;28:237–42. DOI:10.1016/j.bej.2005.11.007.

[16] Ngo TH, Schumpe A. Oxygen absorption into stirred emulsions of n-alkanes. International Journal of Chemical Engineering. 2012;2012:1–7. DOI:10.1155/2012/265603.

[17] Dumont E, Andrès Y, Le Cloirec P. Mass transfer coefficients of styrene and oxygen into silicone oil emulsions in a bubble reactor. Chemical Engineering Science. 2006;61:5612–9. DOI:10.1016/j.ces.2006.04.026.

[18] Kundu A, Dumont E, Duquenne A-M, Delmas H. Mass transfer characteristics in gas-liquid-liquid system. Canadian Journal of Chemical Engineering. 2003;81:640–6. DOI: 10.1002/cjce.5450810341.

[19] Dumont E, Delmas H. Mass transfer enhancement of gas absorption in oil-in-water systems: A review. Chemical Engineering and Processing: Process Intensification. 2003;42:419–38. DOI: 10.1016/S0255-2701(02)00067-3.

[20] Dumont É. Mass transport phenomena in multiphasic gas/water/NAP systems. Advances in Chemical Engineering. 2019;54:1–51. DOI: 10.1016/bs.ache.2018.12.001.

[21] Correia LDC, Aldrich C, Clarke KG. Interfacial gas-liquid transfer area in alkane-aqueous dispersions and its impact on the overall volumetric oxygen transfer coefficient. Biochemical Engineering Journal. March 2010;49: 133–7. DOI:10.1016/j.bej.2009.12.007.

[22] Littel RJ, Versteeg GF, Van Swaaij WPM. Physical absorption of CO2 and propene into toluene/water emulsions. AIChe Journal. 1994;40: 1629–38.

[23] Pal R. Rheology of simple and multiple emulsions. Current Opinion in Colloid and Interface Science. 2011;16: 41–60. DOI:10.1016/j.cocis.2010.10.001.

[24] Pal R. Novel viscosity equations for emulsions of two immiscible liquids. Journal of Rheology. 2001;45:509–20. DOI:10.1122/1.1339249.

[25] Pal R. Viscous behavior of concentrated emulsions of two immiscible Newtonian fluids with interfacial tension. Journal of Colloid and Interface Science. 2003;263:296–305. DOI:10.1016/S0021-9797(03)00125-5.

[26] Ho CS, Ju L-K, Baddour RF. Enhancing penicillin fermentations by increased oxygen solubility through the addition of n-hexadecane. Biotechnology and Bioengineering. 1990;36:1110–8. DOI:10.1002/bit.260361106.

[27] Littlejohns J V., Daugulis AJ. Oxygen transfer in a gas-liquid system containing solids of varying oxygen affinity. Chemical Engineering Journal. 2007;129:67–74. DOI:10.1016/j.cej.2006.11.002.

[28] Gakingo GK, Louw TM, Clarke KG. Modelling of the overall volumetric mass transfer coefficient measured by the dynamic method in gas-liquid-liquid systems. In: 13th International Conference on Heat Transfer, Fluid Mechanics and Thermodynamics; 2017; Portoroz, Slovenia: p. 662–6.

[29] Dumont E, Andrès Y, Cloirec P Le. Mass transfer coefficients of styrene into water/silicone oil mixtures: New interpretation using the "equivalent absorption capacity" concept. Chemical Engineering Journal. 2014; 237:236–41. DOI:10.1016/j.cej.2013.10.021.

[30] Chawla H, Khanna R, Nigam KDP, Adesina AA. A homogeneous model for mass transfer enhancement in gas-liquid-liquid systems. Chemical Engineering Communications. 2008;

195:622–43. DOI:10.1080/00986440701555431.

[31] Wenmakers PWAM, Hoorn JAA, Kuipers JAM, Deen NG. Gas-liquid mass transfer enhancement by catalyst particles, a modelling study. Chemical Engineering Science. 2016;145:233–44. DOI:10.1016/j.ces.2016.01.043.

[32] Bruining WJ, Joosten GEH, Beenackers AACM, Hofman H. Enhancement of gas-liquid mass transfer by a dispersed second liquid phase. Chemical Engineering Science. 1986;41:1873–7. DOI:10.1016/0009-2509(86)87066-X.

[33] Van Ede CJ, Van Houten R, Beenackers AACM. Enhancement of gas to water mass transfer rates by a dispersed organic phase. Chemical Engineering Science. 1995;50:2911–22. DOI:10.1016/0009-2509(95)00133-P.

[34] Svensson FJE, Rasmuson A. PIV measurements in a liquid-liquid system at volume percentages up to 10% dispersed phase. Experiments in Fluids. 2006;41:917–31. DOI:10.1007/s00348-006-0211-0.

[35] Svensson FJE, Rasmuson A. LDA-measurements in a stirred tank with a liquid-liquid system at high volume percentage dispersed phase. Chemical Engineering and Technology. 2004;27:335–9. DOI:10.1002/ceat.200401981.

[36] Laurenzi F, Coroneo M, Montante G, Paglianti A, Magelli F. Experimental and computational analysis of immiscible liquid-liquid dispersions in stirred vessels. Chemical Engineering Research and Design. 2009;87:507–14. DOI:10.1016/j.cherd.2008.12.007.

[37] Zhao Y, Li X, Cheng J, Yang C, Mao ZS. Experimental study on liquid-liquid macromixing in a stirred tank. Industrial and Engineering Chemistry Research. 2011;50:5952–8. DOI:10.1021/ie102270p.

[38] Cheng D, Cheng J, Li X, Wang X, Yang C, Mao Z. Experimental study on gas-liquid-liquid macro-mixing in a stirred tank. Chemical Engineering Science. 2012;75:256–66. DOI:10.1016/j.ces.2012.03.035.

[39] Balachandar S, Eaton JK. Turbulent Dispersed Multiphase Flow. Annual Review of Fluid Mechanics. 2010;42:111–33. DOI:10.1146/annurev.fluid.010908.165243.

[40] Mathai V, Lohse D, Sun C. Bubbly and Buoyant Particle-Laden Turbulent Flows. Annual Review of Condensed Matter Physics. 2020;11:529–59. DOI:10.1146/annurev-conmatphys-031119-050637.

[41] Poelma C, Ooms G. Particle-Turbulence Interaction in a Homogeneous, Isotropic Turbulent Suspension. Applied Mechanics Reviews. 2006;59:78–89. DOI:10.1115/1.2130361.

[42] Gai G, Hadjadj A, Kudriakov S, Thomine O. Particles-induced turbulence: A critical review of physical concepts, numerical modelings and experimental investigations. Theoretical and Applied Mechanics Letters. 2020;10:241–8. DOI:10.1016/j.taml.2020.01.026.

[43] Gore RA, Crowe CT. Effect of particle size on modulating turbulent intensity. International Journal of Multiphase Flow. 1989;15:279–85. DOI:10.1016/0301-9322(89)90076-1.

[44] Hetsroni G. Particles-turbulence interaction. International Journal of Multiphase Flow. 1989;15:735–46. DOI:10.1016/0301-9322(89)90037-2.

[45] Hadinoto K, Jones EN, Yurteri C, Curtis JS. Reynolds number dependence of gas-phase turbulence in gas-particle flows. International Journal of Multiphase Flow. 2005;31:416–34. DOI:10.1016/j.ijmultiphaseflow.2004.11.009.

[46] Ferrante A, Elghobashi S. On the physical mechanisms of two-way coupling in particle-laden isotropic turbulence. Physics of Fluids. 2003;15: 315–29. DOI:10.1063/1.1532731.

[47] Van den Berg TH, Luther S, Lohse D. Energy spectra in microbubbly turbulence. Physics of Fluids. 2006;18: 1–4. DOI:10.1063/1.2185688.

[48] Mazzitelli IM, Lohse D, Toschi F. The effect of microbubbles on developed turbulence. Physics of Fluids. 2003;15:L5–8. DOI:10.1063/ 1.1528619.

[49] Rensen J, Luther S, Lohse D. The effect of bubbles on developed turbulence. Journal of Fluid Mechanics. 2005;538:153–87. DOI:10.1017/ S0022112005005276.

[50] Elghobashi SE, Truesdell GC. On the two-way interaction between homogeneous turbulence and dispersed solid particles. 1: Turbulence modification. Physics of Fluids. 1993;5: 1790–801.

[51] Nielsen DR, Daugulis AJ, McLellan PJ. A novel method of simulating oxygen mass transfer in two-phase partitioning bioreactors. Biotechnology and Bioengineering. 2003;83:735–42. DOI:10.1002/bit.10721.

[52] Versteeg HK, Malalasekera W. An introduction to computational fluid dynamics: the finite volume method. 2nd ed. Essex, England: Pearson Education; .

[53] Joshi JB, Nere NK, Rane C V, Murthy BN, Mathpati CS, Patwardhan AW, Ranade V V. CFD simulation of stirred tanks: Comparison of turbulence models. Part I: Radial flow impellers. Canadian Journal of Chemical Engineering. 2011;89:23–82. DOI: 10.1002/cjce.20446.

[54] Lane GL. Computational modelling of gas-liquid flow in stirred tanks [PhD thesis]. The University of Newcastle, Australia, 2006.

[55] Elghobashi S. Particle-laden turbulent flows: direct simulation and closure models. Applied Scientific Research. 1991;48:301–14. DOI:10.1007/ BF02008202.

[56] Lathouwers D. Modelling and simulation of turbulent bubbly flow [PhD thesis]. Delft University of Technology, The Netherlands, 1999.

[57] ANSYS Academic Research Mechanical and CFD, Release 17.2, Theory Guide 2016.

[58] Lane GL, Schwarz MP, Evans GM. Modelling of the interaction between gas and liquid in stirred vessels. In: 10th European Conference on Mixing; 2000; Delft, The Netherlands: p. 197–204.

[59] Scargiali F, D'Orazio A, Grisafi F, Brucato A. Modelling and simulation of gas–liquid hydrodynamics in mechanically stirred tanks. Chemical Engineering Research and Design. 2007; 85:637–46. DOI:10.1205/CHERD06243.

[60] Zhang D, Deen NG, Kuipers JAM. Numerical simulation of the dynamic flow behavior in a bubble column: A study of closures for turbulence and interface forces. Chemical Engineering Science. 2006;61:7593–608. DOI: 10.1016/j.ces.2006.08.053.

[61] Ishii M, Zuber N. Drag coefficient and relative velocity in bubbly, droplet or particulate flows. AIChE Journal. 1979;25:843–55. DOI:10.1002/ aic.690250513.

[62] Behzadi A, Issa RI, Rusche H. Modelling of dispersed bubble and droplet flow at high phase fractions. Chemical Engineering Science. 2004;59: 759–70. DOI:10.1016/j.ces.2003.11.018.

[63] Simonnet M, Gentric C, Olmos E, Midoux N. Experimental determination of the drag coefficient in a swarm of

bubbles. Chemical Engineering Science. 2007;62:858–66. DOI:10.1016/j.ces.2006.10.012.

[64] Baltussen MW, Seelen LJH, Kuipers JAM, Deen NG. Direct numerical simulations of gas-liquid-solid. Chemical Engineering Science. 2013;100:293–9. DOI:10.1016/j.ces.2013.02.052.

[65] Baltussen MW, Kuipers JAM, Deen NG. Direct numerical simulation of effective drag in dense gas-liquid-solid three-phase flows. Chemical Engineering Science. 2017;158:561–8. DOI:10.1016/j.ces.2016.11.013.

[66] Vaidheeswaran A, Hibiki T. Bubble-induced turbulence modeling for vertical bubbly flows. International Journal of Heat and Mass Transfer. 2017; 115:741–52. DOI:10.1016/j.ijheatmasstransfer.2017.08.075.

[67] Magolan BL. Extending bubble-induced turbulence modeling applicability in CFD through incorporation of DNS understanding [PhD thesis]. Massachusetts Institute of Technology, United States of America, 2018.

[68] Colombo M, Fairweather M. Multiphase turbulence in bubbly flows: RANS simulations. International Journal of Multiphase Flow. 2015;77:222–43. DOI:10.1016/j.ijmultiphaseflow.2015.09.003.

[69] Rzehak R, Krepper E. CFD modeling of bubble-induced turbulence. International Journal of Multiphase Flow. 2013;55:138–55. DOI:10.1016/j.ijmultiphaseflow.2013.04.007.

[70] Mandø M, Lightstone MF, Rosendahl L, Yin C, Sørensen H. Turbulence modulation in dilute particle-laden flow. International Journal of Heat and Fluid Flow. 2009; 30:331–8. DOI:10.1016/j.ijheatfluidflow.2008.12.005.

[71] Horender S, Hardalupas Y. Fluid-particle correlated motion and turbulent energy transfer in a two-dimensional particle-laden shear flow. Chemical Engineering Science. 2010;65:5075–91. DOI:10.1016/j.ces.2010.05.033.

[72] Rols JL, Condoret JS, Fonade C, Goma G. Mechanism of enhanced oxygen transfer in fermentation using emulsified oxygen-vectors. Biotechnology and Bioengineering. February 20, 1990;35:427–35. DOI: 10.1002/bit.260350410.

[73] Mcmillan JD, Wang DIC. Mechanisms of oxygen transfer enhancement during submerged cultivation in perfluorochemical-in-water dispersions. Annals of the New York Academy of Sciences. 1990;589: 283–300. DOI:10.1111/j.1749-6632.1990.tb24253.x.

[74] Lamont J, Scott D. An eddy cell model of mass transfer into the surface of a turbulent liquid. AIChE Journal. 1970;16:513–9. DOI:10.1002/aic.690160403.

[75] Shabalala NZP, Harris M, Leal Filho LS, Deglon DA. Effect of slurry rheology on gas dispersion in a pilot-scale mechanical flotation cell. Minerals Engineering. 2011;24:1448–53. DOI: 10.1016/j.mineng.2011.07.004.

[76] Bakker CW, Meyer CJ, Deglon DA. Numerical modelling of non-Newtonian slurry in a mechanical flotation cell. Minerals Engineering. 2009;22:944–50. DOI:10.1016/j.mineng.2009.03.016.

[77] Kataoka I, Besnard DC, Serizawa A. Basic equation of turbulence and modeling of interfacial transfer terms in gas-liquid two-phase flow. Chemical Engineering Communications. November 1, 1992;118:221–36. DOI: 10.1080/00986449208936095.

[78] Viollet PL, Simonin O. Modelling dispersed two-phase flows: closure,

validation and software development. Appl Mech Rev. June 1, 1994;47:80–4. DOI:10.1115/1.3124445.

[79] Mudde RF, Simonin O. Two- and three-dimensional simulations of a bubble plume using a two-fluid model. Chemical Engineering Science. November 1, 1999;54:5061–9. DOI: 10.1016/S0009-2509(99)00234-1.

[80] Bel F'dhila R, Simonin O. Eulerian prediction of a turbulent bubbly flow downstream of a sudden pipe expansion. In: 6th Workshop on Two-Phase Flow Predictions; 1992; Erlangen, Germany: p. 264–73.

[81] Abufalgha AA, Clarke KG, Pott RWM. Characterisation of bubble diameter and gas hold-up in simulated hydrocarbon-based bioprocesses in a bubble column reactor. Biochemical Engineering Journal. 2020;158. DOI: 10.1016/j.bej.2020.107577.

[82] Abufalgha AA, Pott RWM, Cloete JC, Clarke KG. Gas–liquid interfacial area and its influence on oxygen transfer coefficients in a simulated hydrocarbon bioprocess in a bubble column reactor. Journal of Chemical Technology and Biotechnology. 2021;96:1096–106. DOI: 10.1002/jctb.6625.

Chapter 11

Industrial Design Energy Efficiency and GHG Emission Reduction via Steam and Power Systems Optimization

Mana Al-Owaidh, Abdulrahman Hazazi, Solomon Oji and Abdulaziz Dulaijan

Abstract

The energy supply side for a large oil, gas, refinery, or petrochemical facility is designed to provide the site with sufficient heating, cooling, and power utilities requirements. Reducing capital and operating costs from energy supply side is essential to maximize the value added from the industrial facility. Thus, optimization solutions are often developed to optimize industrial utility design and operation, reduce costs while improving the overall system's efficiency and inevitably reducing CO2 emission. Our topic in the chapter is related to a new methodology that aims to identify the optimum design and operation of the energy supply side of a new industrial facility. One major cause for utility design inefficiency is the fact that in a typical project setup, there are different project teams handling the design of the utility supply side and process design independently. This often results in high capital cost, and lower operating efficiency. The potential improvement expected from the optimum design compared with a typical design case for a new industrial facility is over 15% from base-case life cycle cost. This chapter also covers several examples to explain the concept and expected benefits from applying a new Combined Heat and Power (CHP) optimization solution during new project design.

Keywords: CHP, steam system optimization, system's efficiency, GHG emission reduction, steam and power optimum design, industrial utility system, grassroot facility

1. Introduction

Today, optimizing energy consumption, improving energy efficiency, and reducing GHG emissions are essential for a sustainable operation and lower operating cost of an industrial facility such as Oil, Gas, Refining, and Petrochemical facilities. Every industrial facility depends on more than one form of utilities for its operation. Examples of these utilities include power generation, steam system, instrument and plant air, nitrogen system, hot oil system, etc.

Process streams such as gas and liquid are usually heated or cooled by indirect heat exchange with another fluid: either another process stream or a utility stream such as steam, hot oil, cooling water, or refrigerant. Heating utilities are necessary for proper usage of condensers, distillers, and several other integral types of equipment in the hydrocarbon processing facilities. In hydrocarbon processing plants, steam is the most commonly heat utility used.

Steam is used both as utility and a process fluid (heating agent, diluent to absorb heat of reaction, feedstock, and stripping agent in adsorbers and absorbers). It can be used to drive mechanical drivers such as compressors and pumps, heat exchangers, and ejectors (for producing a vacuum). There are few advantages of using steam as opposed to other methods of process heating. For example, see [1].

In general, the supply-side utility systems for industrial facilities are used to produce the required energy for the facility, and the most common system used is steam system. Other alternatives include hot oil and hot water systems.

Steam system is a better choice for a facility with high power demand and high heating demand required by process and at different level of temperatures. Thus, most of the gas plants, refineries, and petrochemicals that are using steam system often include both boilers and Cogen, as the base-case option. The reason behind using steam system for industrial facilities required both heating and power demand can be summarized as follows:

- Generating steam at high pressure and using steam turbines to recover the energy available in the steam for power generation will improve the overall system efficiency of the supply side to reach a level over 70%.

- Using the lower pressure steam extracted from steam turbines for process heating making use of the available latent heat in the steam for a better heat recovery.

- In addition, steam is a clean service that provides energy and heat for the industrial facility.

2. Background of a combined heat and power (CHP) system

For a majority of process plants, the bulk of the energy required is supplied through the utility system. On most facilities, the required heating is provided by combined heat and power (CHP) systems. A CHP system is a combination of two or more systems that are used in the generation and distribution of steam and power through gas turbine generators (GTGs), heat recovery steam generators (HRSGs), boilers, and steam turbines. The use of CHP plants means that the efficiency of the processing facility in terms of its energy consumption is reliant on efficiency of both the process side and the operation of the utility system. Steam is used for various purposes such as heating, drying, and providing a heat source for air conditioning or motive energy for a power generation via steam turbines.

As illustrated in **Figure 1** (Gas-Turbine-Based CHP Plant), a typical CHP starts with the Bryton Thermodynamic cycle where the combustion of a fuel and air mixture in a gas turbine combustion chamber occurs, the combustion product is then channeled through a series of blades attached to a rotating shaft and a generator, which then generates power and hot flue gas through the gas turbine exhaust system. The exhaust gas, which has a high energy content, becomes the heating

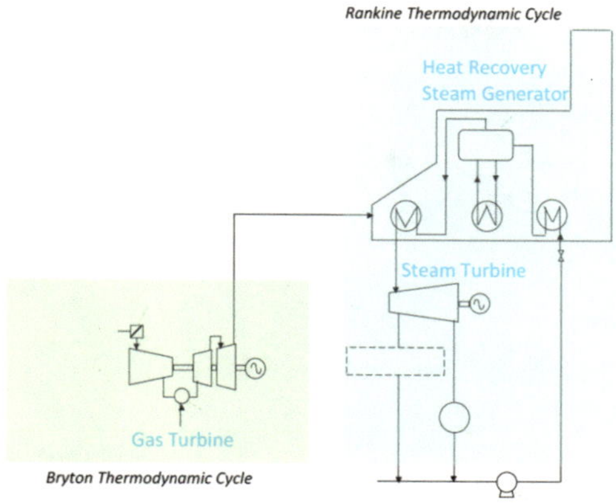

Figure 1.
Gas-turbine-based CHP plant.

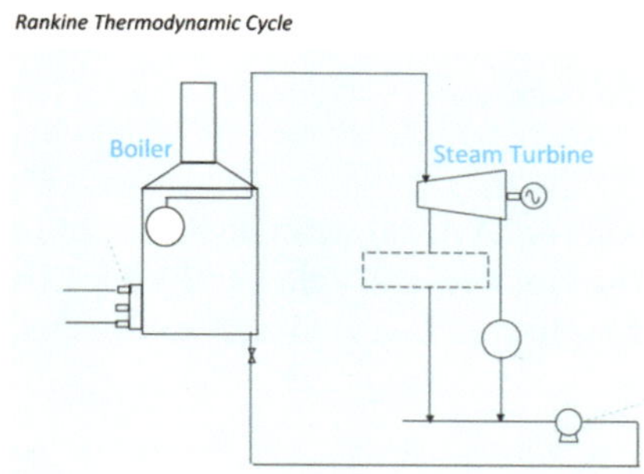

Figure 2.
Boiler-based CHP plant.

medium in the next heat recovery Rankine Cycle of the CHP, the flue gas is channeled through an HRSG to heat up boiler feed water and produce steam. The generated steam is then used to drive a steam turbine generator (known as a combined cycle) or sent to a process plant to be used as a heating medium in a heat exchanger or to mechanically drive rotating equipment such as pumps and compressors (known as cogeneration).

Figure 2 (Boiler-Based CHP Plant) illustrates an alternative type of CHP system. In a boiler-based CHP system, one or more boilers are used to generate steam, and steam turbine generators (STGs) are then used to generate power. In certain configurations, steam is extracted from the steam turbine generator (STG) to be used for process heating via heat exchangers. Most of Saudi Aramco's facilities or plants have a combination of both gas turbine and boiler-based CHP systems. For example, see [2].

3. Toward optimum design of industrial steam and power systems

The concept of simultaneous process and utility design optimization was developed by Saudi Aramco protected by 2-granted patents. **Figure 3** provides an overview of the steps used for the optimization. In Refs [3, 4], the techniques used in optimizing new CHP systems are being derived from unit commitment and economic dispatch power generation concepts. Our optimization problem includes integer (binary), linear, and nonlinear relations between the objective function, variables, and constraints.

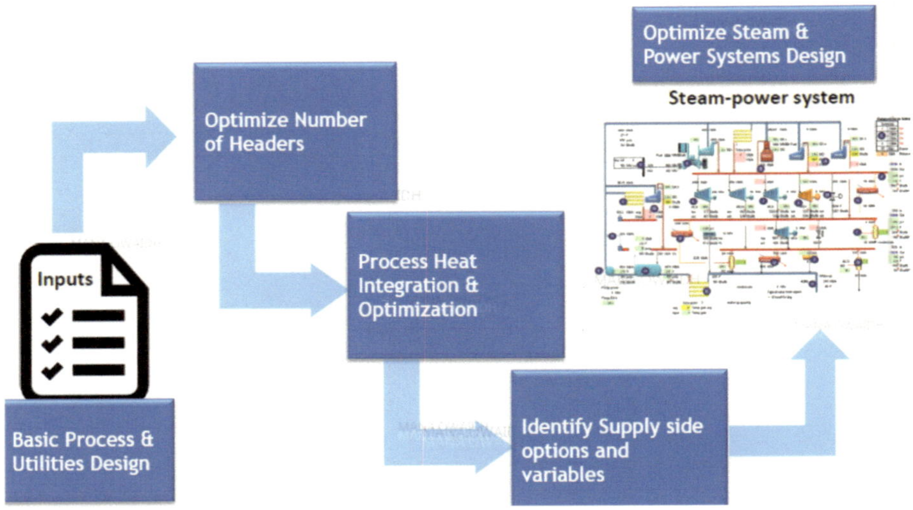

Figure 3.
Key elements of a steam system optimization tool.

The problem formulation for a typical steam system can be summarized as follows:

Objective function is the Net Present Value (NPV) for the new design, and it is a function of capital cost of equipment as well as the expected operating cost of the system configuration.

$$\text{Objective function} = \sum_{i=1}^{n} NPV\left(Capex_i + Opex_i * \frac{Hrs}{yr} * LC\right) \quad (1)$$

Where,

NPV: Net Present Value for the project.

LC: Life Cycle of the new facility, normally used 25 years.

The total capital cost includes major equipment used in the optimization analysis as decision variables

$$= \sum_{i=1}^{n} (Capex_{Blr_i} + Capex_{Cogen_i} + Capex_{STG_i} + Capex_{STi} + Capex_{Motor_i} + Capex_{RO} + Capex_{MED}) \quad (2)$$

Capital cost would be function of number and sizes of major equipment (i.e., decision variables for the optimization algorithm):

- Boilers
- Cogeneration units
- Steam turbine generator
- New steam turbine drivers
- Motors-driven equipment as alternative drives with steam turbines and
- Water desalination facilities and makeup system

The total operating cost is function of the equipment performance and the impact on the energy consumption of the facility. The operating cost includes the following key elements:

- Fuel consumption
- Power export and import tariffs
- Makeup water treatment and chemicals
- CO_2 emissions

In the optimization analysis, there are key constraints that have to be met by the optimizer to confirm the validity of the results. Some of these constraints are related to equipment limitations and others related to systems limitations. Below are some examples of the key constraints used in the optimization analysis:

- Equipment constraint: such as steam generation from boiler should be less than maximum limit and greater than minimum generation limit.
- System constraint: steam production from steam supply equipment shall be greater than or equal to steam demand required.
- System constraint: available steam reserve from boilers should be more than or equal to required steam reserve.
- Steam constraint: input to a steam header should be equal to steam out from steam header
- Mathematical model constraint. Non-negative flows in the steam distribution network

Below is a generic mathematical representation for the steam system:
Steam balance representation includes:

$$\text{Steam balance for a steam header (a)} = \sum_{i=1}^{n} Stm_{in} - Stm_{out} \tag{3}$$

Where i and n are representing the equipment connected to this steam header.
Boiler Feed Water (BFW) Balance is calculated as per the formula below

$$= \sum_{i=1}^{n} Blr_{BFW} + \sum_{i=1}^{n} Cogen_{BFW} - \sum_{i=1}^{n} DSH_{BFW} \tag{4}$$

Makeup water compensates for all loses from steam system, thus makeup water is equal to all loses in the steam system.
BFW makeup balance is calculated as follows:

$$= \sum_{i=1}^{n} Proc_{stm} * (1 - RC\%) + \sum_{i=1}^{n} Blr_{BD} + \sum_{i=1}^{n} Cogen_{BD} + \sum_{i=1}^{n} Vent_{stm} \tag{5}$$

Whereas,
BFW: Boiler feed water.
B_BFW: Boiler feed water to boilers.
COG_BFW: Boiler feed water to Cogen units.
DSH_BFW: De-super heater water into steam network.
BD: Blow down flow.
Bstm: steam generation from boilers.
COG_stm: steam generation from Cogen units.
Steam users.

Case study: grassroot facility.

This section covers the (CHP) optimization assessment to identify the optimum configurations and equipment sizing for the supply side of a new petrochemical complex. References [5–7] include other examples, which can help explaining the concept further.

The assessment for the optimum configuration started with reflecting the utility's initial design data into a newly developed (CHP) for design. The CHP model key input is shown in **Table 1** summary.

For new facilities, 70% (HHV basis) is considered as the minimum efficiency of a site's overall CHP systems thermal efficiency. CHP systems' thermal efficiency for the site can be defined as the ratio between all useful energies generated by the system and total energy input as fuel:

$$CHP\ system\ \ Thermal\ Eff.\% = \frac{Useful\ Energy\ Out}{Energy\ Input} \tag{6}$$

Where:

Useful Energy Out = Total net power generated by Cogen and STGs + total mechanical power recovered in the steam system by STs + total mechanical power driven by GTs + total heat consumed by process at different headers in (MMBtu/hr);

Energy Input = Total fuel consumed by the facility including boilers, Cogeneration units, simple cycle gas turbines, process heaters generating steam, other process heaters, and SEC equivalent fuel for imported power in (MMBtu/hr).

The CHP optimization study evaluated four design scenarios. The CHP analysis and its related economics considered the optimum configuration meeting operational and design requirement. The design requirement accounts for one steam supply unit under T&I and a trip of another unit. The CHP analysis covered the following cases:

Steam Demand		
VHP	101	T/h
HP	978	T/h
MP	936	T/h
LP	413	T/h
Returned Condensate %	65%	

Note: Five different cogeneration frames from different GT manufacturers have been used for the CHP optimization analysis. This is just to give a better understanding and more accurate outcomes from energy efficiency point of view. It is worth highlighting that the analysis for each case is based on the average result of the different frames and not for any specific one.

Table 1.
Process steam headers.

1. Base Case: (5 Cogen Units – 1 standby boiler).

2. Case-1: (4 Cogen Units – 2 boiler Units).

3. Case-2: (3 Cogen Units – 3 boiler Units).

4. Case-3: (2 Cogen Units – 4 boiler Units).

Base Case: The base case composed of five gas turbines each with its heat recovery steam generator and two STGs and with one spare boiler, as shown in **Table 2**.

Table 3 shows that the average CHP model's output from overall supply-side thermal efficiency is in the range of 69%, which is slightly lower than the minimum efficiency requirement of (70%).

Case-1: In this case, the CHP configuration includes four Cogen units, two boiler, and two STGs. The result showed that the average steam system efficiency for the different frames is around 70–73% (**Tables 4** and 5).

Case-2: In this case, the configuration is composed of three Cogen units, three boilers, and two STGs, where the average steam system efficiency is around (74%) (**Tables 6** and 7).

Equipment	Number of units	Avg. size per unit
Cogen	5	260 (MW)
Boiler	1 (standby)	341 (T/h)
STG	2	100 (MW)

Table 2.
Base-case scenario.

Option	System eff.%	Tot. fuel (MMBTU/H)	Tot. STGs (MW)	Net pwr gen (MW)	CO2 emissions reduction (MM ton Co2/year)
Cogen A	71%	10,310	170	1225.2	2.9
Cogen B	70%	10,289	147	1174.9	2.6
Cogen C	69%	10,937	182	1284.9	3.0
Cogen D	68%	14,012	278	1813.7	4.8
Cogen E	69%	11,994	230	1459.0	3.6

Table 3.
Base-case CHP model output.

Equipment	Number of units	Avg. size per unit
Cogen	4	260 (MW)
Boiler	2 (1 standby)	341 (T/h)
STG	2	80 (MW)

Table 4.
Case-1 design basis.

Option	System eff.%	Tot. fuel (MMBTU/H)	Tot. STG (MW)	Net pwr gen (MW)	CO2 emissions reduction (MM ton Co2/year)
Cogen A	73%	8475	113	957.5	2.1
Cogen B	72%	8976	123	1005.2	2.1
Cogen C	70%	11,436	200	1428.2	3.6
Cogen D	71%	9821	161	1144.5	2.6
Cogen E	72%	10,462	178	1297.3	3.2

Table 5.
Case-1 CHP model output.

Equipment	Number of units	Avg. size per unit
Cogen	3	260 (MW)
Boiler	3 (1 standby)	341 (T/h)
STG	2	50 (MW)

Table 6.
Case-2 design basis.

Option	System eff.%	Tot. fuel (MMBTU/H)	Tot. STG (MW)	Net pwr gen (MW)	CO2 emissions reduction (MM ton Co2/year)
Cogen A	75%	7098	95	728.0	1.3
Cogen B	75%	7204	79	740.7	1.3
Cogen C	74%	8860	122	1042.8	2.4
Cogen D	75%	7649	93	830.0	1.7
Cogen E	75%	8129	105	944.6	2.1

Table 7.
Case-2 CHP model output.

Case-3: in this case, the CHP configuration composed of two Cogen units, four boilers, and two STGs resulted in steam system supply-side efficiency around 69%. The reason for having a larger STG in this case is to reduce the power import as much as possible (**Tables 8** and **9**).

To identify the optimum steam and power systems configurations for petrochemical complex, all options have been simulated via CHP optimization model as shown in the previous section.

Equipment	Number of units	Avg. size per unit
Cogen	2	260 (MW)
Boiler	4 (1 standby)	341 (T/h)
STG	2	100 (MW)

Table 8.
Case 3 design basis.

Option	System eff.%	Tot. fuel (MMBTU/H)	Tot. STG (MW)	Net pwr gen (MW)	CO2 emissions reduction (MM ton Co2/year)
Cogen A	67%	7520	218	640.5	0.6
Cogen B	67%	7771	223	664.4	0.6
Cogen C	75%	7154	114	728.0	1.3
Cogen D	69%	8120	236	728.0	0.8
Cogen E	73%	7482	168	728.0	1.1

Table 9.
Case-3 CHP model output.

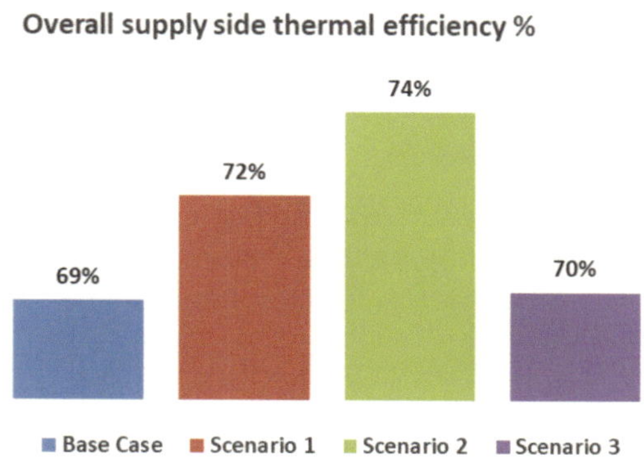

Figure 4.
System efficiency summary of the different design configurations.

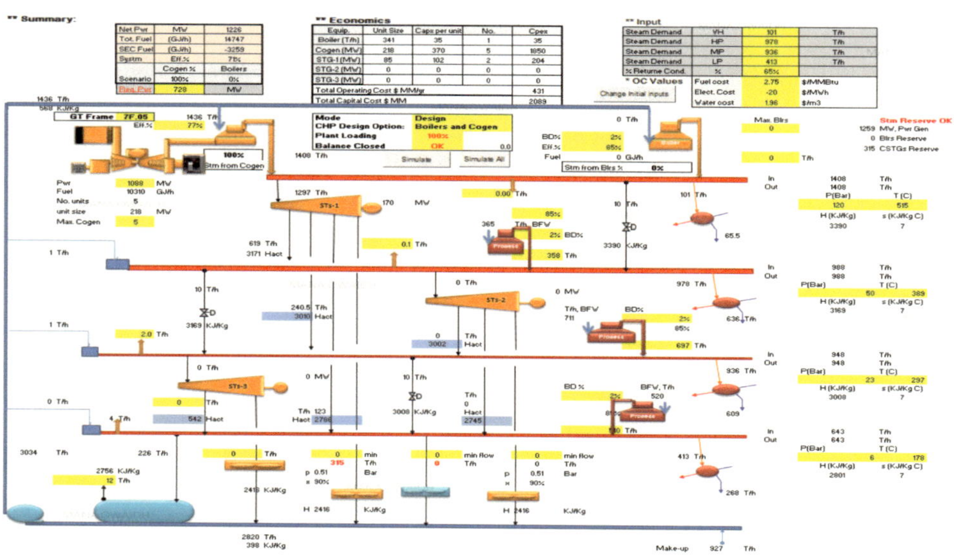

Figure 5.
Base case: Petrochemical complex steam system CHP.

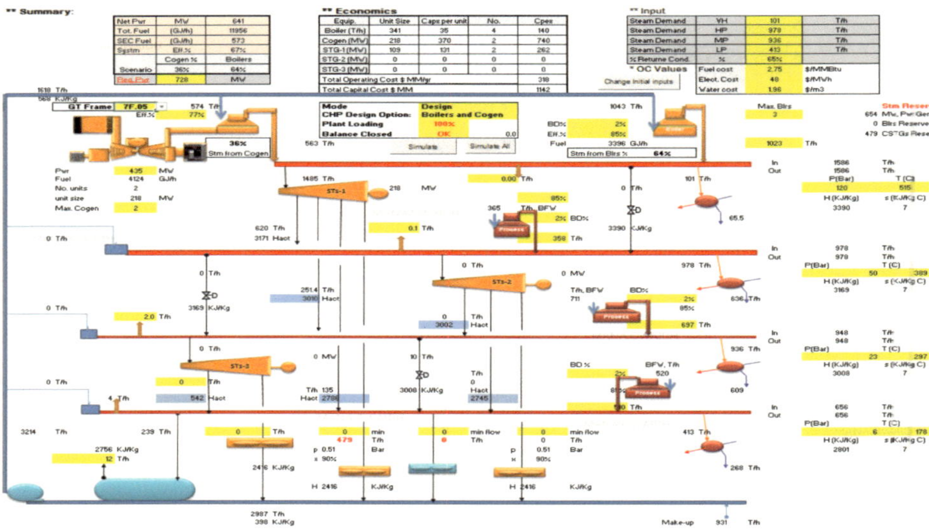

Figure 6.
Case 3: Petrochemical complex steam system CHP.

The study evaluated (4) different cases and compares the outcomes with the base case to identify the best configuration. In summary, in all cases there exist at least two GT frames that can meet the 70% minimum steam system efficiency requirement (**Figures 4–6**).

4. Conclusions

The CHP model includes all the elements involved in the generation and distribution of energy to drive the process and supporting infrastructure. Optimizing such complex system requires a sophisticated model.

Within Saudi Aramco, the CHP model of each operating facility was found to be extremely useful in understanding the interactions between the various utilities' components. The interactions between these components could have been very complex without constructing a reasonably accurate mathematical model. Such tools are used during the design phase of capital projects to optimize cogeneration system sizes and configurations, explore alternatives, and conduct thermal efficiency synthesis to estimate what the final design would look like.

For operating facilities, the CHP model is used to evaluate potential efficiency enhancements, monitor the performance of existing equipment, identify energy saving opportunities, change operation and control based on optimum operational advice, and evaluate the impact of process variations on the CHP utilities system. The features of the model can be summarized in the following points:

- Excel-based model along with Visual Basic (user-friendly)
- Utilize a powerful solver optimization tool
- Include all steam properties
- For real-time asset management, consider connecting the CHP model to a data historian for real-time data
- Account for all system constraints

In summary, having a systematic approach for analyzing operational improvements is viable using CHP optimization solutions. These solutions have been found to be useful with the right features to support plant operation. Following list gives a summary of typical optimization handles from the model for operational modification actions:

1. Minimize steam flow to fin-fan condensers

2. Optimize steam turbine operating load (switch ability of running steam turbines and motors)

3. Boiler load management

4. Maximize cogeneration operation

5. Maintain operation within optimum set points

In addition, optimum design of a new facility will help saving major capital and operating cost. As shown in the example from the case study, the potential benefit from a case and optimum case can exceed 15% in the NPV of the project life cycle.

Author details

Mana Al-Owaidh*, Abdulrahman Hazazi, Solomon Oji and Abdulaziz Dulaijan
Saudi Aramco, Dhahran, Saudi Arabia

*Address all correspondence to: owaidhmm@aramco.com

References

[1] Towler GP, Sinnot R. Chemical Engineering Design: Principles, Practice and Economics of Plant and Process Design. Elsevier; 2012. ISBN: 9780080966595

[2] Oji S, Al-Owaidh MM. Overview of a Combined Heat and Power (CHP) Model. CEP. New York, USA: American Institute of Chemical Engineers; 2019

[3] Al-Owaidh MM, Phung BT. Efficiency-Based Optimization Model for a Large CHP System of an Oil Stabilization Plant. Australia: AUPEC; 2006

[4] Al-Owaidh MM. Energy savings opportunities through CHP optimization models. Bahrain: PEATEM; 2012

[5] Amminudin KA, Al-Owaidh MM, Najjar AA, Al-Dossary IS, Faifi YH. Aramco team plots energy savings at Berri gas plant. Oil & Gas Journal. 5 October 2009

[6] Amminudin KA, Al-Owaidh MM, Barri ZS. Process Utility Interaction Analysis for a True Energy Saving Value Determination. Tampa, Florida: AIChE Spring Meeting; 2009

[7] Al-Owaidh MM, Dhaifullah UA, Ken GR, Ghazal AH. Optimum Operation of Complex Combined Heat and Power Systems of Parallel gas Facilities. Dhahran, Saudi Arabia: Energy Forum; 2013

Section 3

Environmental Impact Assessment

Chapter 12

Improve Energy Efficiency in Surface Mines Using Artificial Intelligence

Ali Soofastaei and Milad Fouladgar

Abstract

This chapter demonstrates the practical application of artificial intelligence (AI) to improve energy efficiency in surface mines. The suggested AI approach has been applied in two different mine sites in Australia and Iran, and the achieved results have been promising. Mobile equipment in mine sites consumes a massive amount of energy, and the main part of this energy is provided by diesel. The critical diesel consumers in surface mines are haul trucks, the huge machines that move mine materials in the mine sites. There are many effective parameters on haul trucks' fuel consumption. AI models can help mine managers to predict and minimize haul truck energy consumption and consequently reduce the greenhouse gas emission generated by these trucks. This chapter presents a practical and validated AI approach to optimize three key parameters, including truck speed and payload and the total haul road resistance to minimize haul truck fuel consumption in surface mines. The results of the developed AI model for two mine sites have been presented in this chapter. The model increased the energy efficiency of mostly used trucks in surface mining, Caterpillar 793D and Komatsu HD785. The results show the trucks' fuel consumption reduction between 9 and 12%.

Keywords: artificial intelligence, energy efficiency, fuel consumption, haul trucks, prediction, optimization, mining engineering

1. Introduction

Climate change, energy security, water scarcity, land degradation, and dwindling biodiversity put pressure on communities, requiring more excellent environmental knowledge and resource-conscious economic practices. As a response to these genuine difficulties, both mining and industrial activities have adopted environmental plans.

The global accord, which 125 countries have signed, aims to reduce global glasshouse gas (GHG) emissions by 80% by 2050 to achieve a low-carbon society. Thus far, the agreement has significantly impacted energy-related laws, such as carbon taxes and energy pricing.

However, following the Paris agreement, the energy costs in the mining industry have risen substantially in respect of overall operating costs. Six years ago, energy accounted for 10% of mining companies' operational costs; now, it is pushing close to 20%. This increases the cost base of companies significantly.

Mining is critical to our national security, economy, and the lives of individual citizens. Millions of tons of resources should be mined each year for each individual to maintain his or her quality of living [1]. In addition, the mining sector is a critical component of the world economy, supplying crucial raw materials such as coal, metals, minerals, sand, and gravel to manufacturers, utilities, and other enterprises [2]. To put it another way, mining will continue to be an essential part of the global economy for many years.

Mining necessitates much energy. Mining, for example, is one of the few non-manufacturing industrial sectors recognized as energy-intensive by the U.S. Department of Energy [3]. It is also widely acknowledged that the mining industry could enhance its energy efficiency dramatically. Using the United States as an example, the U.S. Department of Energy (DOE) estimates that the U.S. mining sector consumes around 1315 PJ per year and that this annual energy consumption might be reduced to 610 PJ or about 46% of current annual energy usage [3]. According to the most recent data, energy consumption in Australia's mining sector was at 730 petajoules (P.J.) in 2019–2020, up 9% from the previous year [4]. This is slightly greater than the average rate of increase in energy use during the last decade. Mining consumes 175 PJ of energy per year in South Africa and is the largest consumer of electricity at 110.9 PJ per year, according to 2003 figures. The association between rising interest in energy efficiency and energy prices demonstrates increasing energy intensity on mining operating expenses [5, 6]. Given recent governmental moves by various governments to make industry pay for the expenses associated with carbon emissions, such high energy-intensive processes are not sustainable or cost-effective (carbon taxes and similar regulatory costs). As a result, all stakeholders have a vested interest in improving mine energy efficiency.

Since the rise in fuel prices in the 1970s, the importance of reducing energy usage has gradually grown. In addition, because the mining industry's primary energy sources are petroleum products such as electricity, coal, and natural gas, increasing margins through efficiency savings can also save millions of tons of gas emissions.

Mining companies are looking into reducing energy consumption and emissions to cut costs and emissions, especially considering any possible carbon emissions strategy. First, however, businesses must have a comprehensive understanding of their current energy usage, which involves using technology that allows employees to make decisions.

Mining businesses actively review their investment, capital expenditure, and operational plans to ensure that their operations are sustainable and ecologically beneficial. Sustainable practices and capital equipment investments must result in measurable cost savings. Mining businesses are looking to increase their energy efficiency to cut costs and lessen their environmental effect.

Sustainable investments were not thought to produce significant returns on investment in earlier years, but they are becoming more appealing with the quickly changing legislative and economic climate. When all the advantages of new technology and business practices are considered, including direct savings from increased efficiency as well as associated incentives such as carbon tax credits, investments become much more appealing. Furthermore, when considered over a longer time horizon, these same investments in energy savings, for example, become incredibly beneficial.

Data analytics represents a very appropriate approach to pulling together disparate data sources since it is the science of examining raw data to conclude that information. In addition, cost savings, faster and better decision making, and finally, new goods and services are some of the key benefits of data analytics [7].

Data analytics represents a very appropriate approach to pulling together these disparate data sources since it is the science of examining raw data to conclude that information. Cost savings, faster and better decision making, and finally, new goods and services are some of the most significant advantages of data analytics [7]. Data analytics is widely used and can be used in areas many might not have thought about before. One area that sees much potential in data analytics is the mining industry. Data analytics should be considered a necessity, not a luxury, for an industry that does trillions of dollars in business every year.

One of the advanced data analytic techniques discussed in this chapter aims to enhance the crucial issue of mining energy efficiency. The focus will be on open-pit mine haulage activities. This study aims to create a sophisticated data analytics model for assessing the complex connections that affect haul truck energy efficiency in surface mining. The application of Artificial Neural Networks for predictive simulation and Genetic Algorithms (GAs) for optimization in the investigation of energy efficiency is the focus of this study.

2. Mining energy efficiency—Using artificial intelligence

Global resource firms are currently struggling in challenging economic and regulatory environments. However, most companies in the mining business are now disclosing their performance in this area in response to growing social concern about the industry's numerous consequences and the birth of the idea of sustainable development. Many firm sustainability reports include total energy consumption and associated glasshouse gas (GHG) emissions in absolute and relative terms, indicating that energy consumption and its impact on climate change are priorities.

Mining companies are setting goals to improve these metrics, but there is also a global trend towards more complicated and lower-grade orebodies, which require more energy to process. As a result, mining businesses must be more innovative to improve their environmental sustainability and efficiency operations. In addition, companies must consider the specific energy usage of their processes to limit glasshouse gas emissions.

According to Australian government research, the most significant energy use industries in 2013–2014 were transportation, metal manufacturing, oil and gas, and mining. Transportation consumes a quarter of Australia's annual energy. The manufacturing of metal products such as aluminum, steel, nickel, lead, iron, zinc, copper, silver, and gold accounted for over 16% of total energy consumption. The mining industry consumes 10% of all energy used by participants. **Figure 1** shows the other industries that used the most energy in 2019–2020.

Grinding (40%) and materials handling by diesel equipment are the most energy-intensive equipment types in the mining industry (17%) [8].

According to the Australian Energy Statistics, Australian energy consumption has increased by an average of 0.6% a year for the past decade and reached 6171 PJ in 2019–2020.

Energy efficiency can significantly cut energy demand while also assisting in reducing GHG emissions at a low cost to industry and the larger economy. Therefore, it makes commercial and environmental sense to be aware of opportunities to maximize energy efficiency. The glasshouse gas emissions produced by mining companies were calculated using various fuels, including electricity, natural gas, and diesel. The mining companies' energy savings translated to a possible reduction in glasshouse gas emissions.

Data analytics is the science of examining raw data to discover useful information, reach conclusions about the meaning of the data, and support decision-making. The

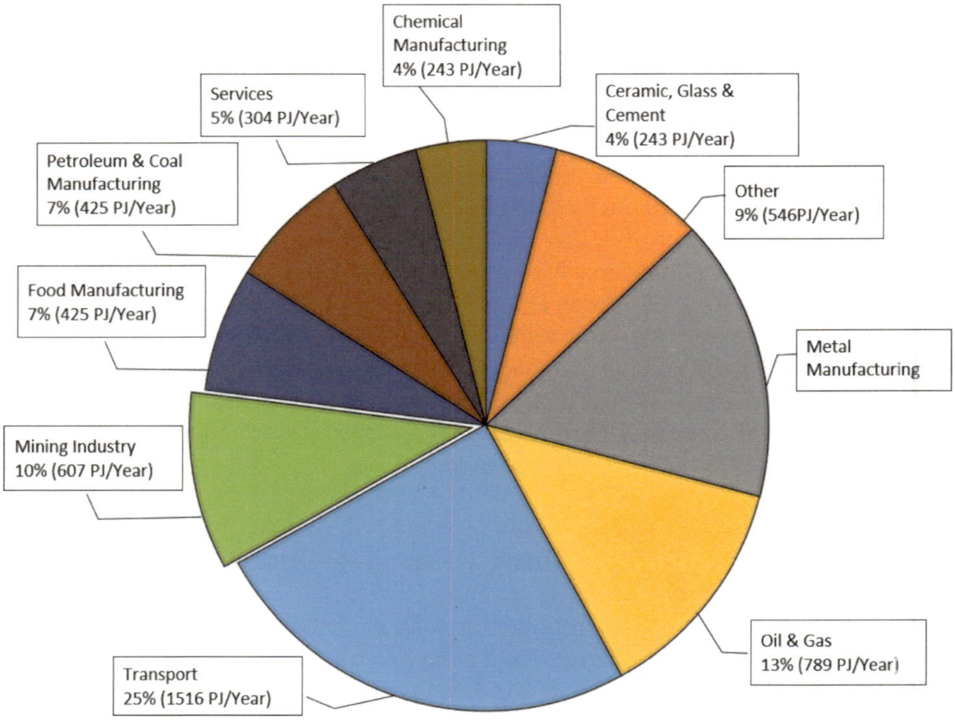

Figure 1.
Top energy users by industry sector 2019–2020 (Total 6069 PJ) [8].

foremost opportunity that data analytics presents for mining is its potential to identify, understand, and then guide the correction of complex root causes of high costs, poor process performance, and adverse maintenance practices. Therefore, data analytics can reduce costs and accelerate better decision-making, which ultimately enables new products and services to be developed and delivered, creating added value for all [7].

Figure 2 illustrates the two dimensions of maturity: a time dimension (over which capability and insights are developed) and a competitive advantage dimension (the value of insights generated). At the lowest levels, analytics are routinely used to produce reports and alerts. These use simple, retrospective processing and reporting tools, such as pie graphs, top-ten histograms, and trend plots. They typically answer the fundamental question: 'what happened and why?' Increasingly, sophisticated analytical tools, capable of working at or near real-time and providing rapid insights for process improvement, can show the user "what just happened" and assist them in understanding "why" as well as the following best action to take. Towards the top end of the comparative advantage scale are predictive models and ultimately optimization tools, with the capability to evaluate 'what will happen and the ability to identify the best available responses—'what is the best that could happen?'

The mining sector and governments have been pushed to perform research on energy consumption reduction due to the potential for energy (and financial) savings. As a result, a significant number of research studies and industrial projects have been conducted worldwide to achieve this in mining operations [8]. As a result, the mining industry might save roughly 37% of its current energy use by fully implementing state-of-the-art technology and installing new technology through research and development expenditure [9]. Furthermore, energy usage is

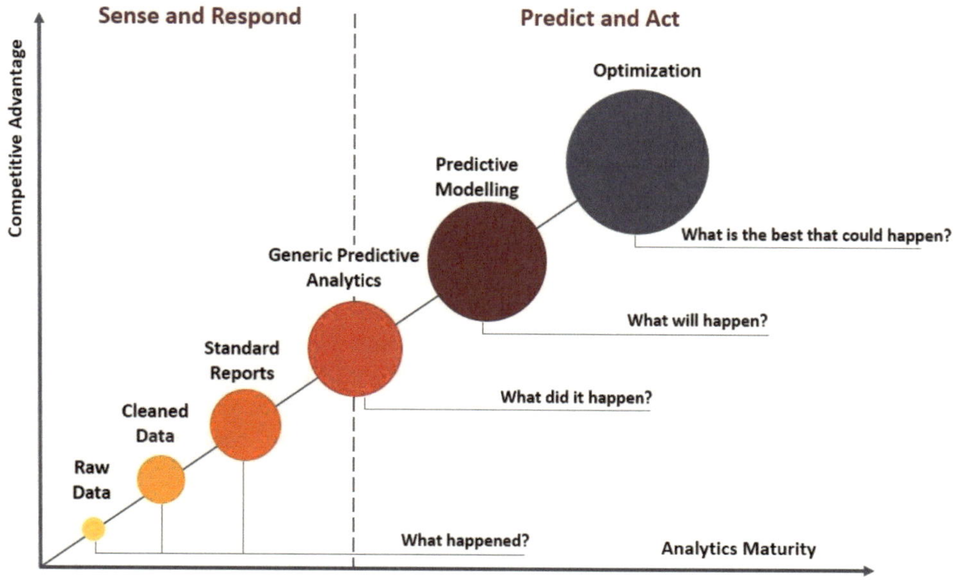

Figure 2.
Data analytics maturity levels [7].

significantly reduced when mining technologies and energy management systems improve. To put it another way, there are substantial further chances to minimize energy use in the mining business.

The four main phases of the mining process that data analytics can use are (1) extraction of ore, (2) materials handling, (3) ore comminution and separation, and (4) mineral processing. The focus of many companies is efficiency improvements in the materials handling phase. For example, the hauling activity at an open-pit mine consumes a significant amount of energy and can be more energy-efficient [10]. The case study presented here- haulage equipment- is one of these potential areas for improving the mining energy efficiency as well as reducing greenhouse gas emissions.

3. Improve haul trucks energy efficiency

In a surface mining operation, truck haulage accounts for most costs. In surface mines, diesel fuel is used as an energy source for haul trucks, which is expensive and has a significant environmental effect. Energy efficiency is widely acknowledged as the easiest and most cost-effective strategy to manage rising energy bills and lower glasshouse gas emissions.

Depending on the production capacity and site layout, haul trucks are utilized in conjunction with other equipment such as excavators, shovels, and loaders. They collaborate to dig ore or waste material out of the pit and carry it to a disposal site, stockpile, or the next step in the mining operation [11].

The pace of energy consumption is determined by various factors that can be evaluated and tweaked to achieve optimal performance levels [12]. The energy efficiency of the mine fleet is affected by a variety of factors, including site production rate, vehicle age and maintenance, payload, speed, cycle time, mine layout, mine plan, idle time, tire wear, rolling resistance, dumpsite design, engine operating parameters, and transmission shift patterns. To improve energy efficiency, this knowledge can be incorporated into mining plan costing and design methods [8].

To assess the prospects for strengthening truck energy efficiency, a comprehensive analytical framework can be built.

We can not only save money each year by improving the energy efficiency of mine haulage systems, but we can also save considerable emissions of glasshouse gases and other air pollutants.

4. Data analytics models

A novel integrated model was proposed to improve haul truck energy usage's three most significant and critical effective characteristics. Payload (P), truck speed (S), and total resistance (R) are the three parameters (T.R.). However, the relationship between energy usage and these characteristics on an actual mining site is complicated. Therefore, to predict and reduce haul truck fuel consumption in surface mines, we apply two AI technologies to develop an advanced data analytic model (**Figure 3**).

In the first step, an artificial neural network (ANN) model was developed to create a Fuel Consumption Index (FC_{Index}) as a function of P, S, and T.R. This index shows how many liters of diesel fuel are consumed to haul 1 ton of mined material in 1 h. In this model, the main parameters used to control the algorithm were R^2 and MSE. After the first step, the optimum values of P, S, and T.R. will be determined using a novel multi-objective GA model. These improved parameters can be utilized to boost haul truck energy efficiency.

The proposed model's methods are all based on actual data obtained from surface mines. Below are the results of utilizing the developed model for two genuine major surface mines in Australia and Iran. On the other hand, the finished methods can be expanded for various mines by substituting the data.

5. Prediction model—Artificial neural network

The artificial neural network (ANN) is a popular AI model and a robust computational tool based on the human brain's organizational structure [13]. ANNs are the

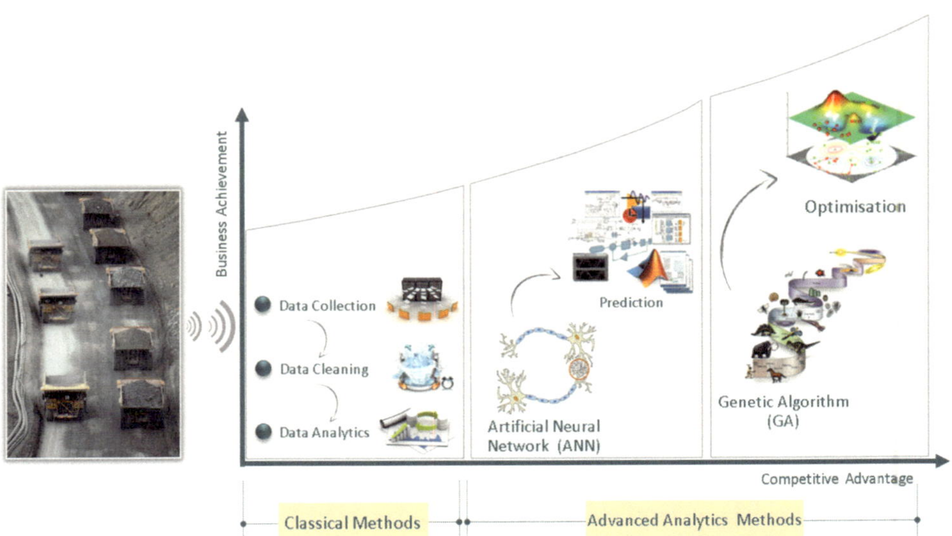

Figure 3.
A schematic of the developed model [8].

representation of methods that the brain uses for learning which are known as neural networks (NNs), simulated neural networks (SNNs), or parallel distributed processing (PDP). ANN simulates the effect of multiple variables on one significant parameter by a fitness function. Thus, ANNs are excellent solutions for complex problems as they can signify the compound relationships between the various parameters involved in a problem.

ANN methods are established as powerful techniques to solve various real-world problems among the different machine intelligence procedures due to ANN's excellent learning capacity in recent decades. The approximate solution by ANN is found to be useful, but it depends upon the ANN model that one considers [14].

Layers are commonly used to organize neural networks. Layers are made from various interconnected "neurons/nodes," which include "activation functions." ANN processes information to solve problems through neurons/nodes in a parallel manner. First, ANN obtains knowledge through learning and is stored within inter-neuron connections' strength, expressed by numerical values called "weights." Then, these weights and biases are combined to calculate output signal values for a new testing input signal value. Next, patterns are provided to the network through the "input layer," which connects to one or more "hidden layers," where the actual processing is completed through a system of weighted "connections." The hidden layers then correlate to an "output layer," which generates the output through the activation functions [Eqs. (1)–(3)].

$$E_k = \sum_{j=1}^{q} \left(w_{ijk} x_j + b_{ik}\right) \qquad k = 1,\ 2,\ \dots,\ m \tag{1}$$

Where i is the input, x is the normalized input variable, w is the weight of that variable, b is the bias, q is the number of input variables, and k is the counter of neural network nodes, and m is the number of neural network nodes in the hidden layer.

In general, the activation functions contain linear and nonlinear equations. The coefficients related to the hidden layer are grouped into matrices w_{ijk} and b_{ik}. Eq. (2) is often used as the activation function between the hidden and output layers, where f is the transfer function.

$$F_k = f(E_k) \tag{2}$$

The output layer calculates the weighted sum of the signals provided by the hidden layer, and the related coefficients are grouped into matrices W_{ok} and b_o. Thus, the network output can be determined by Eq. (3).

$$Out = \left(\sum_{k=1}^{m} w_{ok} F_k\right) + b_o \tag{3}$$

The most significant component of neural network modeling is network training, which can be done in two ways: controlled and uncontrolled. Backpropagation is the most widely used training algorithm, which was established after examining several types of algorithms. A training algorithm modifies the coefficients (weight and bias) of a network to reduce the error between the estimated and actual network outputs.

The Mean Square Error (MSE) and Coefficient of Determination (R^2) were used in this study to investigate the error and performance of the neural network output and determine the appropriate number of nodes in the hidden layer. **Figure 4** depicts the created model's basic structure.

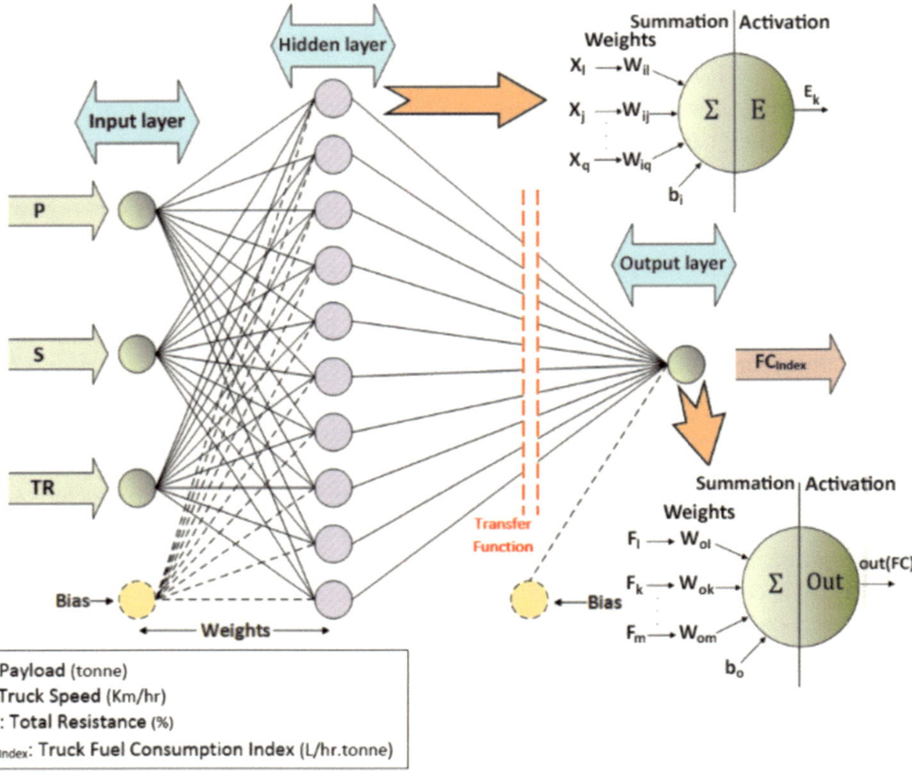

Figure 4.
Structure of artificial neural network [8].

Case study	Mine type	Mine details	Location	Investigated truck
Mine 1	Surface coal mine	The mine contains 877 million tons of coking coal reserves, making it one of Asia's and the world's most significant coal deposits. It can produce 13 million tons of coal per year.	Queensland, Australia	CAT 793D
Mine 2	Surface iron mine	There are 36 million tons of iron deposits in the mine. It has a 15-million-ton ore and waste extraction capacity per year.	Kerman, Iran	Komatsu HD785

Table 1.
Case studies information.

The developed AI model was tested against actual data taken from standard trucks in two surface mines in Australia and Iran. **Table 1** contains some information from these case studies.

For a standard range of loads, **Figures 5** and **6** show the correlation between P, S, T.R., and FC_{Index} created by the constructed ANN model for two types of standard trucks employed in case studies.

The presented graphs show a nonlinear relationship between FC_{Index} and P. The fuel consumption rate increases dramatically with increasing T.R. However, this rate does not change sharply with changing truck speed (S).

The results show good agreement between the estimated and actual values of fuel consumption. **Figure 7** presents sample values for the independent (tested) and the estimated (using the ANN) fuel consumption to highlight the insignificance of the importance of the absolute errors in the analysis for studied mines.

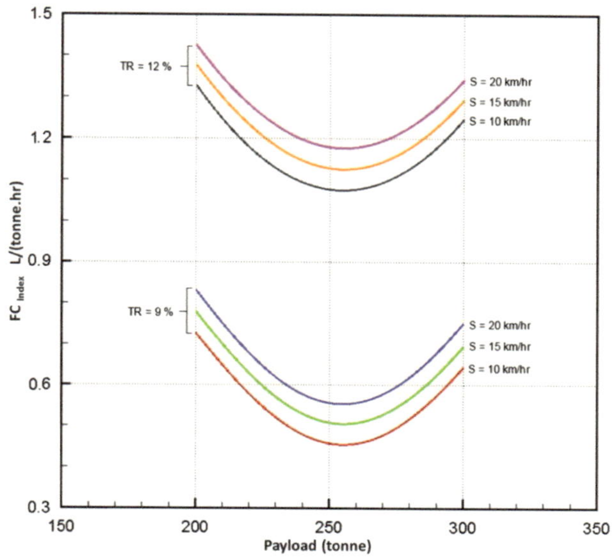

Figure 5.
Correlation between payload, S, T.R., and FC_{Index} based on the developed ANN model for CAT 793D (mine 1).

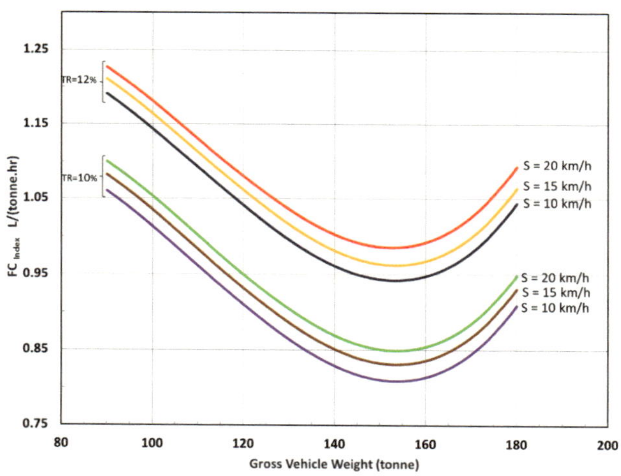

Figure 6.
Correlation between GVW, S, T.R., and FC_{Index} based on the developed ANN model for Komatsu HD785 (mine 2).

6. Optimization model—Genetic algorithm

Optimization is a branch of computational science that shows how to find the best measurable solution to various issues. It is critical to consider the search area and goal function components when solving a specific problem. All the solution's possibilities are investigated in the search area. The objective function is a mathematical function that connects each point in the search area to an actual value that may be used to evaluate all search area members.

Traditional optimization methods are described by the stiffness of their mathematical models and limit their application in presenting dynamic and complex

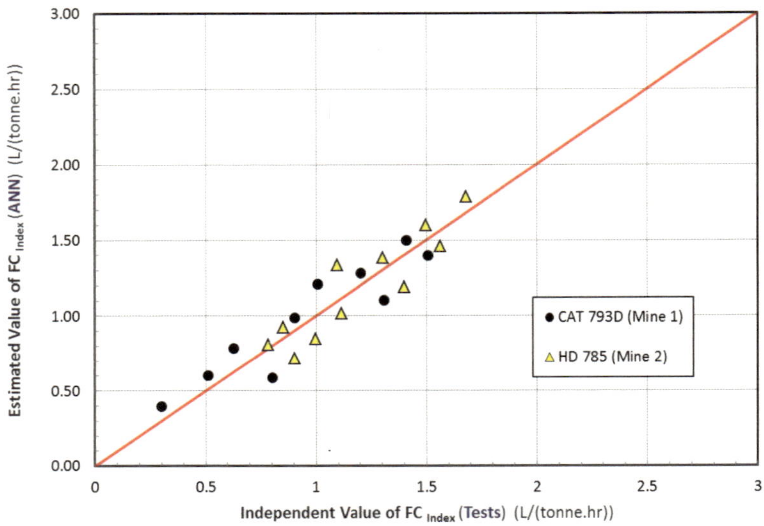

Figure 7.
Sample values for the estimated and the independent fuel consumption index.

situations of "real life." Optimization techniques based on AI, underpinned by heuristic rulings, can reduce the problem of stiffness and are suitable to solve various kinds of engineering problems.

Some heuristic algorithms were developed in the 1950s to replicate biological processes in engineering. When computers were developed in the 1980s, it became possible to employ these algorithms to optimize functions and processes, whereas older methods failed.

During the 1990s, some new heuristic methods were developed by prior algorithms, such as Swarm Algorithms, Simulated Annealing, Ant Colony Optimization, and (GA). GA is one of the most widely used evolutionary optimization algorithms.

GAs were proposed by Holland (1975) based on an abstraction of biological evolution using ideas from natural evolution and genetics to design and implement robust adaptive systems [15]. In optimization methods using the new generation of GA is relatively novel. Moreover, they have good chances to escape from local minimums because of no need for any derivative information. As a result, their application in practical engineering problems can provide more satisfactory solutions than other traditional mathematical methods [16].

GAs are similar to the evolutionary aspects of natural genetics. The individuals are randomly selected from the search area. The fitness of the solutions is determined from the fitness function, subsequently. It is the result of the variable that is to be optimized. The individual that creates the best fitness in the population (a group of possible solutions) has the highest chance to return in the next generation with the opportunity of reproduction by the crossover with another individual, thus producing decedents with both characteristics. The possible solutions will converge to an optimal solution for the proposed problem by correctly developing a GA Crossover, which contributes to the evolution based on selection, reproduction, and mutation.

Due to their potential as optimization techniques for complex functions, GAs have been used in various scientific, engineering, and economic problems [17–20]. There are four significant advantages of using GAs to optimize problems [21]:

- GAs do not have many mathematical requirements for optimization problems.
- They can use many objective functions and constraints (i.e., linear or nonlinear) defined in discrete, continuous, or mixed search spaces.
- They are very efficient at doing global searches due to the periodicity of evolution operators.
- They provide high flexibility to hybridize with domain-dependent heuristics to enable an efficient implementation for a problem.

It is crucial to investigate the impact of particular parameters on GA behavior and performance to determine their relevance to the problem requirements and available resources. Furthermore, the type of problem being addressed determines the impact of each parameter on the algorithm's performance. As a result, determining the best values for these characteristics will necessitate a significant amount of experimentation.

In the GA model, Fitness Function, Individuals, Populations and Generations, Fitness Value, Parents and Children are the main parameters [17]. In addition, the population size impacts global performance and GA efficiency, and the mutation rate ensure that a given position does not remain fixed in value or the search becomes essentially random. **Figure 8** depicts the basic framework of a GA model.

A GA model was created to optimize the significant, influential factors on the energy consumption of haul trucks. **Tables 2** and **3** show the outcomes of utilizing the proposed model for actual case studies with an optimal range of variables.

Using the developed AI models in the two studied mines site shows energy efficiency improvements between 9 and 12%. Reaching the mentioned fuel consumption reduction and energy efficiency is promising when one mostly used truck in the mine site consumes around 110 L of diesel per hour. The haul trucks normally are used 24 h and 7 days per week to move mined materials in the site. Studied mine site had more than 100 trucks in their fleet, and the average price of diesel in those regions was 1.3 dollars per liter. It means that 9–12% energy efficiency improvement equals millions of dollars in saving annually.

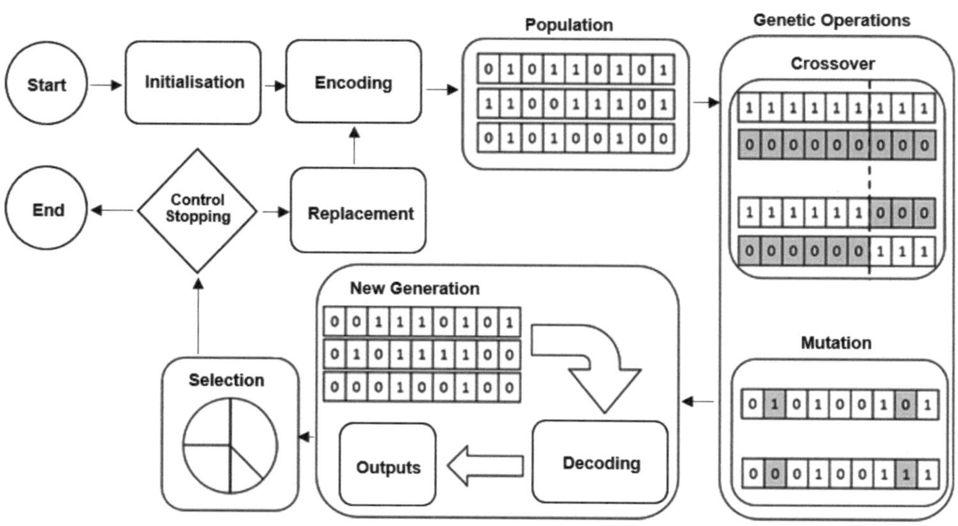

Figure 8.
GA processes (developed model) [8].

Variables	Normal values		Optimized values	
	Minimum	**Maximum**	**Minimum**	**Maximum**
Gross vehicle weight (ton)	150	380	330	370
Total resistance (%)	8	20	8	9
Truck speed (Km/hr.)	5	25	10	15

Table 2.
The result of the GA model for CAT 793D in mine (1).

Variables	Normal values		Optimized values	
	Minimum	**Maximum**	**Minimum**	**Maximum**
Gross vehicle weight (ton)	75	170	150	160
Total resistance (%)	8	15	8	10
Truck speed (Km/hr.)	5	40	10	18

Table 3.
The result of the GA model for HD 785 in mine (2).

7. Conclusion

The purpose of this chapter was to demonstrate the value of modern data analytics models in improving energy efficiency in mining sectors, particularly in haulage operations, which are one of the most energy-intensive activities. However, improving haul truck fuel consumption for actual mining operations based on the link between influential factors, such as P, S, and T.R., was difficult. Thus, two AI methods were utilized to construct a reliable model to assess the problem.

At first, an ANN model was utilized to simulate truck fuel consumption as a function of payload, truck speed, and total resistance. Then, the ANN was generated and tested using the collected accurate mine site datasets, and the results showed good agreement between the actual and estimated values of FC_{Index}.

After that, to improve the energy efficiency in haulage operations, a GA method was developed to determine the optimal value of effective parameters on fuel consumption in haulage trucks. The developed model was used to analyze data for two surface mines in Australia and Iran. This model also can be applied to improve the haul truck fuel consumption for any dataset obtained from actual mine operations.

The results of two successful case studies show plenty of opportunities to use advanced analytics and AI in the mining industry to improve energy efficiency.

Author details

Ali Soofastaei[1*] and Milad Fouladgar[2]

1 AI Program Leader, Vale

2 Department of Mechanical Engineering, Najafabad Branch, Islamic Azad University, Najafabad, Iran

*Address all correspondence to: ali@soofastaei.net

References

[1] Golosinski Tad S. Mining Education in Australia: A Vision for the Future; CIM Bulletin, Canadian Institute of Mining and Metallurgy, Jan 2000.

[2] Zheng S, Bloch H. Australia's mining productivity decline: implications for MFP measurement. Journal of Productivity Analysis. 2014;**41**(2): 201-212

[3] Doe U. Mining Industry Energy Bandwidth Study. Washington DC: BCS; 2007

[4] Allison B et al. Australian Energy Update. Office of the Chief Economist, Department of Industry Innovation and Science: Canberra, Australia; 2016. pp. 20-30

[5] Kecojevic V, Komljenovic D. Haul truck fuel consumption and CO_2 emission under various engine load conditions. Mining Engineering. 2010; **62**(12):44-48

[6] Kecojevic V, Komljenovic D. Impact of bulldozer's engine load factor on fuel consumption, CO_2 emission and cost. American Journal of Environmental Sciences. 2011;**7**(2):125-131

[7] Soofastaei A, Davis J. Advanced data analytic: A new competitive advantage to increase energy efficiency in surface mines. Australian Resources and Investment. 2016;**1**(1):68-69

[8] Soofastaei A. Development of an Advanced Data Analytics Model to Improve the Energy Efficiency of Haul Trucks in Surface Mines, in School of Mechanical and Mining Engineering. Australia: The University of Queensland, School of Mechanical and Mining Engineering; 2016

[9] DOE. Mining Industry Energy Bandwidth Study. Washington, DC: Department of Energy, USA Government; 2012. pp. 26-33

[10] Norgate T, Haque N. Energy and greenhouse gas impacts of mining and mineral processing operations. Journal of Cleaner Production. 2010;**18**(3): 266-274

[11] Darling P. SME mining engineering handbook. SME. 2011; pp. 109-118

[12] EEO. Driving energy efficiency in the mining sector. Australian Government, Department of Resources Energy and Tourism: Canberra, Australia; 2010. pp. 18-22

[13] McCulloch WS, Pitts W. A logical calculus of the ideas immanent in nervous activity. The Bulletin of Mathematical Biophysics. 1943;**5**(4): 115-133

[14] Chakraverty S, Mall S. Artificial Neural Networks for Engineers and Scientists: Solving Ordinary Differential Equations. London: CRC Press; 2017

[15] Sivanandam SN, Deepa SN. Introduction to Genetic Algorithms. Vol. 3. New Delhi, India: Springer; 2008. pp. 246-263

[16] Whitley D, Starkweather T, Bogart C. Genetic algorithms and neural networks: Optimizing connections and connectivity. Parallel Computing. 1990; **14**(3):347-361

[17] Velez-Langs O. Genetic algorithms in the oil industry: An overview. Journal of Petroleum Science and Engineering. 2005;**47**(1–2):15-22

[18] Singh A, Rossi A. A genetic algorithm based exact approach for lifetime maximization of directional sensor networks. Ad Hoc Networks. 2013;**11**(3):1006-1021

[19] Beigmoradi S, Hajabdollahi H, Ramezani A. Multi-objective aeroacoustic optimization of the rear end in a simplified car model by using hybrid robust parameter design, artificial neural networks, and genetic algorithm methods. Computers & Fluids. 2014;**90**:123-132

[20] Sana SS et al. Applying the genetic algorithm to job scheduling under ergonomic constraints in the manufacturing industry. Journal of Ambient Intelligence and Humanized Computing. 2019;**10**(5):2063-2090

[21] Yousefi T et al. Optimizing the free convection from a vertical array of isothermal horizontal elliptic cylinders via genetic algorithm. Journal of Engineering Physics and Thermophysics. 2013;**86**(2):424-430

Chapter 13

Energy, Economic and Environmental (3E) Assessments on Hybrid Renewable Energy Technology Applied in Poultry Farming

Yuanlong Cui, Saffa Riffat, Elmer Theo, Tugba Gurler, Xuan Xue and Zohreh Soleimani

Abstract

This chapter aims to design, construct and test a new and renewable heating system for fulfilling the energy demand and ameliorating the interior environment of poultry farming in the UK. This system consists of a photovoltaic/thermal module attached to the polyethylene heat exchanger integrated with a geothermal copper pipe array and heat pump. The thermal and electrical energy performance of the hybrid renewable heating system is investigated based on a numerical model and experimental test. Moreover, the economic analysis (and environmental assessment are conducted. It is concluded that the electrical energy production from the photovoltaic array could reach 11867 kWh per annum whereas the heat pump thermal output is about 30210 kWh per annum. Meanwhile, the overall gas and electrical cost of the hybrid renewable heating system are £320 and £129, which are much less than that of the gas burners system and could save £763 and £750, respectively, resulting in less than 6-year of payback period. The energy consumption of the hybrid renewable heating system could decrease about 28873 kWh, resulting in a reduction in total CO_2 emission of approximately 8.3 tons, in comparison with the gas burners system.

Keywords: poultry farming, photovoltaic/thermal array, geothermal copper pipe array, energy efficiency, economic and environmental assessments

1. Introduction

The poultry industry is a significant economic part, supplying energy, meat, eggs and livelihoods to an increasing human population. Nevertheless, it is highly exposed to global-scale warming and climate change caused by human activities [1]. The direct impacts on poultry farming involve the growth, breeding, health and welfare whereas the indirect influences are owing to the global warming on the productivity of forage crops, pastures and feeds [2]. Poultry farming makes up a large proportion of the world's entire requirement for family livestock, meanwhile, the global population is going up continuously, which results in a growth in the need for poultry providing over the upcoming decades. To be more specific, compared with 2010, the consumption of poultry meat is anticipated to enhance from 330 to 455 million tons per annum in 2050 [3].

Traditional farming solutions are not the capability of fulfilling this demand, in particular for the broiler breeding. This is because the indoor temperature needs to be controlled with accuracy for achieving optimum growth [4, 5]. And also, the health status of the chicken extremely depends on the ambient temperature inside the poultry shed. In the heating season, it is necessary to sustain indoor air temperature in the range of 21–32°C for broiler birds, while in the cooling season, it should avoid the overheating and heat stress on chicken. What is more, there is major pollutant gas including carbon dioxide (CO_2) and ammonia (NH_3) emitted from poultry facilities that must be maintained underneath the critical concentration levels of $\sim$2500 ppm and $\sim$ 25 ppm, respectively [6].

Although, indoor ambient temperature and harmful gases emission should be controlled effectively, the energy consumption and overall expenses still need to be decreased. Energy is utilized for environmental control including lighting, ventilation cooling and heating, preparation and distribution of feed as well as manure management [7, 8]. Specifically, broiler breeding farming for indoor temperature control makes up 96.3% and 75.5% of the entire thermal and electrical energy demands, respectively. Meanwhile, in laying hen sheds, the power output for indoor condition control and ventilation demand is 58.9% and 43.7%, respectively [9, 10]. Additionally, the electricity expense involving heating, lighting, ventilation and cooling is the biggest part for poultry farmers [11].

Water is a vital input for poultry meat and eggs production and plays an important role in guaranteeing chicken health. The traditional water sources, such as rivers, lakes, rainfall, aquifers and snowmelts are not sufficient to fulfill the minimum water demands. Currently, 61% fresh water is obtained from seawater desalination processes, 21% from brackish water as well as 8% from river water. In general, 90% of fresh water is attained from these three sources and the remaining is extracted from brine, wastewater and other sources. Desalination is regularly utilized to produce freshwater eliminating salts, pollutants and minerals from brackish water [12]. However, desalination technology requires considerably higher energy, cost and greenhouse gas (GHG) emissions compared to traditional water treatment approaches. Currently, two mature technological solutions are employed include thermal and membrane approaches. It is found that the thermal-based approach has much higher energy-intensive compared to membrane-based ones [13, 14]. Specifically, Ahmed et al. [15] found that the desalination capacity across the globe based on the membrane desalination method makes up about 73% whereas the thermal-based solutions account for merely 27% until 2016. Moreover, the membrane approaches require high operating pressure ranging from 55 to 70 bar, by comparison, the normal pressure for the brackish water desalination varies from 15 to 30 bar [16]. Integrating the desalination technologies with renewable energy sources have the potential to produce fresh water for future development. This mainly includes three merits, namely, energy sustainability, future fresh water sustainability and environmental sustainability. Renewable energy technologies like solar photovoltaic (PV), solar photovoltaic/thermal (PV/T) and geothermal heat pumps are state-of-the-art and could become feasible and economically promising for different areas. Nevertheless, when the technologies continue to enhance, the fresh water becomes scarce and fossil fuel energy price increases, thereby renewable energy desalination suits more viable economically. Additionally, the CO_2 emission is the major factor by the operation of desalination processes. It is reported [17, 18] that the CO_2 emission is over 1500% and it is predictable to increase to 2200% by 2040. Herein, the efficiency enhancement of water and electricity, is vital to regulate CO_2 emission to protect the environment.

Hence, to ensure energy sustainable development, decrease cost as well as GHG emission in poultry farming, sustainable energy development in poultry production and reduce cost and GHG emissions, there is a strong incentive to explore energy conservation and deployment of renewable energy technologies for improving energy conversion efficiency for heating and cooling and replacing the utilization of fuel. In comparison with conventional oil and gas energy resources, renewable energy technology shows massive potential owing to its excellent quantity and environmental friendliness.

To be more specific, solar energy technologies including photovoltaic (PV), solar thermal collector and solar photovoltaic/thermal (PV/T) are the ideal solutions for warming poultry shed. This is because they are both inexpensive and efficient to operate and could solve fossil fuel-oriented environmental matters, compared to traditional energy sources such as gas and oil. Gad et al. [19] built a solar energy heating system for poultry shed to assess the energy efficiency and system cost. It is observed that the thermal and electrical efficiencies could achieve 71.6% and 12.5%, respectively. The electricity cost of renewable technology is about 1.12 US $/kg which is less compared to 1.46 US $/kg of the conventional power operating system. Mirzaee Ghaleh et al. [20] built a solar thermal collector system for heating a poultry shed in Iran, and demonstrated that the system could fulfill at least 20% of the energy demand in the heating season. Bazen et al. [21] carried out an economic evaluation of a solar PV system for Tennessee's poultry farm in USA, and concluded that the effects of initial cost, installed expense and tax credit on the net present value could reach 35%, 10.6% and 15%, respectively. Fawaza et al. [22] performed a techno-economic assessment of a solar heating system for broiler breeding in Lebanon. The results illustrated that the hybrid system can achieve 74% of energy-saving and overlay 84% of heating load demand in the heating season. Moreover, annual operating cost saving is approximately $3389, resulting in a 4.6-year' payback period. Chen and Sheng [23] proposed a solar thermal vacuum tube collector system for warming poultry shed, and found that the system can save around 148.6 kg of CO_2 emissions.

Geothermal energy is a potential heat source to provide space heating for a chicken shed owing to the comparatively constant temperature of the soil. And also it has minimal maintenance during the long operation period. As a result, the influences of the GHP poultry shed on indoor temperature control and energy efficiency assessment are investigated in some case studies [24, 25]. Specifically, Kharseh and Nordell [24] developed an integrated solar-geothermal system for evaluating the energy demand for a poultry shed in Syria. It is concluded that this hybrid unit can generate 92 MWh of heating and 13 MWh of cooling, respectively. Choi et al. [25] applied a GHP unit for heating a broiler in Korea. It is demonstrated that the maximum and minimum indoor temperatures could be maintained in the range of 26.4°C-33.5°C and 22.4°C-30.9°C, respectively.

Green poultry shed is a wise choice for resolving basic and applied problems in connection with livestock production in an economic and ecological way. There is still currently a research gap in the area of investigating the energy, economic and environmental (3E) evaluation to study conversion efficiency, economic and GHG emission elements for design and performance estimation of the renewable energy unit in poultry shed. The major novelty of this work is to utilize the techno-economic evaluation approach to predict the annual electrical and thermal energy output and calculate system electrical and thermal energy cost savings, net present value, payback period and GHG emission.

2. System description

Figure 1 illustrates the fundamental design schematic of the hybrid renewable heating system. To be more specific, the hybrid system could produce highly efficient heating by solar photovoltaic/thermal (PV/T) array with a novel category of cheap polyethylene heat exchanger (PHE) loop assisted a heat pump and couple to a low expense geothermal copper pipe array. The PV/T module can simultaneously produce electrical energy for driving the running of heat pump compressor and thermal energy for inputting to the heat pump evaporator along with the geothermal pipe array. In the meantime, geothermal energy can offset the heat source production from the PV/T module for heat pump condenser, such as in the absence of solar radiation or nighttime. The fan coil as a recirculation device is utilized to provide space heating to the poultry shed.

2.1 JWL poultry farm

This hybrid heating renewable system is installed in an actual poultry farm called John Wright Ltd. (JWL), located in Newark-on-Trent of Nottinghamshire in the East Midlands of England, UK. It has four poultry houses with 40,000 chickens in at a time. **Figure 2(a)** presents an actual photo of JWL. The dimension of shed 1 is 62 × 8 × 2 m (L × W × D) which is selected for the hybrid heating system because it is the smallest one on-site and thus needs the least heating demand in winter as given in **Figure 2(b)-(d)**. The photo of young birds, the control unit and 66 kW gas burner are shown in **Figure 2(e)-(g)**, respectively. Additionally, in shed 1, the indoor temperature should be maintained in the range from 32–20°C. The relative humidity (RH) varies from 50–70%. Two 66 kW gas burners are utilized to warm the poultry shed in winter.

2.2 Meteorological data

Meteorological data are crucial for assessing the thermal and electrical energy output and shed heating load. **Figure 3** describes the average ambient temperature,

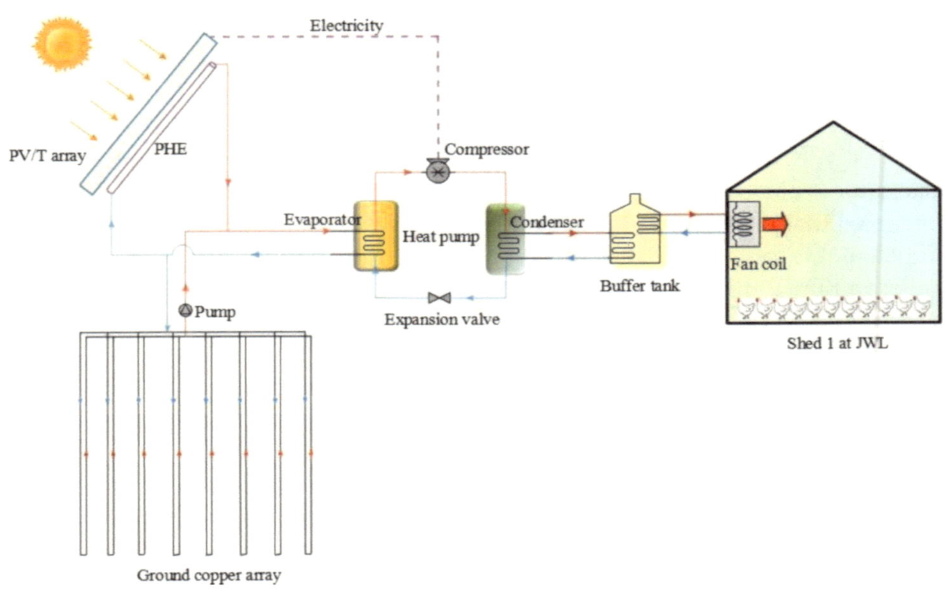

Figure 1.
Schematic diagram of hybrid renewable heating system.

DOI: http://dx.doi.org/10.5772/intechopen.102025

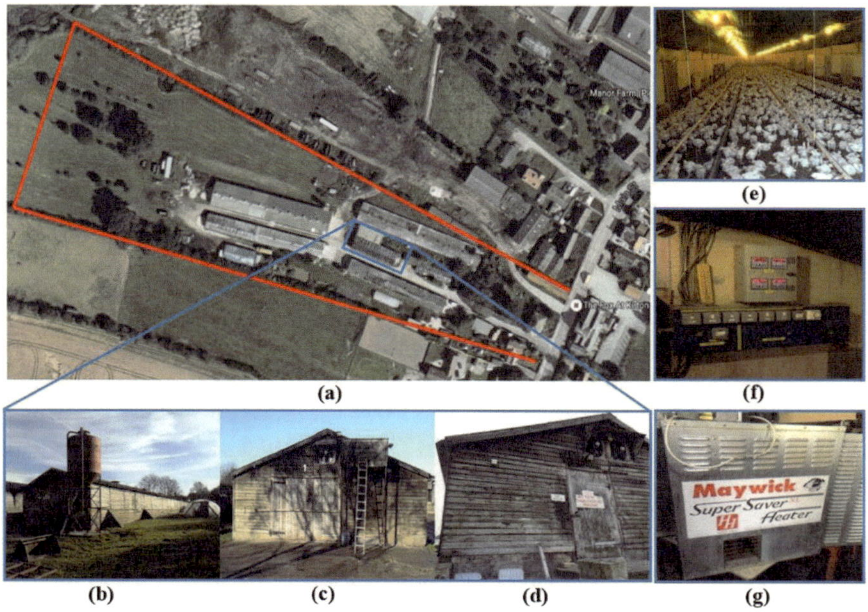

Figure 2.
JWL photos: (a) satellite view; (b) south view; (c) back view; (d) front view; (e) young birds; (f) heating control unit; (g) 66 kW gas burner.

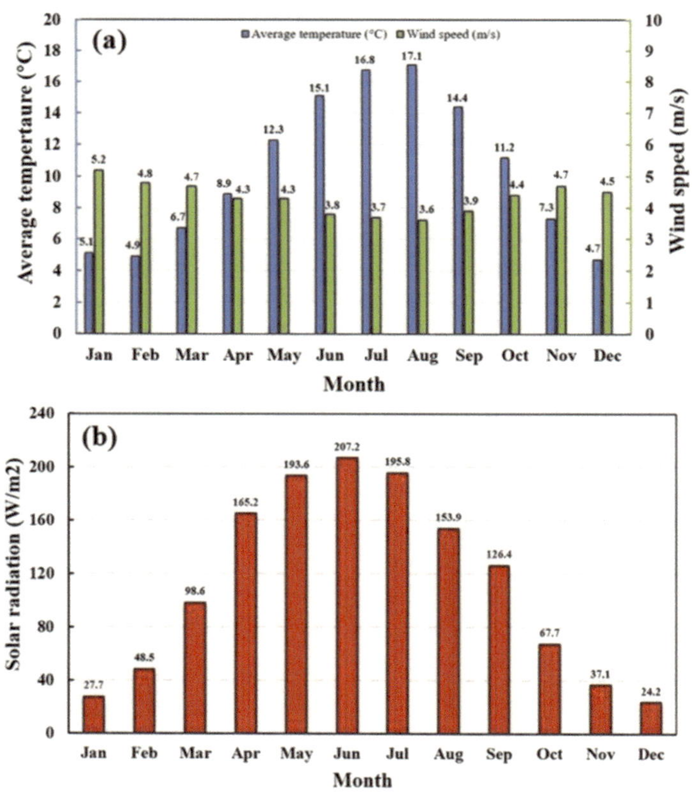

Figure 3.
Weather conditions: (a) average air temperature and wind velocity; (b) solar radiation in Newark over a year.

wind velocity as well as solar radiation [26]. Specifically, the highest average ambient temperature is 17.1°C in August whereas the lowest reaches 4.7°C in December. The monthly wind speed varies from 3.6 m/s to 5.2 m/s on average. Meanwhile, the highest and lowest solar radiation are 207.2 W/m^2 in June and 24.2 W/m^2 in December.

2.3 Experimental description

Figure 4 depicts the layout of the hybrid heating system installation. Because the survey presented that there are not any poultry sheds in JWL which are fit for the installation of the PV/T array. Hence, only the workshop, which is located next to shed 1 as presented in **Figure 4(a)** is fulfilled the requirement of the structural reinforcement. The energy output from the PV/T module is piped down the side of the workshop, under the entrance road and through shed 1 to the plant room as exhibited in **Figure 4(b)**. Moreover, 52 (260Wp) Canadian solar PV panels [27] and four 1 × 12 m PHE mats are mounted on the roof of the workshop with 15° oriented to the south for enhancing solar energy harvesting. The solar PV inverter is placed on the external wall of the workshop as displayed in **Figure 4(c)-(d)**. A 15 kW F1145 NIBE heat pump [28] with R407C refrigerant is designated as it is the biggest capacity the consortium and can afford almost the required thermal energy input from the PV/T and ground copper pipe arrays as given in **Figure 4(e)**. The heat pump system could be employed in reverse to supply cooling in summer if

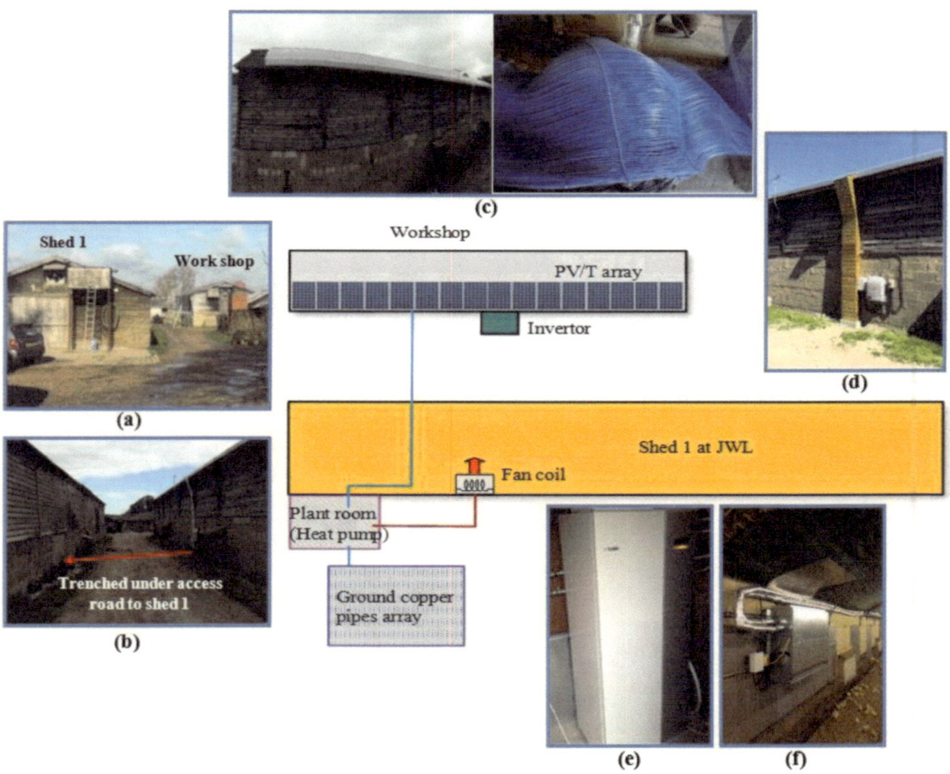

Figure 4.
Hybrid renewable energy heating system installation: (a) workshop and shed 1; (b) trenched under access road to shed 1; (c) PV/T installed on the workshop roof; (d) inverter along with the external wall; (e) NIBE F1145 heat pump; (f) fan coil.

required. Additionally, the fan coil is mounted vertically on the south internal to provide spacing heating for shed 1 as shown in **Figure 4(f)**. Furthermore, there are fifty ground copper pipes with 15 mm diameter and their dimension size is 2.5 × 5 m (length × deep). The overall surface region of the geothermal pipe array is approximately 10 × 10 m. The vertical copper pipes are connected and run back to the plant room.

3. Numerical model

3.1 Energy model

3.1.1 PV/T array

Thermal energy conversion is classified into two processes including solar radiation conversion into thermal energy and transferring collected thermal energy towards PHE. Hence, this dynamic model is expressed by:

$$\frac{\partial Q_t}{\partial t} = m_{PV/T} C_{PV/T} \frac{\partial T_{PV/T}}{\partial t} = Q_{abs} - Q_{PV/T-loss} - Q_{ele} \tag{1}$$

where Q_t is the overall thermal energy (kW); $m_{PV/T}$ is the mass of PV/T (kg); $C_{PV/T}$ is the heat capacity of PV/T (J/ (kg·K)); $T_{PV/T}$ is the temperature of PV/T (°C); t is the time (s); Q_{abs} is the solar energy absorbed (kW); $Q_{PV/T\text{-}loss}$ is the overall thermal loss (kW); Q_{ele} is the power output (kW).

$$Q_{abs} = \tau_c \alpha_{abs} A_{eff} I \tag{2}$$

τ_c is the transmittance of PV/T; α_{abs} is the absorptivity of PV/T; A_{eff} is the effective area of PV/T (m^2); I is the solar radiation (W/m^2);

$$Q_{PV/T-loss} = Q_{conv,c} + Q_{c,sky} + Q_{conv,pl,heo} + Q_{pl,heo} \tag{3}$$

$$Q_{conv,c} = h_{cv}(T_c - T_a) \tag{4}$$

where h_{cv} is the forced convection coefficient ($W/m^2 \cdot K$), which is written as:

$$h_{cv} = 5.7 + 3.8 \cdot V_{wind} \tag{5}$$

$$T_c = 30 + 0.0175 \times (\mathrm{I} - 300) + 1.14 \times (\mathrm{T_a} - 25) \tag{6}$$

where V_{wind} is the wind velocity (m/s); T_c is the PV surface temperature (°C); T_a is the air temperature (°C);

$$Q_{c,sky} = \varepsilon_c \cdot \sigma \cdot \left(T_c^4 - T_s^4\right) \tag{7}$$

$$T_s = 0.037536 \cdot T_a^{1.5} + 0.32 T_a \tag{8}$$

where ε_c is the emissivity of PV/T cover layer; σ is the Stefan-Boltzmann's constant, 5.67×10^{-8} $W/m^2 \cdot K^4$; T_s is the sky temperature (°C).

$$Q_{conv,pl,heo} = h_{air} \cdot \left(T_{pl} - T_{heo}\right) \tag{9}$$

where h_{air} is the convective heat transfer coefficient ($W/m^2 \cdot K$); T_{pl}, T_{heo} are the temperature of the EVA layer and PHE wall temperature, respectively (°C).

$$h_{air} = \frac{N_u \cdot \lambda_{air}}{\delta_{air}} \tag{10}$$

where λ_{air} is the air thermal conductivity (W/m K); δ_{air} is air gap thickness between the surface cover and PV module (m).

Nu is the Nusselt number as expressed:

$$N_u = \left[0.06 - 0.017\left(\frac{\beta_s}{90}\right)\right] \mathrm{Gr}^{1/3} \tag{11}$$

where β_s is the title–angle of PV panels; Gr is the Grashoff number given as:

$$Gr = \frac{g \cdot (\mathrm{T}_{pl} - T_{heo}) \cdot \delta_{air}^3}{\nu_{air}^2 \cdot T_{air}} \tag{12}$$

$$Q_{pl,heo} = \varepsilon_{pl} \cdot \sigma \cdot \left(T_{pl}^4 - T_{heo}^4\right) \tag{13}$$

where ε_{pl} is the emissivity of EVA layer;

$$Q_{ele} = \eta_e A_{eff} I \tag{14}$$

where η_e is the electrical efficiency of PV array (%); A_{eff} is the effective area of PV array (m^2).

The overall heat production is written as:

$$Q_t = A_{eff} \cdot h_t \cdot (T_a - T_w) \tag{15}$$

where h_t is the overall heat transfer coefficient between the water and PV module ($W/m^2 \cdot K$); T_w is the water temperature within the PHE (°C).

$$\eta_t = \frac{Q_t}{A_{eff} \cdot I} \tag{16}$$

where η_t is the thermal efficiency (%).

3.1.2 Ground copper pipe array

To supply an adequate heat source for the evaporator of the heat pump unit, a low expense geothermal copper pipe array is developed. In lights of the heat transfer fields, it is categorized into solid and fluid regions [29].

3.1.2.1 Solid field

The solid field contains soil and pipe wall as presented:

$$\rho_{soil} c_{soil} \frac{\partial T_s}{\partial t} = \frac{\partial}{\partial x}\left(\lambda_{\text{soil}} \frac{\partial T_s}{\partial x}\right) + \frac{\partial}{\partial y}\left(\lambda_{\text{soil}} \frac{\partial T_s}{\partial y}\right) + \frac{\partial}{\partial z}\left(\lambda_{\text{soil}} \frac{\partial T_s}{\partial z}\right) \tag{17}$$

$$\rho_{pipe} c_{pipe} \frac{\partial T_p}{\partial t} = \frac{\partial}{\partial x}\left(\lambda_{pipe} \frac{\partial T_p}{\partial x}\right) + \frac{\partial}{\partial y}\left(\lambda_{pipe} \frac{\partial T_p}{\partial y}\right) + \frac{\partial}{\partial z}\left(\lambda_{pipe} \frac{\partial T_p}{\partial z}\right) \tag{18}$$

where ρ is the density (kg/m^3); c is the thermal capacity (J/kg·°C); λ is the thermal conductivity (W/m·K).

3.1.2.2 Fluid field

The energy balance equation of the inlet pipe is given as:

$$\rho_{fluid}c_{fluid}\frac{\partial T_{inlet}}{\partial t}+(\rho cv)_f\frac{\partial T_{inlet}}{\partial z}=\lambda_{fluid}\frac{\partial^2 T_{inlet}}{\partial z^2}+b_{ig}\left(T_{grout}-T_{inlet}\right) \tag{19}$$

The energy balance equation of the outlet pipe is expressed as:

$$\rho_{fluid}c_{fluid}\frac{\partial T_{outlet}}{\partial t}+(\rho cv)_f\frac{\partial T_{outlet}}{\partial z}=\lambda_{fluid}\frac{\partial^2 T_{outlet}}{\partial z^2}+b_{og}\left(T_{grout}-T_{outlet}\right) \tag{20}$$

3.1.3 Heat pump

The heat source of the heat pump is provided by PV/T and geothermal copper pipe to offer a comfortable climate for the shed in the heating season. Hence, a heat pump model is expressed as [30]:

$$m_r=V_c\omega\rho_{r,suc}\cdot\left[1+C_v\left(1-\frac{\mathrm{P}_{r,cond}}{\mathrm{P}_{r,evap}}\right)^{\frac{1}{n}}\right] \tag{21}$$

$$\Delta\xi_{comp}=\xi_{r,dis}-\xi_{r,suc}=\frac{n}{n-1}\cdot\frac{\mathrm{P}_{r,evap}}{\rho_{r,suc}}\cdot\left[\left(\frac{\mathrm{P}_{r,cond}}{\mathrm{P}_{r,evap}}\right)^{\frac{n-1}{n}}-1\right] \tag{22}$$

$$Q_{el}=\frac{m_r\Delta h_{comp}}{\eta_{comp}} \tag{23}$$

where m_r is the refrigerant mass flow rate (kg/s); V_c is the compressor swept volume (m^3); ω is the compressor rotational speed (rev/s);

The coefficient of performance (COP) is given as:

$$COP_h=\frac{Q_{heating}}{Q_{el}} \tag{24}$$

3.2 Economic model

Economic policy has a vital effect on the life cycle cost (LCC) analysis in lights of the hybrid renewable heating system. Hence, the LCC is given as [31, 32]:

$$LCC=E_{IC}+\sum_{i=1}^{n}\left(E_{SEC}+E_{ME}+E_{PC}+E_{ITS}\right) \tag{25}$$

where LCC is the system lifetime expense (£); E_{IC} is the original expense (£); E_{SEC} is the system energy cost (£); E_{ME} is the maintenance cost (£); E_{PC} is the system periodic cost (£); E_{ITS} is the system income tax savings (£).

The payback period (PBP) is employed to determine the time required to recoup the fund expended in an investment [33, 34].

$$PBP=X+\frac{Y}{Z} \tag{26}$$

where X is the number of years of final recovery (£); Y is the balance amount to be recovered (£); Z is the cash inflow (£).

4. Results and discussion

4.1 Model validation

Before performing the prediction of the annual thermal and electrical energy output, it is necessary to validate the accuracy of the numerical model. Hence, the comparisons between numerical and experimental results, in terms of the PV electrical energy production, PHE and thermal energy output from the heat pump, are analyzed based on the error analysis model from 01/Nov/2016 to 31/Jan/2017.

$$Error = \left| \frac{T_{numerical} - T_{experimental}}{T_{numerical}} \right| \tag{27}$$

4.1.1 Electrical energy production from PV array

Figure 5 displays the comparison of daily electrical energy output from PV array based on the simulation and test results. It is concluded that the error is up to 14.93% occurred at the termination of the operating phase, the mean error reaches 9.26%. Moreover, the experimental data demonstrated that the total electrical energy production could achieve 1125.89 kWh from 01/Nov/2016 to 31/Jan/2017 (228 days), by contrast, the numerical result exhibits close proximity of value, achieving 1247.51 kWh within a 10% error. This means that the numerical result is in very good agreement with experiment data, which validates the reliability of the numerical model.

4.1.2 Thermal energy output from PHE

It can be observed from **Figure 6** that the temperature variation between the experimental data and the numerical result has a similar trend. The highest thermal fluid temperature within the PHE reaches 15.75°C on 10/NOV/2016, by comparison, the lowest one is 0.75°C on 25/JAN/2017. And also, the minimum temperature difference was around 3.29% on 31/DEC/2016, the mean one being 9.11%, while the maximum temperature difference is approximately 14.72% on 14/NOV/2016.

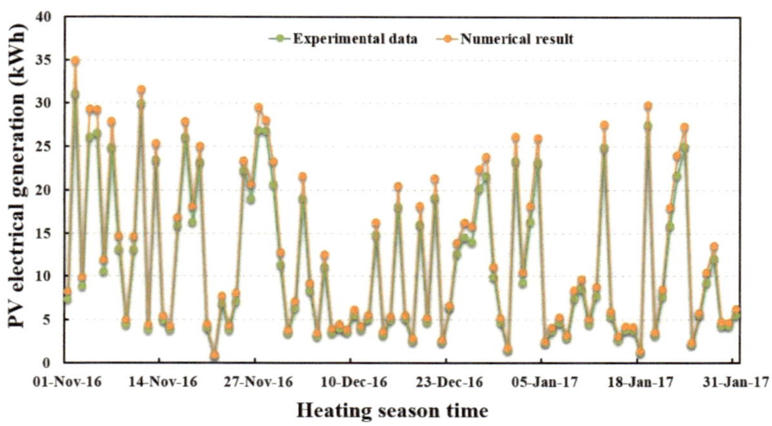

Figure 5.
Electrical energy output from PV array.

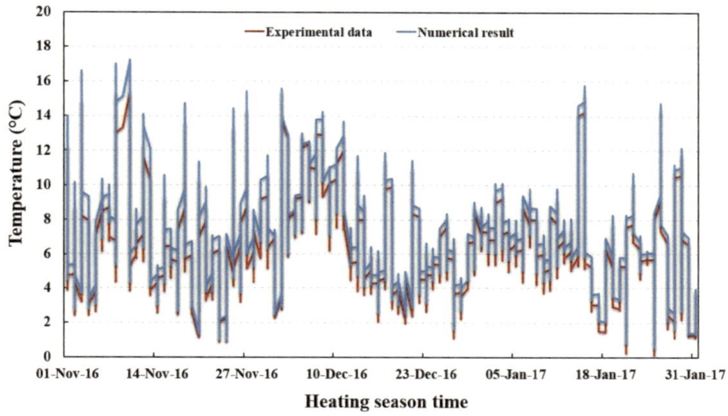

Figure 6.
Thermal energy output from PHE.

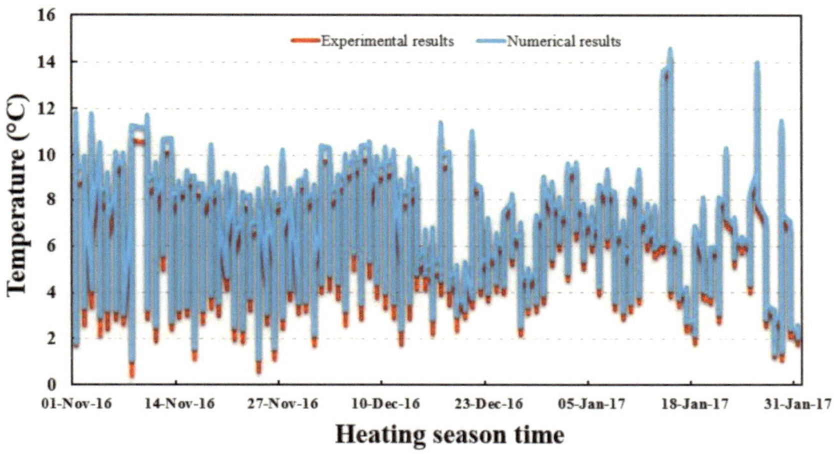

Figure 7.
Heat production from ground copper pipe.

4.1.3 Thermal output from ground copper pipe

Figure 7 displayed a similar temperature change tendency between the simulation and test results. Specifically, the temperature of the test could realize up to 14.23°C on 15/JAN/2017 whereas the lowest temperature could reach 0.89°C on 08/NOV/2016. Additionally, the maximum, minimum and average relative errors are 11.33% on 16/DEC/2016, 2.40% on 15/JAN/2017 and 6.36%, as clarified in **Table 1**.

4.1.4 Thermal energy output from the heat pump

Figure 8 compared the thermal energy output from the heat pump system, and found that the daily maximum and minimum differences are about 9.30% appeared on 16/JAN/2017 and 5.49% occurred on 05/DEC/2016, respectively. Consequently, the numerical model could be employed to assess the annual thermal and electrical energy output of the hybrid renewable heating system over a year.

Table 1 illustrated the relative error analysis of PV electrical, thermal, geothermal thermal and heat pump outputs between simulation results and experimental data. It is found that all error values are less than 15% which fulfill the requirement.

Date	PV electrical output (%)	PHE thermal output (%)	Geothermal thermal output (%)	Heat pump output (%)
03/NOV/2016	14.84	12.39	8.07	7.01
14/NOV/2016	4.15	14.98	7.07	6.33
28/NOV/2016	9.05	11.86	6.73	5.94
15/DEC/2016	8.22	9.75	14.79	5.49
23/DEC/2016	3.48	8.78	6.12	9.01
31/DEC/2016	14.88	3.29	4.98	8.28
08/JAN/2017	14.93	8.50	4.47	8.86
21/JAN/2017	8.45	8.91	2.40	9.30
26/JAN/2017	5.38	3.49	2.57	9.13

Table 1.
Relative error analysis of PV electrical, thermal, geothermal thermal and heat pump outputs.

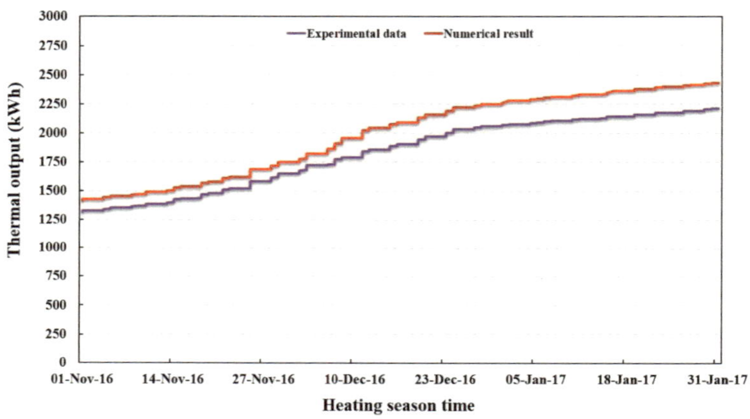

Figure 8.
Thermal energy output from the heat pump system.

4.2 Year-round system performance assessment

Figure 9 depicts the monthly power energy production and efficiency from the PV array. It can be concluded that the minimum and the maximum monthly electrical energy generation are around 335.81 kWh with the lowest efficiency (about 14.85%) in December and 1830.35 kWh with the highest efficiency (about 15.9%) in June, respectively. And also, the overall electrical energy obtained from the PV array could reach 11,867 kWh during a year. This means that it not only can meet the power demand of the poultry shed, but also could supply around 43.5% power requirement of the heat pump compressor operating.

Additionally, diminishing the PV surface temperature contributes to increasing the voltage and electrical efficiency. The PHE under the PV array could help to decrease the PV surface temperature resulting in a PV electrical efficiency enhancement. **Figure 10** exhibits the monthly thermal energy output and COP variation of the heat pump unit. Results show that the highest and the lowest monthly thermal energy output could achieve 3848.77 kWh in July and 1610.77 kWh in February, respectively. And also, the overall thermal energy output is about 30210.98 kWh per annum. This means that some capacity of the gas burners would

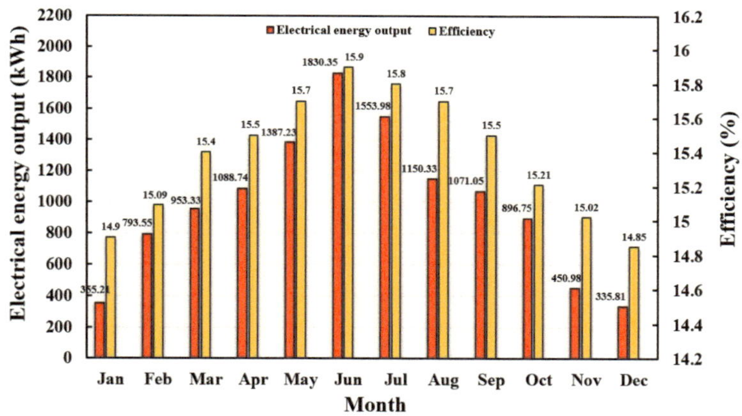

Figure 9.
Monthly PV electrical energy production and efficiency.

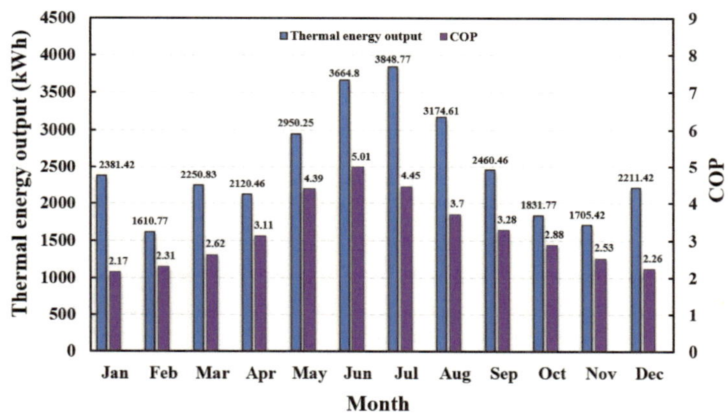

Figure 10.
Monthly thermal energy output and COP of heat pump system.

be needed alongside the heat pump to warm the shed sufficiently, especially from December to February.

Furthermore, the highest and lowest PV/T thermal efficiency could reach 28.3% in June and 7.3% in December. When the PV/T operates in conjunction with the geothermal copper pipe array, the COP of the heat pump could achieve up to 5.01 in June, while a minimum COP of 2.17 can be achieved in December.

4.3 Economic assessment

It can be observed from **Figure 11** that the comparison of gas and electricity cost between the current system and PV/T and heat pump system each period. Notably, the overall cost of the PV/T with heat pump system is lower than the gas burners system. To be more specific, the gas cost of the PV/T with heat pump system is approximately £319.74, which is significantly lower than that of the gas burner system (approximately £1083), saving about £763. Similarly, the electrical cost of the PV/T with heat pump system is approximately £128.52, which is lower compared to the gas burner system (approximately £893), saving about £750. Additionally, the payback period is about 5.5 years.

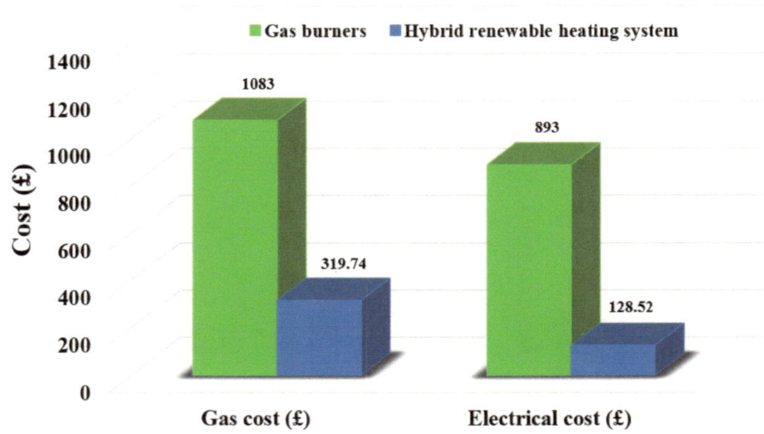

Figure 11.
Comparison of gas and electrical cost between gas burners and hybrid renewable heating systems.

4.4 Environmental evaluation

To perform the environmental evaluation, the CO_2 emissions related to gas and electricity prices are needed, as applied to the UK. Therefore, the values assigned to the parameters in this study are provided [34, 35] including 0.5246 kg CO_2 (e)/ kWe h for electricity and 0.1836 kg CO_2 (e)/kWth h for gas. Additionally, in the UK, the electricity price at Feed-In Tariff (FIT) and the renewable heat incentive (RHI) are £0.1097/kWh and £0.052/kWh, respectively [36, 37]. Results confirmed from **Figure 12** that the gas burners system produces energy consumption of about 39,851 kWh resulting in about 10.27 tons CO_2 emission, while the hybrid renewable heating system has an only energy consumption of 10,978 kWh which is equivalent to 2.026 tons CO_2 emission. This indicates that the novel hybrid system could save about 28,873 kWh energy consumption, making for a reduction of total CO_2 emission of approximately 8.3 tons. **Table 2** describes the calculation processes of energy consumption, CO_2 emission and operating cost of the gas burners and PV/T with heat pump systems.

4.5 Summary

To sum up, the hybrid renewable energy heating system could save 28,873 kWh of thermal and electrical energy consumption, £1528 of operating

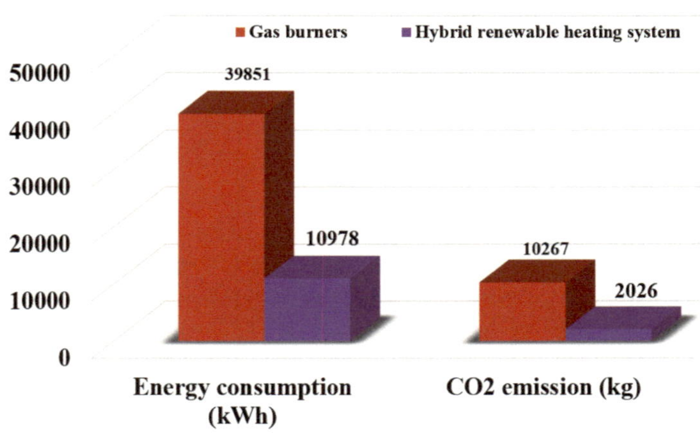

Figure 12.
Comparison of energy consumption and CO_2 emission between gas burners and hybrid renewable heating systems.

System/items	Gas burners system	PV/T with heat pump system	Saving
Total energy consumption, kWh (228 days)	31,199 + 8652 = **39,851**	10946.57 + 31.21 = **10,978**	28,873 kWh
CO_2 emission, tons per (228 days)	31,199 × 0.1836 + 8652 × 0.5246 =10266.98/1000 = **10.27 tons**	10946.57 × 0.1836 + 31.21 × 0.5246 =2026.16 kg/1000 = **2.03tons**	8.3 tons
Operating cost (£) (228 days)	£1083 + £893 = **£1976**	£319.74 + £128.52 = **£448.26**	£1528

Table 2.
Calculation process of energy consumption, CO_2 emission and operating cost between gas burners and PV/T with heat pump systems.

cost with 5.5 years' payback period and 8.3 tons of CO_2 emission. Additionally, the electrical output of the PV/T array could achieve approximately 11,867 kWh per annum whereas the thermal energy output is about 30,210 kWh per annum.

According to previous studies [19, 22, 25, 38], it is observed that the hybrid of solar and geothermal energy systems used in a poultry house is rare. Specifically, Fawaz et al. [22] demonstrated that the solar-assisted localized heating system could save approximately 74% of the energy demand and exhibit a 4.6 years of payback period. To improve the chicken meat and eggs production, Gad et al. [19] concluded that the thermal efficiency of the solar heating system is about 71.6% whereas the PV electrical efficiency is 12.5%. Choi et al. [25] designed, constructed and tested a geothermal heat pump system for ameliorating the interior environment of the poultry shed. It is demonstrated that the average interior air temperature could be kept in the range from 24.8 to 32.2°C whereas the relative humidity varies from 45.2 to 72.6%. Moreover, the GHP poultry house could save about 92% of the overall energy expense in comparison with the normal poultry shed. And also, the concentration of CO_2 in the GHP poultry house could be decreased by 3299 ppm, by comparison, in the conventional shed, it is decreased by 4945 ppm. Uzodinma et al. [38] assessed the performance of a solar thermal collector with a phase change materials system for poultry incubating chamber, and observed that the temperature of the chamber could be kept in the range from 36 to 39°C, meanwhile, an average egg hatchability could reach 62.37%.

Herein, this proposed hybrid PV/T with heat pump system could allow taking the benefit of high solar irradiation rates and soil heat, thus improving system performance, ameliorating the interior environment of poultry shed and boost meat and eggs production.

5. Conclusions

In this chapter, a novel PV/T with PHE array coupled to a low expense geothermal copper pipe array and heat pump system is installed in the JWL poultry farm. A numerical model is established and has a good agreement with experimental data within a 15% error. In the meantime, the yearly system energy output is predicted based on the Newark-on-Trent of Nottinghamshire, UK weather conditions. Moreover, the comparisons of the gas cost, electrical cost and CO_2 emission are investigated between the gas burners and hybrid renewable heating systems. Some significant outcomes are obtained as below:

- The electrical production from the PV array could realize 11,867 kWh per annum. It not only meets the power requirement of the poultry shed, but also supply around 43.5% power needed for the heat pump compressor operation.
- The heat pump thermal energy output is about 30,210 kWh per annum, which indicates that some capacity of the gas burners would be required alongside the heat pump to warm the shed adequately in winter.
- When the PV/T operates in conjunction with the geothermal pipe, the COP of the heat pump could reach up to 5.01 in June, while a minimum COP of 2.17 could be achieved in December.
- The overall gas and electrical cost of the hybrid renewable heating system are £320 and £129, which are much less than that of the gas burners system saving £763 and £750, respectively, resulting in less than 6 years of payback period.
- In comparison with the gas burner, the energy consumption of the hybrid renewable heating system can decrease about 28,873 kWh, making for a decline of total CO_2 emission of approximately 8.3 tons.

Acknowledgements

The authors would like to acknowledge the support from Innovate UK Agri-Tech Catalyst for the Welfare Enhanced Living Conditions for Healthier Chickens (WelChic) project (131897).

Conflict of interest

The authors declare that they have no known competing financial interests or personal relationships that could have appeared to influence the work reported in this paper.

Acronyms and abbreviations

CO_2	carbon dioxide
COP	coefficient of performance
FIT	feed-in tariff
GHG	greenhouse gas
GHP	geothermal heat pump
LCC	life cycle cost
NH3	ammonia
PBP	payback period
PHE	polyethylene heat exchanger
PV	photovoltaic
PV/T	photovoltaic/thermal
RH	relative humidity
RHI	renewable heat incentive
WelChic	welfare enhanced living conditions for healthier chickens
3D	three-dimensional
3E	energy, economic and environmental

Nomenclature

A	area (m^2)
c	thermal capacity (J/kg·K)
H	depth (m)
h	heat transfer coefficient [W/(m·K)]
λ	thermal conductivity [W/(m·K)]
r	radius (m)
T	temperature (°C)

Author details

Yuanlong Cui[1,2]*, Saffa Riffat[1], Elmer Theo[1], Tugba Gurler[1], Xuan Xue[3] and Zohreh Soleimani[2]

1 Department of Architecture and Built Environment, The University of Nottingham, Nottingham, UK

2 School of Built Environment, College of Engineering and Technology, University of Derby, Derby, UK

3 School of Pharmacy, The University of Nottingham, Nottingham, UK

*Address all correspondence to: ezzyc14@yahoo.com

References

[1] Wreford A, Topp CFE. Impacts of climate change on livestock and possible adaptations: A case study of the United Kingdom. Agricultural Systems. 2020; **178**:102737

[2] Rasul G. Twin challenges of COVID-19 pandemic and climate change for agriculture and food security in South Asia. Environmental Challenges. 2021;**2**:100027

[3] Izar-Tenorio J, Jaramillo P, Griffin WM, Small M. Impacts of projected climate change scenarios on heating and cooling demand for industrial broiler chicken farming in the eastern U.S. Journal of Cleaner Production. 2020;**255**:120306

[4] Anastasiadis F, Tsolakis N. Environmental hotspots analysis: A systematic framework for food supply chains and implementation case in the UK poultry industry. Journal of Cleaner Production. 2021;**305**: 126981

[5] Costantino A, Fabrizio E, Ghiggini A, Bariani M. Climate control in broiler houses: A thermal model for the calculation of the energy use and indoor environmental conditions. Energy and Buildings. 2018;**169**:110-126

[6] Du L, Hu C, Yang C, Yang L, Du H, Li Q, et al. Investigation of a preliminary ventilation energy-recovery system for poultry houses. Computers and Electronics in Agriculture. 2020; **175**:105521

[7] Zhao Y, Shepherd TA, Li H, Xin H. Environmental assessment of three egg production systems–part I: Monitoring system and indoor air quality. Poultry Science. 2015;**94**:518-533

[8] Ren G, Lin T, Ying Y, Chowdhary G, Ting KC. Agricultural robotics research applicable to poultry production: A review. Computers and Electronics in Agriculture. 2020;**169**:105216

[9] Costantino A, Fabrizio E, Biglia A, Cornable P, Battaglini L. Energy use for climate control of animal houses: The state of the art in Europe. Energy Procedia. 2016;**101**:184-191

[10] Weindl I et al. Livestock in a changing climate: Production system transitions as an adaptation strategy for agriculture. Environmental Research Letters. 2015;**10**:094021

[11] Bazen EF, Brown MA. Feasibility of solar technology (photovoltaic) adoption: A case study on Tennessee's poultry industry. Renewable Energy. 2009;**34**(3):748-754

[12] Jones E, Qadir M, van Vliet MTH, Smakhtin V, Kang S, mu. The state of desalination and brine production: A global outlook. Science of the Total Environment. 2019;**657**:1343-1356

[13] Tomaszewska B, Akkurt GG, Kaczmarczyk M, Bujakowski W, Keles N, Jarma YA, et al. Utilization of renewable energy sources in desalination of geothermal water for agriculture. Desalination. 2021;**513**: 115151

[14] Al-Karaghouli A, Kazmerski LL. Energy consumption and water production cost of conventional and renewable-energy-powered desalination processes. Renewable and Sustainable Energy Reviews. 2013;**24**:343-356

[15] Ahmed FE, Hashaikeh R, Hilal N. Solar powered desalination-technology, energy and future outlook. Desalination. 2019;**453**:54-76

[16] Dashtpour R, Al-Zubaidy SN. Energy efficient reverse osmosis desalination process. Journal of Membrane Science. 2012;**3**(4):339

[17] Shahzad MW, Burhan M, Ang L, Ng KC. Energy-water-environment nexus underpinning future desalination sustainability. Desalination. 2017;**413**: 52-64

[18] Shahzad MW, Ng KC, Burhan M, Chen Q, Jamil MA, Imtiaz N, et al. Demystifying integrated power and desalination processes evaluation based on standard primary energy approach. Thermal Science and Engineering Progress. 2022;**27**:101153

[19] Gad S, El-Shazly MA, Wasfy KI, Awny A. Utilization of solar energy and climate control systems for enhancing poultry houses productivity. Renewable Energy. 2020;**154**:278-289

[20] Mirzaee-Ghaleh E, Omid M, Keyhani A, Javadikia P. Forecasting the thermal load for implementing solar energy in a model poultry house. Journal of Agricultural Engineering and Biotechnology. 2013;**1**:30-36

[21] Bazen EF, Brown MA. Feasibility of solar technology (photovoltaic) adoption: A case study on Tennessee's poultry industry. Renewable Energy. 2009;**34**:748-754

[22] Fawaza H, Abiad MG, Ghaddar N, Ghali K. Solar-assisted localized ventilation system for poultry brooding. Energy and Buildings. 2014;**71**:142-154

[23] Chen W, Sheng C. Mitigation of carbon dioxide emissions in a warming system for chicks by using solar energy. Life Science Journal. 2013;**10**:1845-1850

[24] Kharseh M, Nordell B. Sustainable heating and cooling systems for agriculture. International Journal of Energy Research. 2014;**35**:415-422

[25] Choi HC, Salim HM, Akter N, Na JC, Kang HK, Kim MJ. Effect of heating system using a geothermal heat pump on the production performance and housing environment of broiler chickens. Poultry Science. 2012;**91**: 275-281

[26] Weather statistics for Newark-On-Trent, England. Available at: Newark-On-Trent, Nottinghamshire, United Kingdom Historical Weather Almanac (worldweatheronline.com)

[27] CS6P-250P (250W) Solar Panel. Available at :CS6P-P_en.cdr (zonnepanelen.net)

[28] Ground Source Heat Pump NIBE™ F1145. Available at: http://www.ecolivinguk.com/media/29491/nibe_product_guide_ f1145.pdf

[29] Cui Y, Zhu J. 3D transient heat transfer numerical analysis of multiple energy piles. Energy and Buildings. 2017;**134**:129-142

[30] Jin H, Spitler JD. A parameter estimation based model of water-to-water heat pumps for use in energy calculation programs. ASHRAE Transactions. 2002;**108**:4493-4510

[31] Gu Y, Zhang X, Myhren JA, Han M, Chen X, Yuan Y. Techno-economic analysis of a solar photovoltaic/thermal (PV/T) concentrator for building application in Sweden using Monte Carlo method. Energy Conversion and Management. 2018;**165**:8-24

[32] Cui Y, Zhu J, Zoras S, Qiao Y, Zhang X. Energy performance and life cycle cost assessments of a photovoltaic/thermal assisted heat pump system. Energy. 2020;**206**:118108

[33] Chandra S, Yadav A. Site selection based on thermo mechanical decay and payback period of solar PV system: Need of present scenario. Materials Today: Proceedings. 2021;**43 Part 1**: 287-292

[34] Herrando M, Markides CN, Hellgardt K. A UK-based assessment of hybrid PV and solar-thermal systems for

domestic heating and power: System performance. Applied Energy. 2014;**122**: 288-309

[35] Amrizal N, Chemisana D, Rosell JI. Hybrid photovoltaic–thermal solar collectors dynamic modeling. Applied Energy. 2013;**101**:797-807

[36] Ofgem. Feed-in Tariff: Guidance for renewable installations. 2016. Available at: 2005_fits_guidance_for_installa tions_v15_clean.pdf (ofgem.gov.uk) [Accessed: June 20, 2016

[37] Ofgem. Tariffs and payments: Domestic RHI. 2016. Available at: Tariff s and payments: Domestic RHI | Ofgem [Accessed: April 15, 2016]

[38] Uzodinma EO, Ojike O, Etoamaihe UJ, Okonkwo WI. Performance study of a solar poultry egg incubator with phase change heat storage subsystem. Case Studies in Thermal Engineering. 2020;**18**:100593

Chapter 14

The Impact of Energy Efficiency Programmes in Ghana

Edwin Kwasi Tamakloe

Abstract

Ghana experienced widespread power shortages due to series of droughts spanning from the 1980s to the 1990s. Energy efficiency programmes were identified to solve these energy supply challenges. Consequently, the residential sector has been recognized as an important target group for energy efficiency programmes in the country. The residential sector in Ghana accounts for 47% of the total final energy use. Reducing the inefficiencies in the residential sector energy use could be an effective way of reducing global energy use and related environmental impacts. Therefore, Ghana enacted four Legislative Instruments to regulate the importation of refrigerating, air conditioning and lighting appliances and also to ensure these appliances meet the minimum energy performance standards (MEPS). The purpose of this paper was to review and establish the impact of the MEPS programmes in Ghana from 2007 to 2020. The content of this desktop review is based on data gathered through a series of reviews of available energy efficiency policy documents from governmental agencies. The results revealed that the implementation of MEPS programmes in Ghana yielded 8317.8 GWh of electricity savings, which translates into carbon emission reduction of 4.60 million tonnes of CO_2 and energy cost savings of USD 832 million in term of electricity bills.

Keywords: energy efficiency, minimum energy performance standards, standards and labelling and energy saving

1. Introduction

Globally, the building sector is accountable for 30% of the total final energy consumption and without action, the energy demand in this sector could increase by another 30% by 2060 [1]. Again, buildings represent 28% of global energy-related CO_2 emissions worldwide [2]. This sector's energy consumption is related to human programmes involving the use of equipment, lighting and other electrical appliances [3]. The household/residential sector, which forms part of the total building stocks, has been identified as one of the target groups for energy efficiency programmes [4, 5] also underscored that households can adopt and implement energy efficiency measures to reduce energy consumption significantly.

The residential sector in Ghana accounts for 47% of the total final energy consumption [6], compared to other countries such as the United States (25%), United Kingdom (30%), Japan (26%), Saudi Arabia (50%), China (15.8%) and Malaysia (15%) [5, 7–9]. Electrical appliances such as refrigerators, freezers and room air conditioners (RACs), which are energy-intensive, are among the most common Ghanaian household appliances. Whilst refrigerators and freezers consume between 25–30% of the total residential energy use, air conditioners account for 6.5% of residential energy use in Ghana [10]. Whilst [11] indicated that most of the markets in sub-Saharan Africa (SSA) countries are inundated with "used electrical appliances" [12], assert that this phenomenon was a result of the proliferation of used or "second-hand" goods from Europe and elsewhere. For instance, it has been estimated that there were over two (2) million used and inefficient refrigerating appliances in Ghana in 2012 that were laden with hydrofluorocarbons (HFCs) [13]. This large number was primarily due to the introduction of standards and labelling in developed countries and the era of energy efficiency programmes in Europe in the early 1970s saw many of these countries disposing of their old and energy inefficient appliances into SSA [14]. The share of the used electrical appliance market in Ghana at the time was about 80%, thereby promoting climate injustice [14]. These used and inefficient refrigerating appliances consume, on average, 1200 kWh per unit per year compared to 250 kWh and 400 kWh per year in Europe and the US respectively [15].

Minimum energy performance standards (MEPS) and the introduction of more efficient appliances through energy efficiency standards and labelling (EES&L) programmes have been identified as potential means of reducing the energy consumption of these inefficient appliances. It has been noted that improving energy efficiency is the best way to simultaneously meet sustainable development goals (SDGs) 7 and 13 in the energy sector [16]. This approach, when used judiciously, frees resources for other projects, helps economies to grow and reduces environmental impacts such as greenhouse gas (GHG) emissions. MEPS programmes have also proven to be effective in stimulating the development of cost-effective, energy-efficient technologies and are said to be the cornerstone of most national energy and climate change mitigation programmes [17].

Ghana's move to improve its energy efficiency is part of a larger energy sector reform programme designed to support the country's long-term economic developmental agenda. In 2002, Ghana, therefore, identified the benefits of MEPS such as EES&L programmes for equipment and other electrical appliances such as deep freezers, RACs, refrigerators, industrial motors and lighting systems [18]. These standards and labelling programmes serve as benchmarks and catalysts in meeting the MEPS objectives.

According to [19], energy efficiency legislations and policies have continued to increase globally through energy efficiency research and developmental programmes [20]. However, several barriers attributable to the low adoption and implementation of MEPS in SSA have to be overcome. These barriers, according to [21, 22], include financial constraints, techno-economic, political-institutional barriers, market barriers, lack of incentives and lack of information (knowledge). As a result of market failures in SSA, [23] suggested in his paper, "The market for lemons: Quality and uncertainty in the market mechanism", that in markets where consumers do not have reliable and adequate information in respect of the quality of the products, it leads to the proliferation of cheaper and low-quality products. The effect of this failure is that more efficient products or appliances are pushed out of the market space.

Notwithstanding these continental constraints, Ghana was able to overcome these barriers through stakeholders' consultative engagements using the ***quadruple helix model*** of policy (government, academia, industry and the media) and marketplace innovations such as standards and labelling [14]. Four (4) Legislative Instruments (L.Is) were subsequently enacted by the Ghanaian Parliament in collaboration with the Ghana Energy Commission and Ghana Standards Authority. They include:

i. Energy efficiency standards and labelling (non-ducted air conditioners and self-ballasted fluorescent lamps) regulations 2005 (L. I 1815);

ii. Energy efficiency (prohibition of manufacture, sale or importation of incandescent filament lamp, used refrigerator, used refrigerator-freezer, used freezer and used air-conditioner) regulations 2008 (L. I 1932);

iii. Energy efficiency standards and labelling (household refrigerating appliances) regulations 2009 (L. I 1958); and

iv. Energy commission (efficiency standards and labelling (light emitting diode and self-ballasted fluorescent lamps) regulations 2017 (L. I 2353).

These standards and labelling initiatives provide a mandatory labelling regime in Ghana, where energy guide labels have to be affixed conspicuously on these appliances to indicate the minimum energy performance levels of these appliances. The indicators on these energy guide labels include annual energy consumption, type of refrigerant, climate class, star rating, manufacturer, model number, fresh and frozen volumes. This mandatory labelling regime is intended to promote energy efficiency, transform the appliances market, reduce energy demand in households and reduce Ghana's energy-related CO_2 and ozone-depleting substances (ODS) emissions. This paper reviews the impact of the first three regulations and other energy efficiency projects between 2007 and 2020.

To commence the enforcement of these regulations, the national refrigerator turn-in and rebate scheme was launched by the Government of Ghana, on the advice of the Ghana Energy Commission in July, 2012. The objective was to recover about fifty thousand (50,000) inefficient refrigerating appliances from homes and encourage individuals to use more energy-efficient ones. So, in 2013, the Ghana Energy Commission commenced the full implementation of these regulations at the ports of entry. It is therefore imperative to review and establish the impact of these regulations between 2007 and 2020.

The rest of the paper is organized as follows: the next section following the introduction looks at the methodology being employed. The concept of energy efficiency is examined in Section 3. Section 4 provides the Global and National overviews of MEPS implementations. Testing and inspection protocols for refrigerating, lighting and air-conditioning equipment in Ghana are considered in Section 5. Section 6 offers the discussion of the results of some of the real impacts due to MEPS whilst Section 7 concludes the paper.

2. Methodology of the study

This section explains the methodology adopted for this work.

2.1 Desktop review

The content of this desktop review is based on energy efficiency appliance import data, policy documents and market transformation data from the Ghana Energy Commission, Energy Foundation, Council for Scientific and Industrial Research—Institute of Industrial Research (CSIR-IIR), Ghana Revenue Authority (GRA-Custom Division) and Ghana Statistical Service (GSS). Other sources of information include the International Energy Agency (IEA) as well as other related web searches. Besides, global and SSA energy efficiency documents were also reviewed by examining secondary data from standards, regulations, protocols, market report series and other available statistical data.

2.2 Calculation of electricity savings

The basic assumption for computing electricity savings is that without MEPS regulation and its accompanying awareness programmes. Eqs. (1)–(4) were used to estimate the annual energy saving per appliance, total energy savings for all the appliances and CO_2 emissions reductions between 2007 and 2020 as a result of MEPS implementation. Therefore, the annual electricity saving, *AES* (kWh) gained from the use of each MEPS-compliant appliance can be estimated by comparing its calculated annual electricity consumption (*AEC*) with the annual consumption of a used and inefficient refrigerator as shown by Eq. (1):

$$AES = AEC_{\text{before MEPS implementation}} - AEC_{\text{after MEPS implementation}} \tag{1}$$

where *AES* is the annual electricity saving for MEPS-compliant refrigerating appliance (kWh/year), and *AEC* is the annual electricity consumption before and after MEPS implementation. In Ghana, used and inefficient refrigerating appliances consumed on average 1200 kWh/year (consumption before MEPS implementation).

The total annual electricity saving, AES_{total} (GWh), of MEPS-compliant appliances was calculated by aggregating the products of electricity saving and the number of units sold (*NUS*) in a particular year, as shown in Eq. (2):

$$AES_{total,year} = \sum (AES \times NUS_{,year}) \tag{2}$$

2.3 Calculation of cost savings

The total cost savings from MEPS implementation are computed in terms of the electricity savings from operating a more efficient appliance. Thus, the annual electricity savings computed using Eq. (2) is used to estimate the total cost savings by multiplying the computed electricity savings with the electricity tariff as shown in Eq. (3):

$$CS = AES_{\text{total,year}} \times ET_{\text{residential}} \tag{3}$$

where $CS_{total,\ year}$ [USD] is the total cost savings in a year and $ET_{\text{residential}}$ [USD/kWh] is the average electricity tariff for the residential sector, set at USD 0.10/kWh [6]. The residential sector tariff was used because MEPS regulations in Ghana is primarily targeted at the residential sector.

2.4 Calculation of carbon emission reduction

The annual carbon emission reduction, $CER_{total,\ year}$ [$MtCO_2eq$] was evaluated based on the total annual electricity saving, $AES_{total,\ year}$ [MWh] with the help of Eq. (4).

$$CER_{\text{total,year}} = AES_{\text{total, year}} \times GEF_{\text{Ghana,year}} \tag{4}$$

where $GEF_{\text{Ghana, year}}$ is the grid emission factor for Ghana [tCO_2eq/MWh].

3. The concept of energy efficiency

Several definitions for energy efficiency have emerged over the years. Energy efficiency may be considered as investing in more energy-efficient technologies or appliances which results in more energy savings. Butler et al. [24] defines energy efficiency as "using energy cautiously and economically to sustain everyday life, live comfortably and support wellbeing". According to [25], energy efficiency is achieving the same service and performance while using technology with less energy use and therefore enhancing the security of the energy supply. For [26], energy efficiency is an effective tool for reducing electricity or energy consumption which limits greenhouse gas (GHG) emission and thereby reducing global warming.

4. Overview of MEPS and EES&L programmes

This section reviews the Global and the Ghanaian perspectives of MEPS and EES&L programmes.

4.1 Global overview of MEPS and EES&L programmes

Both minimum energy performance standards (MEPS) and energy efficiency standards and labelling (EES&L) for appliances are the two known key mitigation strategies for electricity conservation worldwide [27]. MEPS is a technique of eliminating inefficient performing appliances through the prescription of minimum efficiency (or maximum energy consumption) that manufacturers must achieve [28]. The benefits of MEPS, according to [29, 30] include the following:

i. It allows manufacturers and suppliers to increase appliance efficiency since less-efficient appliances will no longer be tolerated for sale in the regulated market;

ii. It encourages manufacturers to explore innovative and efficient technologies to gain a competitive edge;

iii. It provides market consistency and certainty, thus creating economies of scale;

iv. Consumers enjoy electricity cost savings over the lifetime of the appliance as these appliances on the market now consume less energy to operate; and

v. Appliance purchase prices are largely falling in real terms in many countries with MEPS regulations due to competition, economies of scale and marketplace innovations.

The main objectives of the EES&L regime are to [31];

i. Prevent the influx of substandard appliances on the market;

ii. Provide the consumer an appliance the needed information to make an informed choice;

iii. Provide information regarding the running cost of the appliance; and

iv. Fulfil environmental treaty commitments such as the Paris Agreement, Kyoto Accord, Kigali Amendment, Montreal Protocol, Rotterdam Convention and Basel Convention.

EES&L programmes, introduced in the 1970s, are now being implemented in over eighty (80) countries including Ghana. The programmes cover more than fifty (50) types of appliances and equipment [32]. Testing protocols are used to determine appliance performance relating to energy efficiency. These protocols are periodically revised to ensure they keep up with trends and advances in technology.

4.2 Overview of Ghana's energy efficiency (MEPS) programme

Over the past decades, Ghana has made significant progress in its energy energy efficiency programmes [33]. These programmes put Ghana on the world map as a pioneer in SSA as a result of extensive collaborations of stakeholders and institutions such as the Ghana Energy Commission, Ministry of Energy, Ghana Standards Authority, Ghana Energy Foundation and CLASP [14]. Ghana experienced widespread power shortages due to series of droughts spanning from the 1980s to the 1990s. Energy efficiency programmes were identified to solve these power shortages as it delivers benefits faster than building new generating power plants [14]. The Government of Ghana, therefore, decided to support energy efficiency standards, policies and programmes. Consequently, the Ghana Electrical Appliance Labelling and Standards Programme (GEALSP) was launched to help transform the country's appliance market. An initial assessment to determine the energy savings potential was then carried out from energy efficiency projects [10].

A comprehensive national household survey on demand-side management (DSM) was then conducted by the Ghana Energy Commission between 2003 and 2006. The results from the survey revealed that residential energy consumption, for example, was 50% of the total national energy use and refrigerating appliances accounted for nearly 59% of the residential use [15]. These used refrigerating appliances consume on average 1200 kWh per annum compared to 250 kWh and 400 kWh per annum in Europe and the US respectively [15] as illustrated in **Figure 1** for comparative energy use basis.

Energy efficiency programmes for household appliances was crucial and therefore needed to be adopted and implemented without further delay to bring Ghana's consumption down (proposed standard—the green line in **Figure 1**). According to [33], energy efficiency is considered a "***low-hanging fruit***" and "***first fuel***" of the clean energy transition due to the low marginal cost of its implementation. However, prior to the adoption and implementation of the MEPS in Ghana, the barriers identified in Section 1, needed to be addressed. The ***quadruple helix model and marketplace innovations*** were adopted in resolving these issues [14]. **Table 1** lists some of the barriers and measures implemented by Ghana to overcome these barriers [14].

Four (4) Legislative Instruments (L.Is) were subsequently enacted by the Ghanaian Parliament in collaboration with the Ghana Energy Commission and Ghana Standards Authority. The first, energy efficiency regulations for non-ducted room air conditioners (RACs) and compact fluorescent lights (L.I. 1815) was developed in 2005 [34], which mandates that all RACs imported into Ghana must meet the minimum energy efficiency ratio (EER) of 2.8 W of cooling per watt of power input, equivalent to a 1-star rating. The second energy efficiency regulations (L.I. 1932) was also enacted in 2008 to help prevent the importation or sale of used

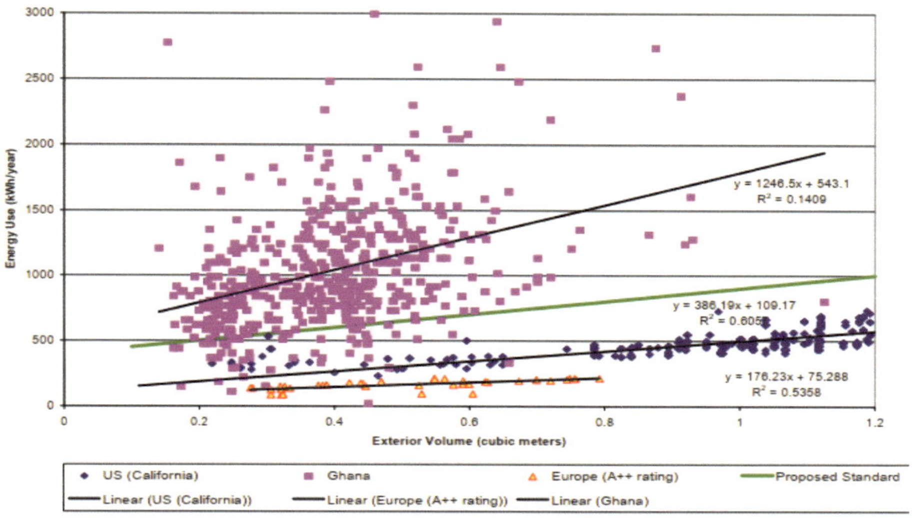

Figure 1.
Refrigerator Electricity use for Ghana, Europe and US in 2007 [14].

No	Types of barriers	Measures
1.	Financial	Implementation of tax incentives and national programmes. For instance, between 2003 and 2016, import duties and taxes were removed from the importation and sale of compact fluorescent lamps (CFLs).
2.	Market	Import duties and value-added tax (VAT) on compact fluorescent lamps (CFLs) were removed by government in April 2003, thus making them available and affordable to consumers.
3.	Lack of information (knowledge)	Organization of national energy efficiency campaigns throughout the country using both print and electronic media. Also, leaflets, brochures and flyers were printed and distributed, including town hall meetings where relevant stakeholders were engaged.
4.	Institutional	Holding stakeholders' consultations and institutional reforms. For instance, Public Utility Regulatory Commission (PURC) is mandated to set electricity tariffs independent of governmental influence and to reward efficient users and punish inefficient users of electricity. Also, the Ghana Energy Commission came in with the labelling programme in 2005 to help in transforming the appliance market.
5.	Technical	Technical cooperation with developmental partners was encouraged to offer training to Ghanaians. For instance, between 2012 and 2013, over 600 technicians were trained to carry out installation, repairs and maintenance of the new and efficient refrigerating appliances under the Ghana Energy Foundation and UNDP project to transform the appliance market.
6.	Political	Independent regulatory bodies were created to deal with the energy crisis. For instance, the Ghana Energy Commission and Public Utility Regulatory Commission were mandated to carry out technical and economic regulations of the power sector respectively. The Ghana Grid Company was also responsible for the national network, separated from the distribution operator. All these programmes were moved from the Ministry of Energy to prevent any political interference.

Table 1.
Types of energy efficiency barriers and measures to overcome them.

inefficient refrigerating and air conditioning appliances in Ghana. To ensure that only energy efficient refrigerating appliances are imported and sold in Ghana, the energy efficiency standards and labelling regulation for household refrigerating

L.I.	Regulation	Scope & targets	Year passed	Year implemented
1815	Energy efficiency standards and labelling (non-ducted air conditioners and self-ballasted fluorescent lamps) regulations.	Gives legal backing to the use of energy-efficient non-ducted air conditioners and fluorescent lamps.	2005	2014
1932	Energy efficiency (prohibition of manufacture, sale or importation of incandescent filament lamp, used refrigerator, used refrigerator-freezer, used freezer and used air-conditioner) regulations.	Places total ban on the importation and sale of incandescent filament lamp, used refrigerator, used refrigerator-freezer, used freezer and used air conditioners effective January 2012.	2008	2013
1958	Energy efficiency standards and labelling (household refrigerating appliances) regulations.	Provides for the enforcement of minimum energy efficiency and labelling for household refrigerating appliances.	2009	2013
2353	Energy commission (efficiency standards and labelling (light emitting diode and self-ballasted fluorescent lamps) regulations.	Provides for the enforcement of minimum energy efficiency and labelling for light emitting diode and self-ballasted fluorescent lamps.	2017	2020

Table 2.
Main end-use policies and regulations.

appliances (L.I. 1958) was enacted in 2009. Finally, in 2017, the energy efficiency standards and labelling for light-emitting diode and self-ballasted fluorescent regulations (L.I. 2353) was enacted. The purpose of the fourth regulation is to enforce the standards for minimum energy efficiency for self-ballasted fluorescent lamps and light-emitting diode lamps imported or manufactured in Ghana.

Details of these four regulations and standard protocols for determining the energy efficiency star ratings for (non-ducted air-conditioners, household refrigerating appliances and lamps and CFLs), categories of household refrigerators and initial luminous efficacy of the lamps are available and can be assessed under Legislative Instruments (list of L.I's for energy efficiency) at [35].

In conclusion, Ghana has developed and implemented four (4) main end-use policies, regulations and standards to promote demand-side management (DSM). Seventeen (17) other regulations are currently being developed for other electrical appliances such television (TV) sets, electric motors, washing machines, blenders, etc. **Table 2** lists the four (4) regulations, its scope and targets, the year they were enacted by the Ghanaian Parliament and their implementation dates.

5. Testing and inspection procedures

To achieve the full potential of the MEPS programmes, Ghana adopted rigorous testing, approval and inspection procedures for importing these regulated appliances into the country. Currently, Ghana does not manufacture these appliances, so testing is done by third-party accredited laboratories such as Vkan Certification & Testing Co., Ltd. (CVC), Intertek, TUV-Rhineland, DEKRA Product Testing & Certification, General Society of Surveillance (SGS) and Bureau Veritis (BVAC). These testing facilities are recommended and designated by the Ghana Standards Authority and Ghana Energy Commission. A performance test report, which details the performance of the appliance in terms of energy consumption, approved refrigerant, climate class, star rating, etc. from these facilities, is then submitted to the Ghana Energy Commission by the importer concerning a particular model, for evaluation. Certificate of approval (COA), containing parameters of the said model, is issued to the importer for model(s) that meet the MEPS requirements as laid down in the energy efficiency guidelines [31].

The testing, approval and inspection procedures of a model are summarized in **Figures 2** and **3**.

Having reviewed all the available documents, regulations, standards, procedures and processes, the next section provides some of the real impacts resulting from MEPS implementation in Ghana since 2007.

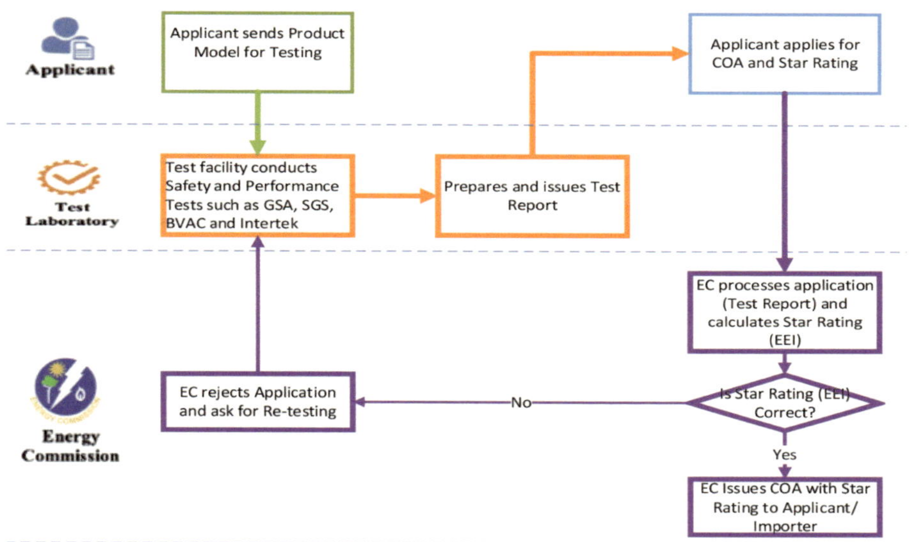

Figure 2.
Testing and approval processes for model(s) to be imported into Ghana.

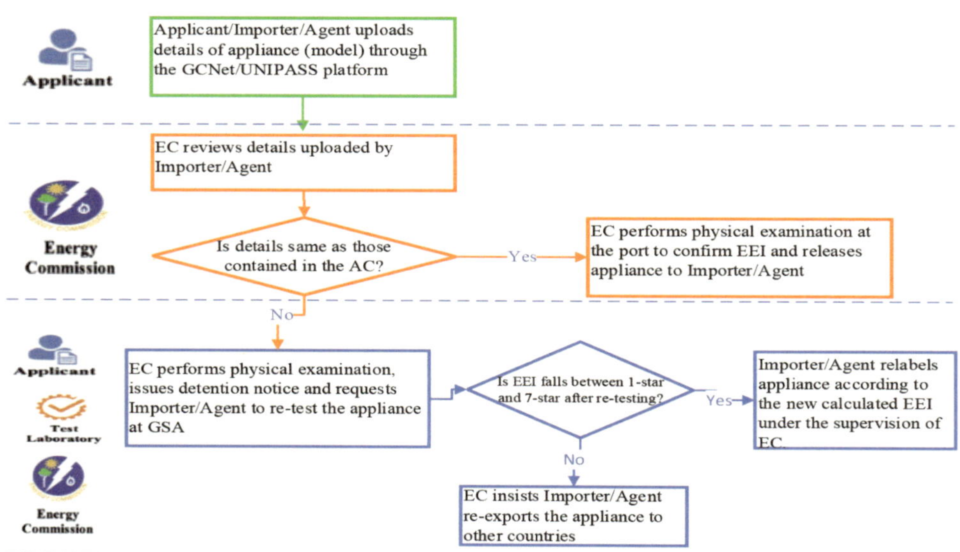

Figure 3.
Physical inspection procedures of a model at the ports of entry.

6. Results and discussion

This section discusses the impact of the major energy efficiency programmes resulting from MEPS implementation in Ghana based on Eqs. (1)–(4) between 2007 and 2020.

6.1 Ghana's efficient lighting project

In 2007, Ghana implemented an efficient lighting project (CFL exchange programme) regarding its policy directions in the area of energy efficiency. On the advice of the Ghana Energy Commission, the Government procured and distributed over six (6) million compact fluorescent lamps (CFLs) to replace the estimated six million incandescent lamps at no cost to the beneficiaries [36]. As a result of this project, incandescent lamp usage in households has reduced from 58–3% while CFLs penetration increased from 20% in 2007 to 79% in 2009 [36]. The country's peak electricity demand was accordingly reduced by 124 MW and peak electricity consumption by 72.8 GWh per year due to this policy implementation. This resulted in an energy cost saving of about US$ 39.5 million per year and carbon dioxide (CO_2) savings estimated at 105,000 tonnes per year [36]. Consequently, there was a delay in the generation expansion of thermal energy investment of US$ 105 million. At US$ 120/bbl, energy cost saving would amount to US$ 39.5 million per year [37]. The project received a Global Energy Efficiency Award in 2010 organized by the Energy Efficiency Global Forum in Brussels, Belgium (12–14 April 2011) [36].

6.2 Ghana's refrigerator turn-in and rebate scheme

Reports available at the Ghana Energy Commission indicated that the Government of Ghana, in September 2012, through the Ghana Energy Commission, launched the national refrigerator turn-in and rebate scheme with the support of the United Nation Development Programme (UNDP), Global Environment Facility (GEF) and Multilateral Fund of the Montreal Protocol (MFMP). The scheme, which encouraged consumers to exchange their old refrigerators for new and efficient ones, available at a discounted price, was to recover about 50,000 inefficient refrigerating appliances from homes and promote the use of more energy-efficient ones and transform the refrigerating appliances market in the country. By mid-June, 2016, a total of 10,472 units of old energy-inefficient appliances have been replaced across the country with new energy-efficient ones [38]. Customers who participated in the project had their consumption reduced from 1200 kWh per year to 385 kWh per year, resulting in a saving of about 400 GWh of electricity, 1.1 million tonnes of carbon dioxide (CO_2) and about 1500 kg of Chlorofluorocarbon (CFC) recovered [38]. This translated into a household income saving of about US$ 140 per year [38].

6.3 Enforcement of legislative instrument 1932

The enforcement of L.I. 1932 at Ghana's ports of entry by the Ghana Energy Commission prohibited the importation of an estimated number of 4,854,864 units of used refrigerating appliances between 2013 and 2020. A total of 5825.84 GWh of electricity would have been consumed with over 2.33 million tonnes of CO_2 released into the atmosphere if the ban was not enforced. However, data available at the Ghana Energy Commission shows that a total of 46,666 used refrigerators and 11,003 used RACs were imported through illegal means by some recalcitrant importers. They were subsequently confiscated by the Ghana Energy Commission inspectors positioned at the ports and evacuated for e-disposal [39]. **Figure 4** shows the yearly trend in the importation of used refrigerators, particularly the downward trend between 2013 and 2020 during MEPS implementation.

The average annual energy consumptions of a used refrigerator and used RAC are 1200 kWh per year and 4000 kWh per year respectively [39]. The rigorous enforcement of L.I. 1932 since 2013, yielded a total of 100 GWh of electricity and 40,000 tonnes of CO_2 savings for those confiscated used appliances based on Eqs. (2) and (3).

6.4 Enforcement of legislative instrument 1958

Enforcing L.I. 1958 resulted in the importation of 2,378,432 new and efficient refrigerating appliances into the country between 2013 and 2020, thus preventing the importation of used and inefficient ones [39]. Data analysis indicated that 92% of all refrigerating appliances imported between 2013 and 2020 were new and efficient in accordance with L.I. 1958. About 74.8% of these appliances were 2- to 5-star rated with 87.2% of all imports laden with R600a refrigerant (hydrocarbon) [39], which has both low global warming potential (GWP) and low ozone-depleting potential (ODP) and therefore more energy-efficient [40]. The remaining 8%, which were used refrigerators were confiscated. The rise in the importation and sale of new refrigerating appliances is primarily due to strict regulations, procedures and controls implemented at the ports of entry, regular market surveillance and stringent compliance monitoring. These measures help to ensure that only appliances that meet MEPS are permitted into the Ghanaian market. **Figure 5** shows how the refrigerating appliance market in Ghana has evolved/transformed over the years (2005–2020) from being a completely used and inefficient refrigerator market (88.9% inefficient in 2005) to new and efficient ones (99.1% efficient in 2020) as a result of MEPS implementation.

The average annual energy consumption of these new and efficient appliances has dropped drastically due to MEPS implementation compared with the used refrigerators. Consumption values reduced from about 1400 kWh per unit per year to 340 kWh per unit per year [39]. **Figure 6** shows the trend in the average annual energy consumption patterns of refrigerating appliances over the years especially that during the implementation period.

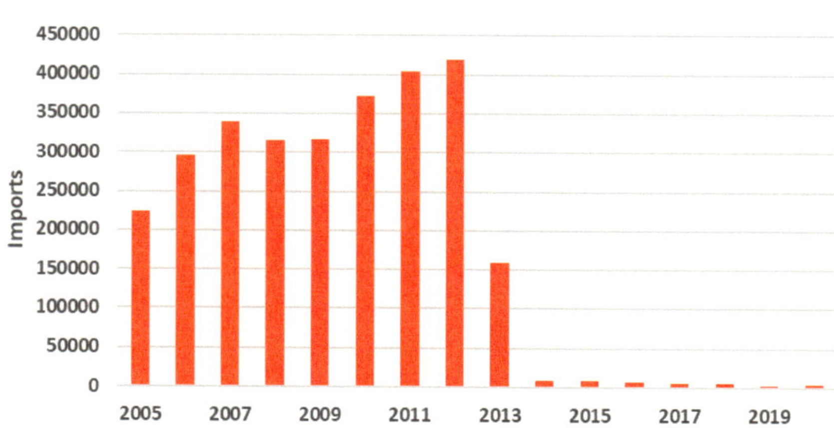

Figure 4.
Yearly imported used refrigerating appliances to Ghana from 2005 to 2020.

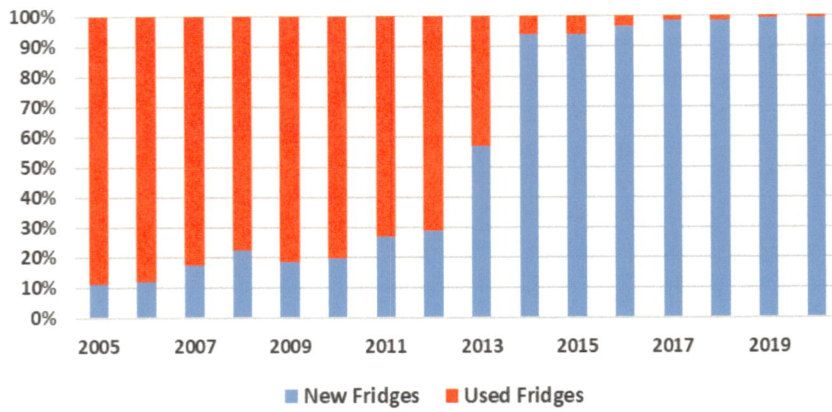

Figure 5.
Evidence of transformed market through MEPS from 2005 to 2020.

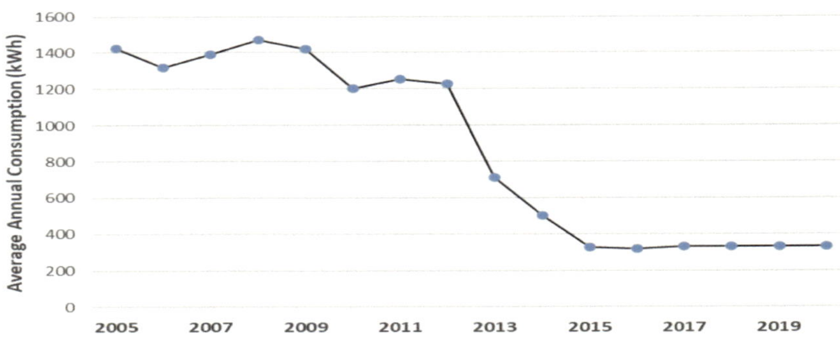

Figure 6.
Average annual energy consumption patterns for new fridges.

Assuming that 90% of the new refrigerating appliances were sold between 2013 and 2020. **Figure 7** presents the analysis of the total electricity and CO_2 emission savings based on Eqs. (1)–(3). From the analysis, a total of 5845 GWh electricity has been saved with a corresponding 2.56 million tonnes of CO_2 emission savings resulting from MEPS due to L.I. 1958 enforcement. This is equivalent to more than 3.4% of the total ***thermal electricity generated*** in 2015 [6], thereby further preventing the construction of a 667 MW power plant capacity.

6.5 Enforcement of legislative instrument 1815

A total of 904,923 new RACs were imported and inspected at the Port of Tema since 2014 [39]. About 54.6% of the RACs on the Ghanaian market are 1-star rated with an average EER of 2.87, which is above the minimum EER of 2.80 with an average annual rated power consumption of 3347.4 kWh [37]. Also, 56% of the RACs were laden with R410a refrigerant which is more energy-efficient than R22 [40]. The average annual energy consumption of a used RAC is 4000 kWh per year [39]. Therefore, 652.6 kWh of electricity has been saved per unit per year due to MEPS. Assuming that 90% of the new RACs were sold between 2014 and 2020. **Figure 8** presents the analysis of the total electricity and CO_2 emission savings using Eqs. (1)–(3). From the analysis, about 1900 GWh electricity has been saved with a corresponding 783,000 tonnes of CO_2 emission savings resulting from MEPS due to L.I. 1815 enforcement.

Year	Imports	Sales	Years Fridge in Use	Energy Savings per Unit (kWh/yr)	Total Energy savings (GWh)	Grid Emission Factor (tCO2eq/MWh) [6]	Cummulative CO2 Savings (MtCO2)
2013	209,824	188,842	8	488	737.24	0.46	0.34
2014	142,764	128,488	7	699	628.69	0.36	0.23
2015	128,697	115,827	6	800	555.97	0.31	0.17
2016	218,073	196,266	5	860	843.94	0.43	0.36
2017	390,686	351,617	4	860	1209.56	0.47	0.57
2018	373,656	336,290	3	860	867.63	0.53	0.46
2019	380,298	342,268	2	860	588.70	0.45	0.26
2020	534,434	480,991	1	860	413.65	0.40	0.17
Total	**2,378,432**	**2,140,589**			**5,845.39**		**2.56**

Figure 7.
Analysis of total electricity and CO_2 emission savings for refrigerating appliances.

Year	Imports	Sales	Years RAC in Use	Energy Savings per Unit (kWh/yr)	Total Energy savings (GWh)	Grid Emission Factor (tCO2eq/MWh)[6]	Cummulative CO2 Savings (ktCO2)
2014	104,205	93,785	7	652.6	428.4	0.36	154.23
2015	101,455	91,310	6	652.6	357.5	0.31	110.83
2016	77,137	69,423	5	652.6	226.5	0.43	97.41
2017	148,583	133,725	4	652.6	349.1	0.47	164.07
2018	139,781	125,803	3	652.6	246.3	0.53	130.54
2019	163,290	146,961	2	652.6	191.8	0.45	86.32
2020	170,472	153,425	1	652.6	100.1	0.40	40.05
Total	**904,923**	**814,431**			**1,899.80**		**783.44**

Figure 8.
Analysis of total electricity and CO_2 emission savings for RACs.

6.6 Compliance levels of refrigerating appliance and RACs

Analysis of some of the reports at the Ghana Energy Commission indicated that the compliance levels of the refrigerating appliances and RACs have been increasing over the years. Between 2017 and 2020, the compliance level of the imported refrigerating appliance increased from 92.6% to 97.0% whilst that of RACs also saw an upsurge from 79.2% to 96.8% [37, 39]. **Figure 9** shows the trends in the compliance levels of these appliances from 2017 to 2020.

The analysis further revealed that the following factors contributed to the high compliance levels in an attempt to transform the appliance market in Ghana [37, 39]:

i. Submission of performance test report by importers from third-party accredited laboratories to the Ghana Energy Commission for evaluation and approval or otherwise;

ii. Establishment of import appliance database/register;

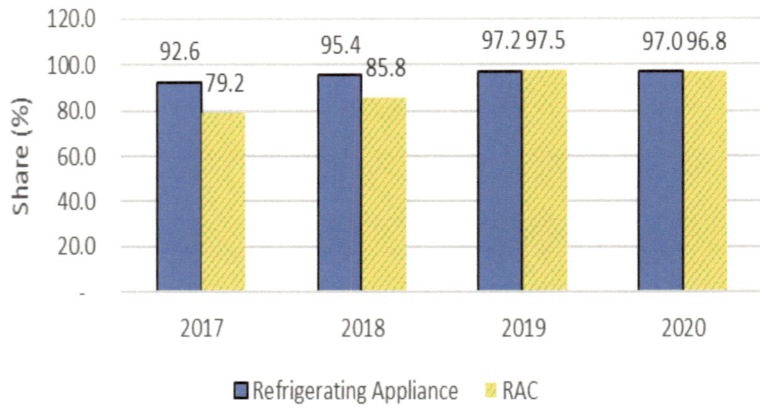

Figure 9.
Trends in compliance levels for refrigerating appliances and RACs from 2017 to 2020.

iii. Operationalization of the GCNet/ICUMS digital portals for approving only appliances that meet MEPS;

iv. Rigorous physical examination procedures/protocols put in place at the ports of entry;

v. Regular and consistent market surveillance;

vi. Stringent compliance monitoring;

vii. Removal of non-compliant appliances from showrooms for testing and re-labelling;

viii. Payment of enforcement fees for non-compliance;

ix. Verification and challenge testing procedures; and

x. Development of a Certified Appliances Mobile Application (APP) which contains all the approved appliances, nearby shops, tips on how to save energy. This APP also helps consumers to make an informed purchase decision. To download the APP, retailers and consumers are required to search for ***CERTIFIED APPLIANCE APP*** on the google play store. iPhone users will have to wait for a while as the APP is being developed for those on the iOS/iPadOS platforms.

7. Conclusion

Ghana's drive to transform the appliance market from the used and inefficient appliances to new and efficient ones has been highly recommended over the years. From this review or study, Ghana developed and implemented MEPS successfully for refrigerating appliances, RACs and lighting systems. This was done through the engagement of relevant stakeholders with complementary financial and technical assistance from development partners. Through MEPS implementation, annual average energy consumptions of refrigerating appliances and RACs have decreased drastically. The implementation of MEPS programmes in Ghana during the period under review yielded 8317.8 GWh (8.32 TWh) of electricity savings, while at the same time reducing fossil CO_2 emissions by 4.60 million tonnes. This figure roughly corresponds to the total thermal electricity generated (8424 GWh) in 2017 [6]. At 10 US cents per kWh, about USD 832 million has been saved on electricity bills. This enormous financial saving and environmental benefit resulting from deferred electricity consumption amplifies the positive implication of energy efficiency and MEPS programmes. In conclusion, the implementation of energy efficiency programmes delayed the construction of a 950 MW power plant capacity. The 17 other regulations that are currently being developed for other electrical equipment such as television sets, electric motors, washing machines, blenders, etc. must be concluded in good time to enable further energy savings for the country.

Acknowledgements

I would like to thank the following organizations that contributed valuable information on the subject: Ghana Energy Commission, Energy Foundation, Council for Scientific and Industrial Research—Institute of Industrial Research, Ghana Revenue Authority, International Energy Agency (IEA), European Union (EU), Collaborative Labelling and Appliance Standards Program (CLASP) and Lawrence Berkeley National Laboratory (LBNL).

Conflict of interest

The author, Edwin Kwasi Tamakloe is an employee of the Ghana Energy Commission which is one of the stakeholder institutions discussed in the article. This is indicated transparently in the author affiliations.

A. Appendices and nomenclature

See **Figure A1**.

Figure A1.
Sample of a three (3) star rated energy guide label for refrigerating appliance in Ghana [41].

Author details

Edwin Kwasi Tamakloe
Energy Statistician, Energy Efficiency Inspection and Enforcement Directorate, Ghana Energy Commission, Accra, Ghana

*Address all correspondence to: edwintamakloe@yahoo.com

References

[1] International Energy Agency. Energy Efficiency in Building [Internet]. 2017. Available from: https://elearning.iea.org/courses/course-v1:IEA+BUILDINGS1+Open/courseware/e9bce3dd7e1c4dc291718549aff3e641/c2c1535f6e814152b668e559f3bc11d2/?child=first%20pdf [Accessed: 31 May 2021]

[2] Global Alliance for Buildings and Construction, International Energy Agency, the United Nations Environment Programme. Global status report for buildings and construction [Internet]. 2019. Available from: https://www.unep.org/resources/publication/2019-global-status-report-buildings-and-construction-sector.pdf [Accessed: 20 May 2021]

[3] Lu W. Potential energy savings and environmental impact by implementing energy efficiency standards for household refrigerators in China. Energy Policy. 2006;**34**(13):583-1589. DOI: 10.1016/j.enpol.2004.12.012

[4] Brounen D, Kok N, Quigley JM. Residential energy use and conservation: Economics and demographics. European Economic Review. 2012;**56**(5):931-945. DOI: 10.1016/j.euroecorev.2012.02.007

[5] Wang Z, Zhang B, Yin Y, Zhang Y. Determinants and policy implications for household electricity-saving behaviour: evidence from Beijing, China. Energy Policy. 2011;**39**(1):3550-3557. DOI: 10.1016/j.enpol.2011.03.055

[6] Ghana Energy Commission. National Energy Statistics [Internet]. 2021. Available from: http://www.energycom.gov.gh/files/National%20Energy%20Statistics%202021.pdf [Accessed: 02 May 2021]

[7] Druckman A, Jackson T. Household energy consumption in the UK: A highly geographically and socio-economically disaggregated model. Energy Policy. 2008;**36**(8):3177-3192. DOI: 10.1016/j.enpol.2008.03.021

[8] Saidur R, Masjuki HH, Jamaluddin MY. An application of energy and exergy analysis in the residential sector of Malaysia. Energy Policy. 2007;**35**(2):1050-1063. DOI: 10.1016/j.enpol.2006.02.006

[9] Azlina AA, Engku Abdullah ES, Kamaludin MA, Radam AL. Energy conservation of residential sector in Malaysia. Journal of Business and Social Development. 2015;**3**(2):51-62

[10] Constantine S, Denver A, Hakim S, McMahon JE, Rosenquist G. Ghana Residential Energy Use and Appliance Ownership Survey: Final Report on the Potential Impact of Appliance Performance Standards in Ghana (No. LBNL-43069). Berkeley, CA, US: Ernest Orlando Lawrence Berkeley National Laboratory; 1999

[11] CLASP. Africa air conditioner market scoping study. Market Survey Report. 2018. pp. 1-43

[12] Adom PK, Adams S. Energy savings in Nigeria. Is there a way of escape from energy inefficiency? Renewable and Sustainable Energy Reviews. 2018;**81**(1):2421-2430. DOI: 10.1016/j.rser.2017.06.048

[13] Ghana Energy Commission. Strategic National Energy Plan (SNEP II): Energy Demand Projections for the Economy of Ghana [Internet]. 2019. Available from: http://energycom.gov.gh/files/SNEP%20Demand%20Oct2019_SNEP2030_Final.pdf [Accessed: 02 May 2021]

[14] Agyarko KA, Opoku R, Van Buskirk R. Removing barriers and promoting demand-side energy

efficiency in households in Sub-Saharan Africa: A case study in Ghana. Energy Policy. 2020;**137**(1):1-11. DOI: 10.1016/j.enpol.2019-111149

[15] Hagan EB, Van Buskirk R, Ahenkorah AO, McNeil MA. Refrigerator efficiency in Ghana: Tailoring an appliance market transformation program design for Africa. Energy Policy. 2007;**35**(4):2401-2411. DOI: 10.1016/j.enpol.2006.08.017

[16] Charter E. Policies That Work: Introducing Energy Efficiency Standards and Labels for Appliances and Equipment [Internet]. 2009. Available from: https://www.energycharter.org/fileadmin/DocumentsMedia/Thematic/EE_Standards_and_Labels_2009_en.pdf [Accessed: 28 June 2021]

[17] Ellis M, Pilven Z, Evans C, McAndrew L. Compliance Counts: A Practitioner's Guidebook on Best Practice Monitoring, Verification, and Enforcement for Appliance Standards & Labelling. Washington DC: CLASP [Internet]; 2010. Available from: https://storage.googleapis.com/clasp-siteattachments/2010-09_MVEGuidebookSingle.pdf [Accessed: 16 May 2021]

[18] Ofosu-Ahenkorah AK. Transforming the West African Market for Energy Efficiency: Ghana Leads the Way with Mandatory Standards and Labels [Internet]. 2002. Available from: https://storage.googleapis.com/clasp-siteattachments/2002_CLASP_GhanaSL.pdf [Accessed: 10 May 2021]

[19] Guo Q, Wu Y, Ding Y, Feng W, Zhu N. Measures to enforce mandatory civil building energy efficiency codes in China. Journal of Cleaner Production. 2016;**15**(119):152-166. DOI: 10.1016/j.jclepro.2016.02.002

[20] Du M, Wang B, Zhang N. National research funding and energy efficiency: Evidence from the National Science Foundation of China. Energy Policy. 2018;**120**(1):335-346. DOI: 10.1016/j.enpol.2018.05.058

[21] Langlois-Bertrand S, Benhaddadi M, Jegen M, Pineau PO. Political-institutional barriers to energy efficiency. Energy Strategy Reviews. 2015;**8**(1):30-38. DOI: 10.1016/j.esr.2015.08.001

[22] Kenfack J, Bossou OV, Tchaptchet E. How can we promote renewable energy and energy efficiency in Central Africa? A Cameroon case study. Renewable and Sustainable Energy Reviews. 2017;**75**(1):1217-1224. DOI: 10.1016/j.rser.2016.11.108

[23] Akerlof GA. The market for lemons: Quality uncertainty and the market mechanism. The Quarterly Journal of Economics. 1970;**84**(3):488-500. DOI: 10.1016/B978-0-12-214850-7.50022-X

[24] Butler K, Gordon R, Roggeveen K, Waitt G, Cooper P. Social marketing and value in behaviour? The perceived value of using energy efficiently among low-income older citizens. Journal of Social Marketing. 2016;**6**(2):144-168. DOI: 10.1108/jsocm-07-2015-0045

[25] European Commission. Communication from the Commission to the European Parliament, the Council, the European Economic and Social Committee and the Committee of the Regions [Internet]. 2011. Available from: https://eurlex.europa.eu/LexUriServ/LexUriServ.do?uri=COM:2011:0109:FIN:EN:PDF [Accessed: 18 June 2021]

[26] Craig CA, Feng S. Exploring utility organization electricity generation, residential electricity consumption, and energy efficiency: A climatic approach. Applied Energy. 2017;**185**(1):779-790. DOI: 10.1016/j.apenergy.2016-10-101

[27] Wiel S, McMahon JE. Energy Efficiency Labels and Standards: A Guide for Appliances, Equipment and Lighting. 2nd ed. Washington, D.C., USA: Collaborative Labelling and

Appliance Standards Program (CLASP); 2005. Available from: https://escholarship.org/content/qt01d3r8jg/qt01d3r8jg.pdf [Accessed: 17 June 2021]

[28] McMahon JE, Wiel S. Governments should implement energy-efficiency standards and labels cautiously. Energy Policy. 2003;**31**(13):1403-1415. DOI: 10.1016/S0301-4215(02)00199-4

[29] Nadel S, DeLaski A. Appliance Standards: Comparing Predicted and Observed Prices [Internet]. 2013. Available from: https://appliancestandards.org/sites/default/files/Appliance_Standards_Comparing_Predicted_Expected_Prices.pdf [Accessed: 16 May 2021]

[30] Van Buskirk RD, Kantner CLS, Gerke BF, Chu S. A retrospective investigation of energy efficiency standards: policies may have accelerated long term declines in appliance costs. Environmental Research Letters. 2014; **9**(11):114010. DOI: 10.1088/1748-9326/9/11/114010

[31] Ghana Energy Commission. Energy Efficiency Guidelines. 2020

[32] International Energy Agency. Achievements of Appliance Energy Efficiency Standards and Labelling Programs: A Global Assessment in 2016, IEA Technology Collaboration Programme [Internet]. 2016. Available from: https://www.iea-4e.org/document/387/achievements-of-appliance-energy-efficiency.pdf [Accessed: 13 May 2021]

[33] Gyamfi S, Diawuo FA, Kumi EN, Sika F, Modjinou M. The energy efficiency situation in Ghana. Renewable and Sustainable Energy Review. 2018;**82**(1):1415-1423. DOI: 10.1016/j.rser.2017.05.007

[34] McMahon J, Van Buskirk R. Standards and labels: Transforming the market for energy-efficient appliances. Clean Energy Solutions Center Webinar Series. 2012

[35] Ghana Energy Commission. Regulation and Codes (List of L.I's for Energy Efficiency) [Internet]. 2021. Available from: http://www.energycom.gov.gh/regulation/regulation-and-codes.pdf [Accessed: 01 May 2021]

[36] Agyarko K. Towards Efficient Lighting Market: The case of Ghana [Internet]. 2011. Available from: http://www.ecreee.org/sites/default/files/event-att/k.agyarko-ouaga_ecreee_presentation.pdf [Accessed: 31 May 2021]

[37] Ghana Energy Commission. Regulated Appliances Market Performance Report [Internet]. 2021. Available from: http://www.energycom.gov.gh/files/Regulated%20Appliances%20Market%20Performance%20Report%20for%20Ghana%20-%202020.pdf [Accessed: 04 July 2021]

[38] Ghana Energy Commission. The success story of the Ghana Refrigerator Efficiency Project [Internet]. 2016. Available from: http://www.energycom.gov.gh/efficiency/the-success-story-of-the-eerp.pdf [Accessed: 13 June 2021]

[39] Ghana Energy Commission. Enforcement of Energy Efficiency Legislative Instruments (1815, 1932 and 1958) at Ports of Entry (with Relevant Indicators/Statistics). 2020. Available from: http://www.energycom.gov.gh/files/Enforcement%20of%20Energy%20Efficiency%20Legislative%20Instruments.pdf [Accessed: 04 July 2021]

[40] Prajapat U. CO_2 as a refrigerant in supermarket refrigeration systems: A Review. International Research Journal of Engineering and Technology. 2019; **6**(6):2472-2489

[41] Ghana Energy Commission. Regulation and Codes (List of L.I's for Energy Efficiency) [Internet]. 2021. Available from: http://www.energycom.gov.gh/files/LI1958.pdf [Accessed: 01 May 2021]

Chapter 15

Optimized Energy Efficiency in a Telecommunication Company: Machine Learning Approach

Ngang Bassey Ngang

Abstract

Energy efficiency is the use of technology that requires less energy to perform the same task. It was considered to introduce optimized energy efficiency by using machine learning to reduce power consumption at communication base station (BTS) sites. This process involves reviewing relevant work to identify defects, characterizing and determining the power consumption of the cell site under investigation, developing a SIMULINK model for the cell site under investigation, and identifying the module. It also includes optimizing high power consumption; design a machine learning rule base to monitor the power consumption of the module. Train artificial neural network (ANN) on machine learning rules designed to reduce cell power consumption, thereby improving network performance. The next step is developing an algorithm to implement it, and finally, to design a power consumption model for the network under investigation. The results obtained after a large simulation show that the traditional maximum power consumed at the cell site is 5746 kW, while the power when machine learning is injected into the system is 4733 kW. Integrating machine learning into the system resulted in 4731 kW, an 8.9% performance improvement.

Keywords: optimized, energy efficiency, reduction of power consumption, telecommunication base transceiver station, machine learning

1. Introduction

The trouble of power efficiency is one of the main challenges dealing with wireless cell community vendors around the world today. In Nigeria, some cell network vendors have been affected, resulting in the closing down of websites due to the trouble of energy efficiency. To handle this situation, which offers power consumption reduction in a base station (BS), several processes have been adopted that led to the introduction of green conversation techniques. Power amplifier (PA) improvement has attracted a good deal of attention because it devours the greatest proportion of the strength consumption of BSs. In cellular communications, the energy amplifier in a macro phone BS consumes the most energy, as a great deal as 65% of the complete power bumps off via all BS elements. The trouble is that the strength efficiency of PA is doable solely with interior equipment, but now not with external prerequisites such as not knowing the wide variety of users asking for getting entry to the BS at a time. This changes the electricity effectivity done through the internal equipment. It is normal that excessive bit error fees result in poor conversation overall performance [1]. On the different hand, Akbari et al. [2] certainly emphasizes the need to integrate ultracapacitors into the device for energy efficiency. The author of egalitarianism [3] emphasized the need to redefine wi-fi conversation to enhance its effectiveness. Bazzi et al. [4] reiterated that greater throughput is a core function of multiradio efficiency. Optimization issues seem to be for most or minimum values that a character can take. In the absolute extremes section, we noticed how to resolve sure optimization problems. Here we have located the maximum and minimum values that the function takes in the interval. Machine learning is a synthetic intelligence (AI) software that provides systems with the ability to automatically learn and improve. Machine getting to know teaches computer systems what people take for granted, that is, what they learn from experience. Machine getting to know algorithms use computational methods to "learn" statistics without delay from data, except relying on specific equations as a model. The algorithm adaptively improves overall performance as the wide variety of samples on hand for education increases. The development and modifications that have taken vicinity in the enterprise lately have entered a new phase in parallel with the improvement of computer technology, fuzzy logic, and, ultimately, a completely new subject of synthetic intelligence, the study [5] reveals that artificial talent (AI) is growing because of its manageable to be predictive and sourcing. Renewable strength sources such as wind and solar are very useful and clean sources; their indepleteable houses help in enhancing or smoothing performance in aggregate with different sources, such as biomass, in particular in rural areas. The methodology to achieve the intention of this work is the adherence to the mentioned research goals, which has to do with the tabulation of the gathered data and characterization of the current telecommunication primary based transceiver underneath learn about [6].

1.1 Aim of the work

This paper is about the optimization of energy efficiency through reduction of power consumption in a telecommunication base transceiver station (BTS).

1.2 The study objectives

The high consumption of power by modules of a cell site has impacted on the operations of the telecommunication company. This has necessitated the introduction of optimized energy efficiency through reduction of power consumption in a telecommunication base transceiver station (BTS) site using machine learning. The specific objectives are stated thus to

1. Determine the power consumption of the cell modules to be optimized from the collected data.

2. Perform optimization of the established high power consumed by the modules of the cell to a minimal value.

3. Design a machine learning rules for a reduced power consumption in the cell to enhance its performance.

4. Train the ANN in the designed machine learning rules for reduced power consumption in the cell site.

5. Develop an algorithm that will implement the sequence.

6. Develop a power consumption model for the network under study based on the result obtained when the algorithm is integrated in it.

Validate and justify the percentage improvement of energy efficiency in the cell site with and without the application of machine learning.

2. A closer look at previous related works

Introduction of renewable strength sources can enhance the diesel generators used in the base stations of all Nigerian carriers, in wind farms, the place the source of electricity is stochastic, the inefficiency of enhancing the percentage efficiency of the mills to raise the production capacity of the industries that fully rely on the generator for their each day manufacturing is addressed with the aid of strength efficiency upgrades of doubly fed induction generator machines, the use of adaptive manipulate approach [7]. In our verbal exchange network, the lengthen in sending data from the transmit point to the get hold of factor is a very massive problem, so the power provides desires to be reliable; the minimization in electricity supply in the United States has induced financial troubles to small-scale industries; one principal hassle attributed to this is inadequate planning mechanism that will forecast the required amount of electricity that will be needed to feed the complete population [8]. During the current period, the renewable energy source has dramatically extended both qualitative and quantitative enhancements with growing pressure to overcome environmental and financial crises; taking awareness of the reality that passing information or transfer of data from one factor to the other has ended up a chronic problem in our communication industry, energy efficiency techniques should be adopted such as laptop; learning this ought to properly improve the robustness of information network, the usage of adaptive modulation technique [9]. Recently, ways to improve energy efficiency have been delivered to improve demand-side management of strength distribution systems. Measures brought in the work of [10] place optimized genetic algorithm (OGA), which was used to improve epileptic electricity provided from the country-wide grid due to instability that has been a problem to strengthen consumers. This instability in strength furnished skilled in energy distribution network ought to be minimized with the aid of introducing optimized genetic algorithm (OGA). Recently, voltage adjustment primarily based on reinforcement gaining knowledge of and distribution evaluation is additionally gaining popularity; renewable energy penetration into the power mix in mild of the developing global demand for strength has elevated the distribution and power satisfaction considerably. Renewable strength systems (RESs) had been hastily developed due to ecological, social, economic, and environmental elements such as the extensively established photovoltaic structures (PVs) and wind turbine systems (WTs) [11]. In learning about [12], Nigeria electricity gadget is confronted with a sequence of technical challenges due to long, radial, weak, and aging transmission network; this paper introduces the idea of electricity conservation and related technologies, as properly as choices to help users achieve the benefits of electricity efficiency improvements. The work includes using computer learning to improve power first-class and decreasing strength consumption as in the case of the telecommunication base transceiver station (BTS).

Research has been carried out on strength effectivity enhancement methods for the motive of reduction in electricity consumption and environmental pollution from unburnt hydrocarbons. Today's standard wireless get admission to networks consumes more than 50% of the total power consumption of mobile communications networks, which excludes the energy fed on via the cell stations (user terminals) whose more than 50% of electricity consumption is without delay attributed to the base station (BTS) equipment. However, a discount on the electricity consumption of cellular networks is of remarkable importance from within your budget (cost reduction), environmental (decreased CO_2 emissions), and efficiency perspectives. Hence, each reduction in strength consumption and CO_2 emission are key drivers for the future of the ICT industry. In the latest file by way of the International Telecommunications Union (ITU) and Alliance for Telecommunications Industry Solutions (ATIS), a quantity of energy environment-friendly practices and strategies for consideration by way of agencies seeking to gain larger efficiencies within their wi-fi networks have been outlined.

There are lively lookup works on energy consumption, reduction, and efficiency in wireless get entry to networks, but issues touching on to the implementation of desktop learning approach had not been utilized considerably and explicitly addressed. This paper investigates power consumption of base transceivers stations (BTS), schemes that may want to doubtlessly reduce the power consumption have been described, and the management of reusing the conserved energy barring compromising first-class of the carrier (QoS) of the community explored. The research additionally investigates the importance of deploying optimization methods on strength efficiency.

We know that base transceiver station (BTS) is a transceiver that acts as an interface between the mobile stations (MS) to the network. A BTS will have between 1 and 16 transceivers (TRX) depending on the geography and demand for the provider of an area. Each TRX represents one ARFCN (absolute radio-frequency channel number). However, relying on geography, carrier demand, and operator's network method and architecture, a BTS can also host up to two, three, or six sectors, or a cell might also be serviced using various BTSs with redundant sector coverage. Each area is protected by a quarter antenna, which is a directional antenna. **Figure 1** indicates the typical macro BTS we found today. A range of remarkable documentations of hooked-up research techniques and philosophy have been mentioned extensively. Unfortunately, little comparison and integration throughout studies exist. In this article, a frequent appreciation of computer mastering and sensible agents' research and its implementation was undertaken. This paper does not extensively discuss electricity effectivity technologies but is looking to utilize modern-day mathematical strategies to successfully minimize strength consumption to retailer price in the industries. A dialogue on the framework protecting the literature on AI and ML research is restricted to energy effectivity techniques. Rather, it attempts to supply a beginning factor for integrating understanding throughout research in this area and suggests paths for future research. It explores research in certain novel disciplines: environmental pollution, medicine, maintenance, manufacturing, etc. Further lookup is wished to lengthen the current boundary of know-how in computer learning and optimization approaches. Utilizing machine gaining knowledge of disciplines into the current AI frameworks should through greater light maximize the beneficial properties of this strategy. This paper provides precious thoughts and perspectives for undergoing research on AI and ML. The closing aim was once to comprehend a reduction in electricity consumption using power efficiency means. The work offers a basis for future implementation of intelligent agents and machine learning strategies to achieve power savings that would finally translate to costs savings.

3. Materials and methods

To get the desired results and achieve our purpose, there is a need to follow the stated objectives sequentially and observe the procedure.

Characterize and determine the power consumption of the modules of the cell site under study; to characterize the cell site and determine its power consumption, the type of base station (BS), configuration model, transceiver, and power models were inspected. The cell site or base station or base transceiver station (BTS) is a microcell managed by IHS Towers West Africa Limited, with site No. IHS-EN-T4670—2G/3G/4G networks (Indoor/Outdoor Site), housing MTN Nig Ltd and Airtel Nig Ltd base station equipment at Mount Street by Idaw River Layout Awkunanaw, Enugu. The site control (hop) about 30 other MTN/Airtel base stations (Terminal and Fiber sites) within its coverage area, it handles transmission (TX) and reception (RX) of voice, data, and streaming services.

A period of 27 days was used to monitor and carry out the measurements. The days include the morning period (peak), afternoon (off-peak), and evening/night (main peak). The readings are shown in sub-section 3.2.

3.1 Obtaining the data from the company under study and determining the power consumption of the cell modules to be optimized

The method used for data collection was a time series method of measurement for 27 days at the cell site under study.

In each day, measurements were taken at a period of 2 h, with an interval of every 15 min for 8 times. In the end, an average for the eight intervals was taken for each day for all equipment.

For instance, on Day 1, the 2G BTS Airtel with the current of 25 Amps as the average of eight intervals for every 15 min in the 2 h has the following current readings of 25.4 A; 24.6 A; 25.8 A; 24.3 A; 25.6 A; 24.4 A; 24.7 A; and 25.4 A for the intervals. The average is

$$\begin{aligned} Average &= \frac{25.4 + 24.6 + 25.8 + 24.3 + 25.6 + 24.4 + 24.7 + 25.4}{8} \\ &= 25.03\text{Amps} \cong 25\text{Amps} \end{aligned}$$

An example of the 2G BTS Airtel measurement process is shown in Table C1 in Appendix C. The measurement was first performed at the BTS equipment booth, which is the backbone of the cell site that houses the transmitter and receiver modules. BTS is also connected to equipment on the tower for broadcast and hopping activities with other cell sites linked to the tower via RF and microwave antennas. At the end of the measurements performed, a 27-day summary was calculated.

3.2 To determine the module to go on sleep mode and its power requirement

$$\text{Power consumed} = \text{P}_{\text{Con}} = \text{V}_{\text{Aver}} \times \text{I}_{\text{T}}\ (\text{Watt}) \quad (1)$$

where P_{Con} is the power consumed in (Watt); V_{Aver} is the average voltage calculated from each day measurement (Volt); I_{T} is the average total current consumed by the equipment in the cell site (Amps).

- Day 1 at 13:30–15:30 h on 3rd September 2019

$$\begin{aligned} &\text{Total current} = \text{I}_{\text{T}} = 109.8\ \text{Amps} \\ &\text{Average voltage} = \text{V}_{\text{Aver}} = 52.5\ \text{Volts} \\ &\text{Power consumed} = \text{P}_{\text{Con}} = \text{I}_{\text{T}} \times \text{V}_{\text{Aver}} = 109.8 \times 52.5 \end{aligned}$$

$= 5764.50$ Watts

- Day 2 at 11:00–13:00 h on 4th September 2019

Total current $= I_T = 102.4$ Amps
Average voltage $= V_{Aver} = 50.7$ Volts
Power consumed $= PCon = I_T \times V_{Aver} = 102.4 \times 50.7$
$= 5191.68$ Watts

- Day 3 at 15:00–17:00 h on 5th September 2019

Total current $= I_T = 110.8$ Amps
Average voltage $= V_{Aver} = 52$ Volts
Power consumed $= P_{Con} = I_T \times V_{Aver} = 110.8 \times 52$
$= 5761.60$ Watts

- Day 4 at 10:00–12:00 h on 6th September 2019

Total current $= I_T = 94.9$ Amps
Average voltage $= V_{Aver} = 52.8$ Volts
Power consumed $= P_{Con} = I_T \times V_{Aver} = 94.9 \times 52.8$
$= 5010.72$ Watts

- Day 5 at 14:00–16:00 h on 9th September 2019

Total current $= I_T = 105.6$ Amps
Average voltage $= V_{Aver} = 51.3$ Volts
Power consumed $= P_{Con} = I_T \times V_{Aver} = 105.6 \times 51.3$
$= 5417.28$ Watts

- Day 6 at 10:30–12:30 h on 11th September 2019

Total current $= I_T = 98.5$ Amps
Average voltage $= V_{Aver} = 52.3$ Volts
Power consumed $= P_{Con} = I_T \times V_{Aver} = 98.5 \times 52.3$
$= 5151.55$ Watts

- Day 7 at 16:00–18:00 h on 12th September 2019

Total current $= I_T = 107.9$ Amps
Average voltage $= V_{Aver} = 52.9$ Volts
Power consumed $= P_{Con} = I_T \times V_{Aver} = 107.9 \times 52.9$
$= 5707.91$ Watts

- Day 8 at 12:00–14:00 h on 15th September 2019

Total current $= I_T = 97.5$ Amps
Average voltage $= V_{Aver} = 52.7$ Volts
Power consumed $= P_{Con} = I_T \times V_{Aver} = 97.5 \times 52.7$
$= 5138.25$ Watts

- Day 9 at 15:30–17:30 h on 17th September 2019

Total current $= I_T = 108.3$ Amps
Average voltage $= V_{Aver} = 52.4$ Volts
Power consumed $= P_{Con} = I_T \times V_{Aver} = 108.3 \times 52.4$
$= 5674.92$ Watts

- Day 10 at 14:30–16:30 h on 19th September 2019

Total current $= I_T = 106.9$ Amps
Average voltage $= V_{Aver} = 53.2$ Volts
Power consumed $= P_{Con} = I_T \times V_{Aver} = 106.9 \times 53.2$
$= 5687.08$ Watts

- Day 11 at 13:00–15:00 h on 20th September 2019

Total current $= I_T = 104.7$ Amps
Average voltage $= V_{Aver} = 52.7$ Volts
Power consumed $= P_{Con} = I_T \times V_{Aver} = 104.7 \times 52.7$
$= 5517.69$ Watts

- Day 12 at 19:00–21:00 h on 23rd September 2019

Total current $= I_T = 160.8$ Amps
Average voltage $= V_{Aver} = 53.7$ Volts
Power consumed $= P_{Con} = I_T \times V_{Aver} = 160.8 \times 53.7$
$= 8634.96$ Watts

- Day 13 at 18:00–20:00 h on 26th September 2019

Total current $= I_T = 151.1$ Amps
Average voltage $= V_{Aver} = 53.7$ Volts
Power consumed $= P_{Con} = I_T \times V_{Aver} = 151.1 \times 53.7$
$= 8114.07$ Watts

- Day 14 at 20:00–22:00 h on 27th September 2019

Total current $= I_T = 144.3$ Amps
Average voltage $= V_{Aver} = 52.4$ Volts
Power consumed $= P_{Con} = I_T \times V_{Aver} = 141.3 \times 52.4$
$= 7404.12$ Watts

- Day 15 at 18:30–20:30 h on 28th September 2019

Total current $= I_T = 151.7$ Amps
Average voltage $= V_{Aver} = 53.3$ Volts
Power consumed $= P_{Con} = I_T \times V_{Aver} = 151.7 \times 53.3$
$= 8085.61$ Watts

- Day 16 at 19:30–21:30 h on 2nd October 2019

Total current $= I_T = 153$ Amps
Average voltage $= V_{Aver} = 53.4$ Volts
Power consumed $= P_{Con} = I_T \times V_{Aver} = 153 \times 53.4$
$= 8170.20$ Watts

- Day 17 at 06:30–08:30 h on 3rd October 2019

Total current $= I_T = 131.4$ Amps
Average voltage $= V_{Aver} = 52.8$ Volts
Power consumed $= P_{Con} = I_T \times V_{Aver} = 131.4 \times 52.8$
$= 6937.92$ Watts

- Day 18 at 07:00–09:00 h on 6th October 2019

Total current $= I_T = 125$ Amps
Average voltage $= V_{Aver} = 53.3$ Volts
Power consumed $= P_{Con} = I_T \times V_{Aver} = 125 \times 53.3$
$= 6662.50$ Watts

- Day 19 at 08:00–10:00 h on 7th October 2019

Total current $= I_T = 121.2$ Amps
Average voltage $= V_{Aver} = 52.4$ Volts
Power consumed $= P_{Con} = I_T \times V_{Aver} = 121.2 \times 52.4$
$= 6350.88$ Watts
$= 5138.25$ Watts

- Day 9 at 15:30–17:30 h on 17th October 2019

Total current $= I_T = 108.3$ Amps
Average voltage $= V_{Aver} = 52.4$ Volts
Power consumed $= P_{Con} = I_T \times V_{Aver} = 108.3 \times 52.4$
$= 5674.92$ Watts

- Day 10 at 14:30–16:30 h on 19th October 2019

Total current $= I_T = 106.9$ Amps
Average voltage $= V_{Aver} = 53.2$ Volts
Power consumed $= P_{Con} = I_T \times V_{Aver} = 106.9 \times 53.2$
$= 5687.08$ Watts

- Day 11 at 13:00–15:00 h on 20th October 2019

Total current $= I_T = 104.7$ Amps
Average voltage $= V_{Aver} = 52.7$ Volts
Power consumed $= P_{Con} = I_T \times V_{Aver} = 104.7 \times 52.7$
$= 5517.69$ Watts

- Day 12 at 19:00–21:00 h on 23rd October 2019

Total current $= I_T = 160.8$ Amps
Average voltage $= V_{Aver} = 53.7$ Volts

Power consumed = $P_{Con} = I_T \times V_{Aver} = 160.8 \times 53.7$
= 8634.96 Watts

- Day 13 at 18:00–20:00 h on 26th October 2019

Total current = I_T = 151.1 Amps
Average voltage = V_{Aver} = 53.7 Volts
Power consumed = $P_{Con} = I_T \times V_{Aver} = 151.1 \times 53.7$
= 8114.07 Watts

- Day 14 at 20:00–22:00 h on 27th October 2019

Total current = I_T = 144.3 Amps
Average voltage = V_{Aver} = 52.4 Volts
Power consumed = $P_{Con} = I_T \times V_{Aver} = 141.3 \times 52.4$
= 7404.12 Watts

- Day 15 at 18:30–20:30 h on 28th October 2019

Total current = I_T = 151.7 Amps
Average voltage = V_{Aver} = 53.3 Volts
Power consumed = $P_{Con} = I_T \times V_{Aver} = 151.7 \times 53.3$
= 8085.61 Watts

- Day 16 at 19:30–21:30 h on 2nd November 2019

Total current = I_T = 153 Amps
Average voltage = V_{Aver} = 53.4 Volts
Power consumed = $P_{Con} = I_T \times V_{Aver} = 153 \times 53.4$
= 8170.20 Watts

- Day 17 at 06:30–08:30 h on 3rd November 2019

Total current = I_T = 131.4 Amps
Average voltage = V_{Aver} = 52.8 Volts
Power consumed = $P_{Con} = I_T \times V_{Aver} = 131.4 \times 52.8$
= 6937.92 Watts

- Day 18 at 07:00–09:00 h on 6th November 2019

Total current = I_T = 125 Amps
Average voltage = V_{Aver} = 53.3 Volts
Power consumed = $P_{Con} = I_T \times V_{Aver} = 125 \times 53.3$
= 6662.50 Watts

- Day 19 at 08:00–10:00 h on 7th November 2019

Total current = I_T = 121.2 Amps
Average voltage = V_{Aver} = 52.4 Volts
Power consumed = $P_{Con} = I_T \times V_{Aver} = 121.2 \times 52.4$
= 6350.88 Watts

- Day 20 at 07:30–07:30 h on 9th November 2019

Total current = I_T = 124 Amps
Average voltage = V_{Aver} = 53.6 Volts
Power consumed = $P_{Con} = I_T \times V_{Aver} = 124 \times 53.6$
= 6646.40 Watts

- Day 21 at 06:45–08:45 h on 16th November 2019

Total current = I_T = 122.3 Amps
Average voltage = V_{Aver} = 52.2 Volts
Power consumed = $P_{Con} = I_T \times V_{Aver} = 122.3 \times 52.2$
= 6384.06 Watts

- Day 22 at 06:00–08:00 h on 17th November 2019

Total current = I_T = 127.8 Amps
Average voltage = V_{Aver} = 53.4 Volts
Power consumed = $P_{Con} = I_T \times V_{Aver} = 127.8 \times 53.4$
= 6824.52 Watts

- Day 23 at 07:45–09:45 h on 20th November 2019

Total current = I_T = 120.7 Amps
Average voltage = V_{Aver} = 52.7 Volts
Power consumed = $P_{Con} = I_T \times V_{Aver} = 120.7 \times 52.7$
= 6360.89 Watts

- Day 24 at 08:15–10:15 h on 22nd November 2019

Total current = I_T = 117.5 Amps
Average voltage = V_{Aver} = 52.7 Volts
Power consumed = $P_{Con} = I_T \times V_{Aver} = 117.5 \times 52.7$
= 6192.25 Watts

- Day 25 at 08:15–10:15 h on 26th November 2019

Total current = I_T = 128.8 Amps
Average voltage = V_{Aver} = 53.3 Volts
Power consumed = $P_{Con} = I_T \times V_{Aver} = 128.8 \times 53.3$
= 6865.04 Watts

- Day 26 at 08:30–10:30 h on 27th November 2019

Total current = I_T = 122.1 Amps
Average voltage = V_{Aver} = 53.1 Volts
Power consumed = $P_{Con} = I_T \times V_{Aver} = 122.1 \times 53.1$
= 6483.51 Watts

- Day 27 at 06:15–08:15 h on 28th November 2019

Total current = I_T = 125.4 Amps

Average voltage $= V_{Aver} = 53.1$ Volts
Power consumed $= P_{Con} = I_T \times V_{Aver} = 128.8 \times 53.3$
$= 6658.74$ Watts
Total $= 170098.78$ Watts $= 170.09878$ kW $= 170.1$ kW (approximation)
Number of hours for 27 days $= 27 \times 2 = 54$ h
KWH $= 170.1 \times 54 = 9185.4$ kWh
#60 = 1 kWh
9185.4 kWh = #9185.4 × 60 = #551124

3.3 Optimization of the established high power consumed by the modules of the cell site to a minimum value

Minimize
$P = X + 13.3Y$
SUBJECT TO
$X + 13.3Y \leq 5764.50$
$7X + 16Y \leq 5707.91$
where P is the minimum power consumed by the cell site; X is the day the power is consumed in the cell site; Y is the hour the power is consumed in the cell site.

>> % OPTIMIZING ENERGY EFFICIENCY THROUGH REDUCTION OF POWER CONSUMPTION IN A TELECOMMUNICATION BASE TRANCEIVER STATION (BTS) SITE USING MACHINE LEARNING

```
%
% Minimize P = X + 13.3Y
%         Subject to
%      X + 13.3Y ≤ 5764.50
%   10X + 14. 3Y≤ 5687.08
%
% Where
% P is the minimum power consumed by the cell site
%X is the day the power is consumed in the cell site
%Yis the hour the power is consumed in the cell site
```

Days	Time (h)	Power consumption (Watts)
1	13:30	5764.50
2	11:00	5191.68
3	15:00	5761.60
4	10:00	5010.72
5	14:00	5417.28
6	10:30	5151.55
7	16:00	5707.91
8	12:00	5138.25
9	15:30	5674.92
10	14:30	5687.08

Table 1.
Power consumed from the characterized cell site under study.

```
    f=[-1;-13.3];
    A=[1 13.3;7 16];
    b=[5764.50;5707.91];
Aeq=[0 0];
beq=[0];
    LB=[0 0];
    UB=[inf inf];
    [X,FVAL,EXITFLAG]=linprog(f,A,b,Aeq,beq,LB,UB)

Optimization terminated.

X =

   0.0000
 356.7444

FVAL =

-4.7447e+003
EXITFLAG = 1

>>
```

3.4 Designing a machine learning rule base that will monitor the power consumed on the modules and minimize it if high

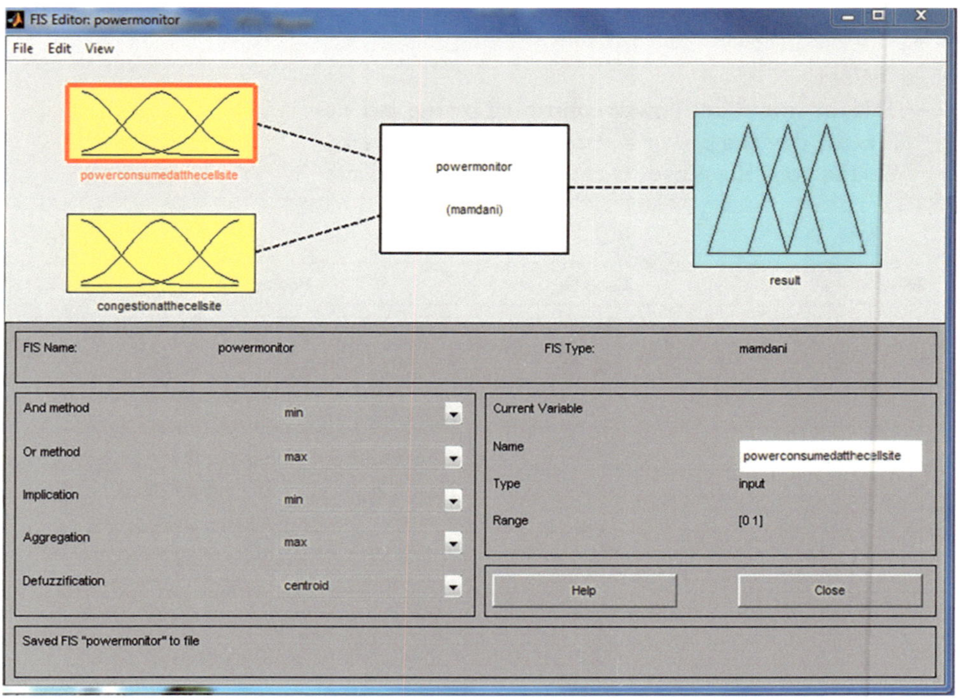

Figure 1.
Designed machine learning fuzzy inference system that will monitor the power consumed on the modules and minimize it if high.

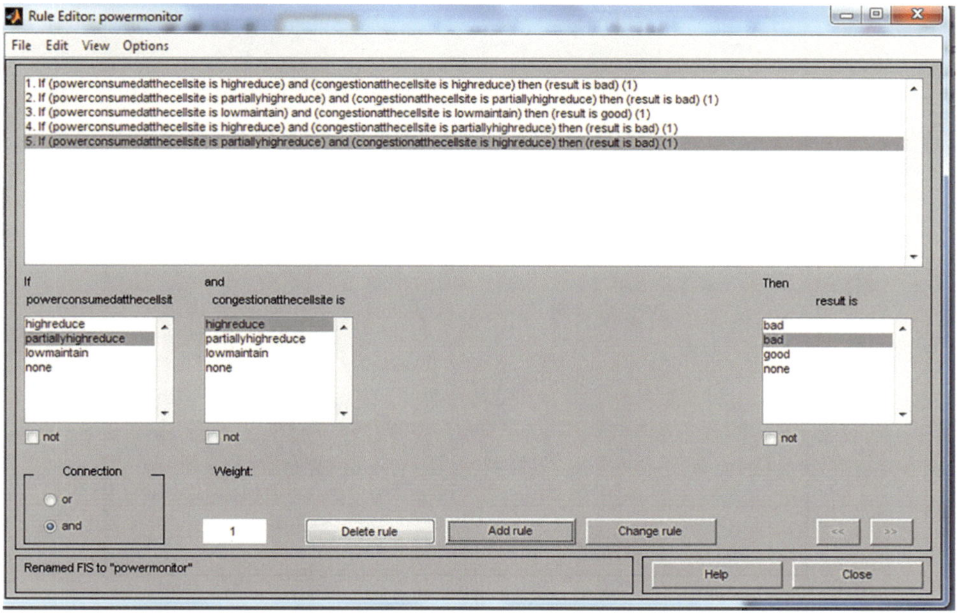

Figure 2.
Designed machine learning rule base that will monitor the power consumed on the modules and minimize it if high.

1	If power consumed at the cell site is high reduce	And congestion at the cell site is high reduce	Then, result is bad
2	If power consumed at the cell site is partially high reduce	And congestion at the cell site is partially high reduce	Then, result is bad
3	If power consumed at the cell site is low maintain	And congestion at the cell site is low maintain	Then, result is good
4	If power consumed at the cell site is high reduce	And congestion at the cell site is partially high reduce	Then, result is bad
5	If power consumed at the cell site is partially high reduce	And congestion at the cell site is high reduce	Then, result is bad

Table 2.
Details of designed machine learning rule base that will monitor the power consumed on the modules and minimize it if high.

3.5 Training ANN in the designed machine learning rules for reduced power consumption in the cell site, thereby enhancing its network performance

Figures 3 and **4** will be implemented in the machine learning to enhance its proper functioning to minimize the power consumption in the cell site to save costs.

3.6 Developing an algorithm that will implement 4, 5, and 6

1. Identify the much power consumed by the module of cell site.

2. Optimize the identified much power consumed by the module of the cell site to a minimal.

3. Apply designed machine learning rule base that will monitor the power consumed on the modules and minimize it if high.

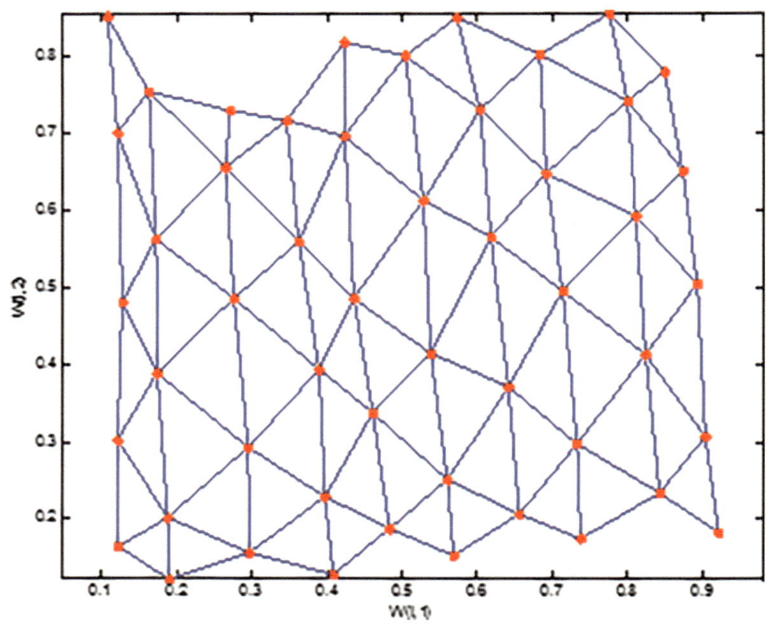

Figure 3.
Trained ANN in the designed machine learning rules for reduced power consumption in the cell site thereby enhancing its network performance.

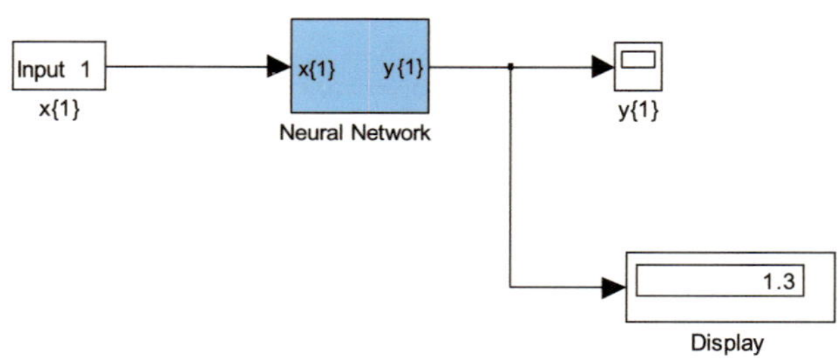

Figure 4.
Model that resulted in the training.

4. Apply the trained ANN in 3 to retain minimal power consumption in the module of the cell site.

5. Does the power consumption at the module of the cell site minimized after the application of 4?

6. No, go to 4.

7. Yes, go to 9.

8. Minimized power consumption by the module of the cell site.

9. Stop.

10. End.

11. To develop a power consumption model for the network under study based on results obtained.

3.7 Developing power consumption model for the network under study based on results obtained

The power consumption model is shown in **Figure 5** reflecting all the simulations obtained as depicted in **Figures 6–9**.

4. Results and discussion

The results obtained using machine learning to minimize power consumption are presented and discussed. In **Figure 1**, we have two inputs of power consumed at the cell site and congestion. It also has an output of results (**Table 1**).

Figure 2 is a designed machine learning rule base that will monitor the power consumed on the modules and minimize it if high. It monitors the power consumed by the modulus of the cell site and minimizes it when detected high. A comprehensive analysis of the rules is as shown in **Table 2** where details of designed machine learning rule base that will monitor the power consumed on the modules and minimize it if high are tabulated. In **Figure 3**, ANN was trained 10 times in the machine learning five rules to give 50 neurons that look like human brain 10 × 5 = 50. These neurons mimic human intelligence and do what it is instructed to do.

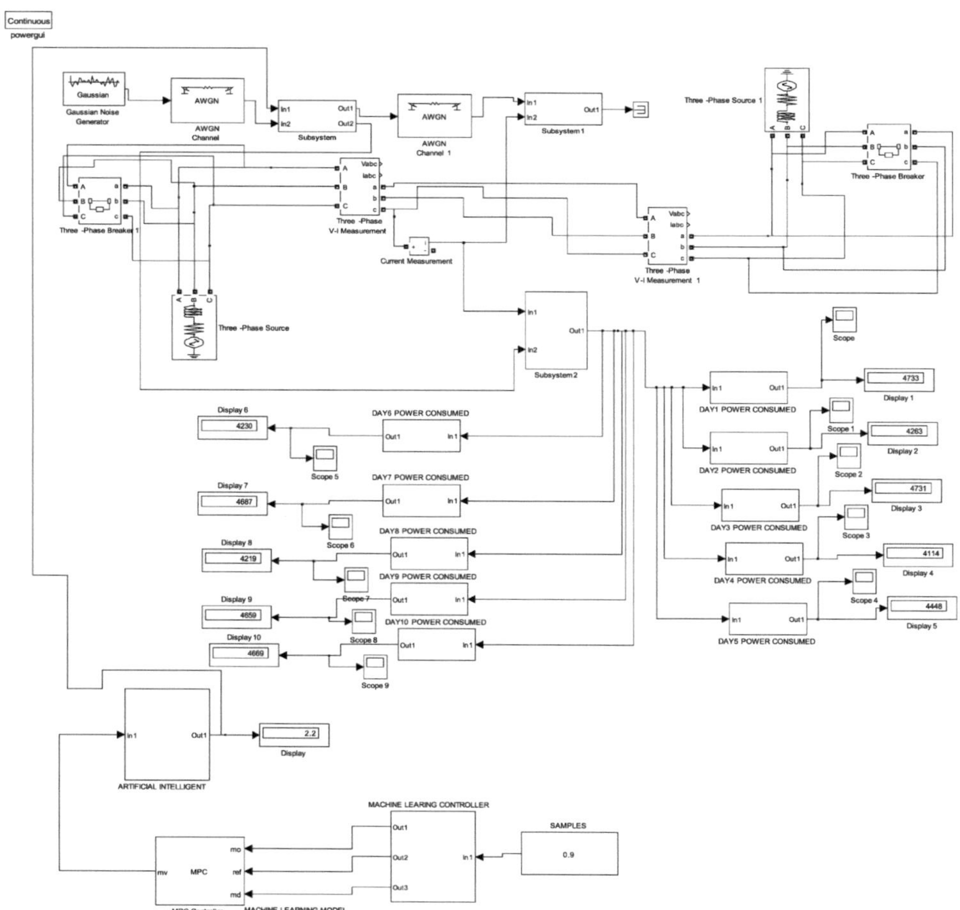

Figure 5.
Developed power consumption model for the network under study based on results obtained. The results obtained after simulation are as shown in ***Figures 6–9****.*

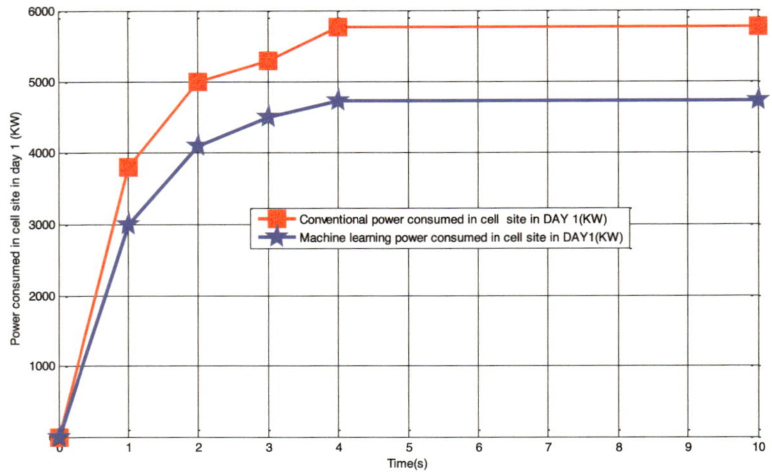

Figure 6.
Comparing conventional and machine learning power consumed in cell site in day.

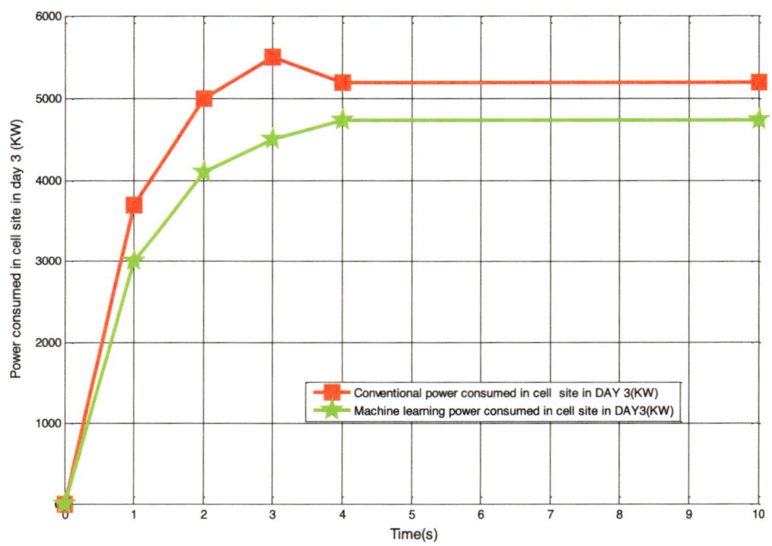

Figure 7.
Comparing conventional and machine learning power consumed in cell site in day 3.

Figure 4 is incorporated in the machine learning to enhance its efficacy in terms of reducing the power consumed in the cell site, thereby enhancing the financial status of the site. **Figure 5** is the developed power consumption model for the network under study based on results obtained. The results obtained after simulation are as shown in **Figures 6–9**. **Figure 6** shows the comparison between conventional and machine learning power consumed in cell site in day 1 (**Table 3**). In **Figure 6**, the highest conventional power consumed by the cell site is 5764 kW while that when machine learning is inculcated in the system is 4733 kW (**Table 4**). With these results, it signifies that the percentage improvement in the reduction of power consumed in the cell site when machine learning is incorporated in the system in day 1 is 17.9%.

Figure 7 shows that the highest conventional power consumed in the cell site in day 3 is 5191 kW while that when machine learning is integrated in the system is 4731 kW. This clearly showed that the percentage improvement in power consumption reduction in the cell site when machine learning technique is imbibed in the system in day 3 is 8.9%.

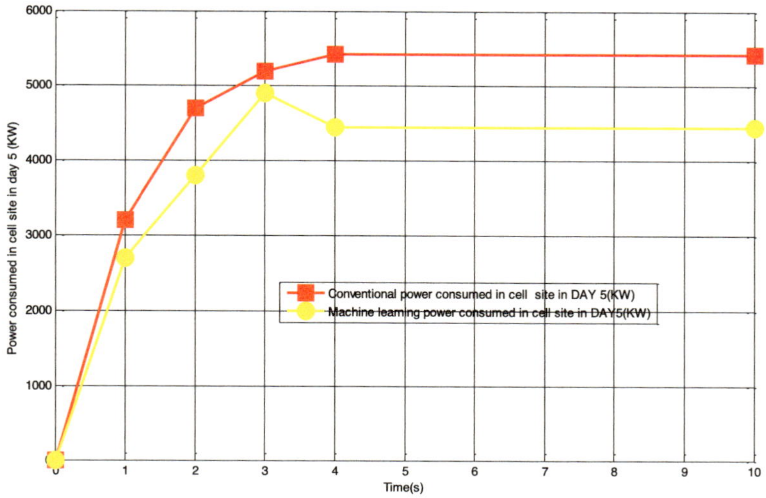

Figure 8.
Comparing conventional and machine learning power consumed in cell site in day 5.

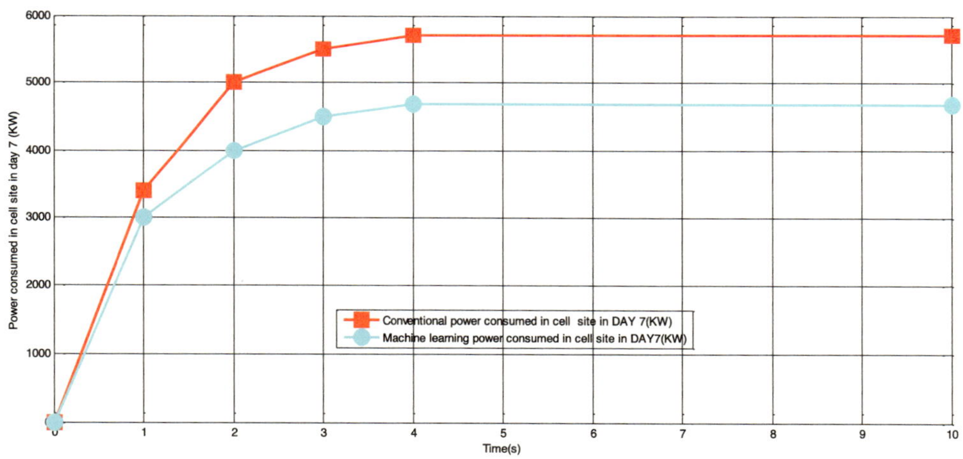

Figure 9.
Comparing conventional and machine learning power consumed in cell site in day 7.

Time (s)	Conventional power consumed in cell site in DAY 1 (kW)	Machine learning power consumed in cell site in DAY1 (kW)
0	0	0
1	3800	3000
2	5000	4100
3	5300	4500
4	5764	4733
10	5764	4733

Table 3.
Comparison between conventional and machine learning power consumed in cell site in day 1.

Time (s)	Conventional power consumed in cell site in DAY3 (kW)	Machine learning power consumed in cell site in DAY3 (kW)
0	0	0
1	3700	3000
2	5000	4100
3	5500	4500
4	5191	4731
10	5191	4731

Table 4.
Comparison between conventional and machine learning power consumed in cell site in day.

Time (s)	Conventional power consumed in cell site in DAY5 (kW)	Machine learning power consumed in cell site in DAY5 (kW)
0	0	0
1	3200	2700
2	4700	3800
3	5200	4900
4	5417	4448
10	5417	4448

Table 5.
Comparison between conventional and machine learning power consumed in cell site in day 5.

Time (s)	Conventional power consumed in cell site in DAY7 (kW)	Machine learning power consumed in cell site in DAY7 (kW)
0	0	0
1	3400	3000
2	5000	4000
3	5500	4500
4	5708	4687
10	5708	4687

Table 6.
Comparison between conventional and machine learning power consumption in day 7.

In **Figure 8**, it is obvious that the highest conventional power consumed in the cell site is 5417 kW. On the other hand, when machine learning is integrated in the system, it reduced drastically to 4448 kW, which is 17.9% power consumed by the cell site reduction (**Table 5**). **Figure 9** symbolizes that the highest conventional power consumed by the cell site is 5708 kW while that when machine learning is introduced into the system is 4687 kW, which is 17.9% better that the conventional approach as regards power consumption reduction in the cell (**Table 6**).

5. Conclusion

Sustainable high power consumption at cell sites has reduced the financial status of some of these cell sites. This ugly situation of high cell site power consumption is mitigated by introducing optimized energy efficiency by using machine learning to reduce power consumption at communication base station (BTS) sites. To achieve this enthusiastically, in this process, the related work is checked to find out its shortcomings, and the power consumption of the inspected cell site module is characterized, determined, and inspected. SIMULINK model is developed, specified, and optimized. Machine learning rule base that minimizes the high power consumption of the cell site module monitors the power consumed by the module and minimizes it at high power. Design and train ANNs with designed machine learning rules to reduce power consumption improve its network performance at base stations. Next, we will develop an algorithm that implements it. Finally, based on the results obtained when the algorithm was integrated into the network, we developed a power consumption model for the network under investigation and improved energy efficiency at the cell site with and without machine learning, validated and justified the rate. The results of extensive simulation show that the conventional maximum power consumption of the cell site is 5746 kW, while the maximum machine learning power consumption of the system is 4733 kW. From these results, the improvement rate of power consumption reduction of cell sites by integrating machine learning into the system is 17.9% on the first day, while the maximum power consumption of conventional cell sites is 3 days of machine learning. You can see that it is 5191 kW by eye. Learning is integrated into the system and is 4731 kW. From these results, it can be seen that the improvement rate of the power consumption reduction of the cell site when the machine learning technology is incorporated into the system on the third day is 8.9%, and the conventional maximum power consumed by the cell site is 5417 kW. I understand, on the other hand, when machine learning is integrated into the system, the reduction of cell sites significantly reduces the power consumption to 4448 kW, which is 17.9% of the power consumption, and the conventional maximum power consumption of cell sites is 5708 kW. The learning built into the system is 4687 kW, which is 17.9 superior to the traditional approach in terms of cell site power savings.

Acknowledgements

I appreciate the effort of my son Junior N. Bassey for the sustained interest in assisting to read the manuscript.

Conflict of interest

There is no conflict of interest.

Author details

Ngang Bassey Ngang
Faculty of Engineering, Veritas University, Abuja, Nigeria

*Address all correspondence to: ngangn@veritas.edu.ng

References

[1] Tanko AL. Evolution of mobile telecommunication with respect to 1G 2G 3G and 4G. In: The Communicator. Nigeria Communication Commission; 2018 2019 Quarter 3/4 edition. Issue 25

[2] Akbari A, Imran MA, Tafazolli R, Dianati M. Energy efficiency contours for single-carrier downlink channels. IEEE Communications Letters. 2011; **15**(12):1307-1309

[3] Akhilesh Kumar P, Ompal S. 5G technology—Redefining wireless communication in upcoming years. International Journal of Computer Science and Management Research. 2012;**1**(1):1-8

[4] Bazzi A, Pasolini G, Andrisano O. Multi-radio resource management: Parallel transmission for higher through put. EURASIP Journal of Advances in Signal Process. 2008;**9**:120-131. Article ID 763264. DOI: 10.1155/2008/763264

[5] Ugwu KI et al. Industrial processing water control for an increased output using intelligent agent. International Journal of Sciences and Business. 2021; **5**(8):214-221. DOI: 10.5281/zenodo.5070521

[6] Ude K et al. Reducing costs and pollutant emissions using renewable energy sources in Eket town, Nigeria. International Journal of Sciences and Applied Research. 2021;**8**(7):11-12

[7] Bakare K, Ngang NB. Energy efficiency improvement of doubly fed induction generator machines using adaptive control technique. International Journal of Research in Engineering and Science. 2021;**3**(7): 661-669. DOI: 10.35629/5252-0307661669

[8] Akaninyene J et al. Optimized electric power generation expansion planning using decomposition technique. International Journal of Science and Business. 2021;**5**:94-116

[9] Anthony LA, Uko O, Ngang NB. Enhancement of data network robustness in Nigeria, using adaptive modulation technique. International Journal of Advances in Engineering and Management. 2021;**3**(6):1470-1478. DOI: 10.35629/5252-030614701478

[10] Aneke NE et al. Improving loss minimization in 33 kV power distribution network using optimized genetic algorithm. International Journal of Systems Engineering. 2021;**5**(1):34-42

[11] Bakare K et al. Optimization of a micro grid operation under uncertainty using model predictive controller. International Journal of Research and Innovation in Applied Science. 2021;**6**(6):81-92

[12] Ngang NB, Bakare K. Improving frequency stability of the Nigerian 330 kV transmission network using fuzzy controller. American Journal of Electrical Power and Energy Systems. 2021;**10**(3):43-50